Lecture Notes in Mathematics 1509

J. Aguadé M. Castellet F. R. Cohen (Eds.)

Algebraic Topology

Homotopy and Group Cohomology

Proceedings of the 1990 Barcelona Conference on Algebraic Topology, held in S. Feliu de Guíxols, Spain, June 6–12, 1990

Springer-Verlag
Berlin Heidelberg New York
London Paris Tokyo
Hong Kong Barcelona
Budapest

Editors

Jaume Aguadé
Manuel Castellet
Departament de Matemàtiques
Unversitat Autònoma de Barcelona
E-08193 Bellaterra, Spain

Frederick Ronald Cohen
Department of Mathematics
University of Rochester
Rochester, New York 14620, USA

Mathematics Subject Classification (1991): 55-06, 55Pxx, 20J06, 20J05

ISBN 3-540-55195-6 Springer-Verlag Berlin Heidelberg New York
ISBN 0-387-55195-6 Springer-Verlag New York Berlin Heidelberg

Printed in Germany

Typesetting: Camera ready by author
Printing and binding: Druckhaus Beltz, Hemsbach/Bergstr.
46/3140-543210 - Printed on acid-free paper

INTRODUCTION

This volume represents the Proceedings of the Third Barcelona Conference on Algebraic Topology. The conference was held June 6-12, 1990, in Sant Feliu de Guíxols and was organized by the Centre de Recerca Matemàtica, a research institute sponsored by the Institut d'Estudis Catalans and located at the Universitat Autònoma de Barcelona. Financial support from the DGICYT and the CIRIT of the Generalitat de Catalunya made this Conference possible by bringing together mathematicians from around the world.

The scientific program consisted of 21 plenary lectures which are listed below. The Proceedings contains papers which were, for the most part, submitted by conference participants. All papers were refereed. We would like to take this opportunity to thank the participants for their contributions. The editors would also like to thank the referees for conscientious and thorough efforts.

We thank Carles Broto, Carles Casacuberta, and Laia Saumell for their generous help with the organization of the conference. We warmly thank the secretaries Consol Roca and Sylvia Hoemke. Most of the papers included in these Proceedings are prints obtained from TEX files submitted by the authors, suitably modified to uniformize headings and margins. We thank Maria Julià for her fine job in typing some of the manuscripts and for her help in solving TEX problems. Finally, we would like to thank the mayor of Sant Feliu de Guíxols for his help in stimulating an excellent mathematical environment.

Jaume Aguadé
Manuel Castellet
Fred Cohen

HOTEL MURLA

1
2
3
4
5
6
7
8
9
10
11
12
13
14
15
16
17
18
19
20
21
22
23
24
25
26
27
28
29
30
31
32
33
34
35
36
37
38
39
40
41
42
43
44
45
46
47
48
49
50
51
52
53
54

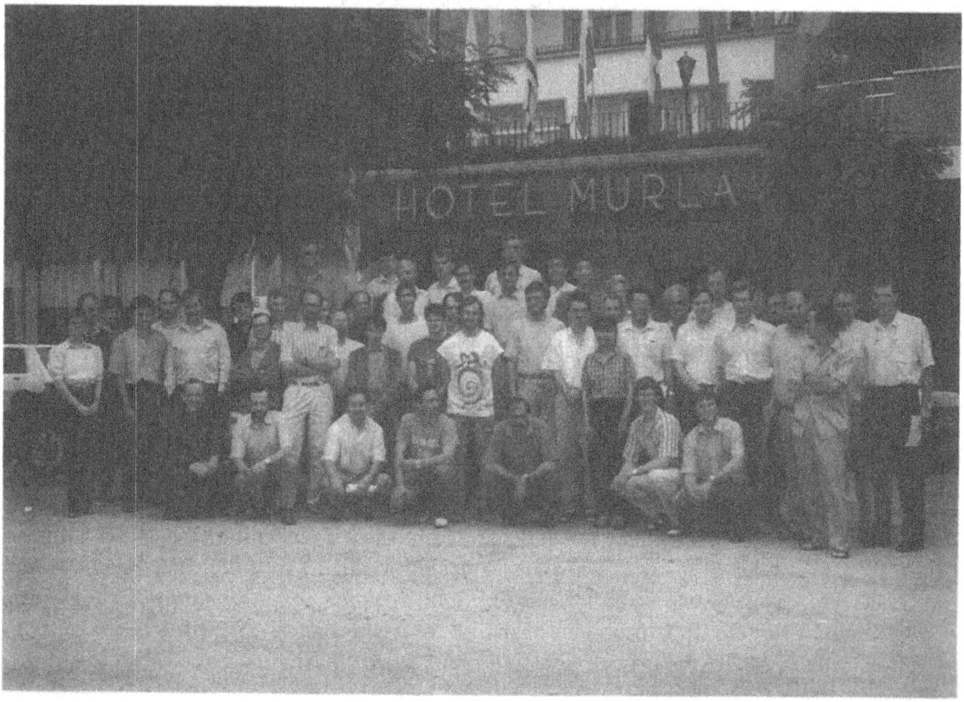

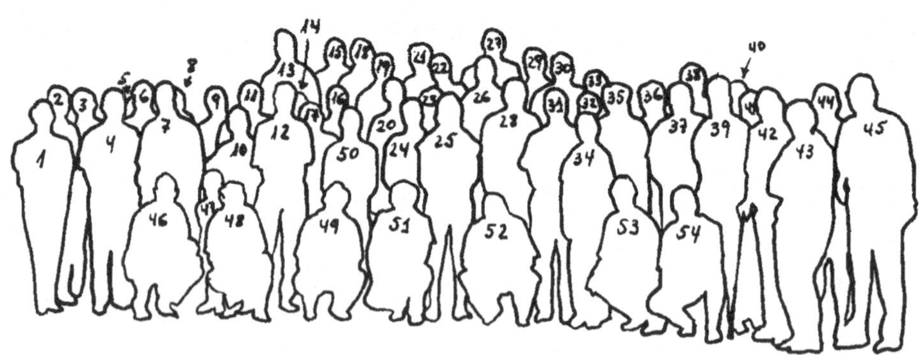

List of participants

The numbers in brackets refer to the group picture

A. Adem (9)
J. Aguadé (48)
N. Álamo
D. Benson (25)
J. Berrick (27)
M. Boileau
I. Bokor (16)
C. Broto (10)
J. Cabeza
C. Casacuberta (7)
M. Castellet (45)
A. Cavicchioli (3)
B. Cenkl (11)
F.R. Cohen (42)
W. Dicks
E. Dror Farjoun (43)
C. Elvira
J.I. Extremiana
H. Glover (33)
F. Gómez (22)
K. Hansen
J.C. Harris (53)
J.C. Hausmann
C. Hayat-Legrand
F. Hegenbarth (2)
H.-W. Henn (54)
L.J. Hernández (49)
J. Hubbuck (14)
K. Irie (40)
K. Ishiguro (35)
M. Izydorek (18)
S. Jackowski
A. Jeanneret (21)
B. Jessup (24)
R. Kane (13)
A. Kono (30)
J. Lannes (50)
I. Llerena (1)
L. Lomonaco (51)
H. Marcum (39)
J.R. Martino (37)
C.A. McGibbon (19)
G. Mislin
A. Murillo (20)
J.L. Navarro (52)
D. Notbohm (38)
B. Oliver
E. Ossa (17)
G. Peschke (6)
Ch. Peterson (34)
M. Pfenniger (41)
S. Priddy (46)
M. Raußen (28)
D.C. Ravenel (12)
M.T. Rivas
C. Safont
L. Saumell
R. Schon
X. Shen (29)
L. Smith (32)
U. Suter (36)
R. Thompson
K. Varadarajan (8)
E. Ventura (4)
A. Vidal (15)
A. Vučic (31)
Z. Wojtkowiak (44)
X. Zarati (26)

1990 Barcelona Conference
on Algebraic Topology.

Titles of Talks

(in chronological order)

K. Ishiguro	Classifying spaces of compact simple Lie groups and p-tori.
D. Notbohm	Homotopy uniqueness of $BU(n)$.
J. Lannes	Representing modulo p homology.
A. Kono	Homology of the Kac-Moody groups.
A. Adem	The geometry and cohomology of some sporadic simple groups.
J.R. Martino	The complete stable splitting for classifying spaces of finite groups.
S. Priddy	On the dimension theory of summands of BG.
G. Mislin	Cohomologically central elements in groups.
C. Broto	On Sub-$\mathcal{A}_p^*$-algebras of H^*V.
H.-W. Henn	Refining Quillen's description of $H^*(BG; F_p)$.
J. Hubbuck	On Miller's stable splitting of $SU(n)$.
C.A. McGibbon	On spaces with the same n-type for all n.
L. Smith	Fake Lie groups and maximal tori.
D. Benson	Resolutions and Poincaré duality for finite groups.
R. Thompson	The fiber of secondary suspension.
J. Harris	Lannes' T functor on summands of $H^*(BZ/p)^s$.
D. Ravenel	Morava K-theory of BG and generalized group characters.
E. Dror Farjoun	Good spaces, bad spaces.
B. Oliver	Approximations of classifying spaces.
S. Jackowski	Vanishing theorems for higher inverse limits.

Table of Contents

1990 Barcelona Conference
on Algebraic Topology.

ON THE GEOMETRY AND COHOMOLOGY OF FINITE SIMPLE GROUPS

ALEJANDRO ADEM*

§0. INTRODUCTION

Let G be a finite group and k a field of characteristic $p \mid |G|$. The cohomology of G with k-coefficients, $H^*(G,k)$, can be defined from two different points of view:

(1) $H^*(G,k) = H^*(BG;k)$, the singular cohomology of the classifying space BG

(2) $H^*(G,k) = Ext^*_{kG}(k,k)$.

From (1) it is clear that group cohomology contains important information about G-bundles and the rôle of G as a transformation group. From (2) we can deduce that $H^*(G,k)$ contains characteristic classes for representations over k, and indeed is the key tool to understanding modular representations via the method of cohomological varieties introduced by Quillen.

The importance of classifying spaces in topology is perhaps most evident in the method of "finite models". Roughly speaking, the classifying spaces of certain finite groups can be assembled to yield geometric objects of fundamental importance. For example, let Σ_m denote the symmetric group on m letters; then the natural pairing $\Sigma_n \times \Sigma_m \to \Sigma_{n+m}$ induces an associative monoid structure on $C(S^\circ) = \coprod_{n\geq 0} B\Sigma_n$. We have the result due to Dyer and Lashof [D-L]:

THEOREM

$Q(S^\circ) = \Omega^\infty S^\infty$ is the homotopy group completion of $C(S^\circ)$ (using the loop sum), and hence

$$H_*(Q(S^\circ),\mathbf{F}_2) \cong \varinjlim H_*(\Sigma_m,\mathbf{F}_2) \otimes \mathbf{F}_2[\mathbf{Z}]\,. \qquad \blacksquare$$

Given that $\pi_i(Q(S^\circ)) = \pi_i^S(S^\circ)$ (the stable homotopy groups of spheres), $Q(S^\circ)$ is the "ground ring" for stable homotopy; and the result above indicates that it can be constructed from classifying spaces of finite groups.

A similar approach (due to Quillen [Q1]) uses the $\mathrm{GL}_n(\mathbf{F}_q)$ to construct the space Im J, which essentially splits off $Q(S^\circ)$ in the natural way. From this space we obtain the algebraic K-theory of the field $\mathbf{F}_q$.

Now from the point of view of cohomology of groups, very few noteworthy examples have been calculated. Most strikingly, the important computations have been for groups involved in the geometric scheme outlined above. The understanding of the cohomology

* Partially supported by an NSF grant.

of the symmetric and alternating groups (only additively) is linked to the geometry of $Q(S^\circ)$; indeed this space can be identified with $B\Sigma_\infty^+$ (the plus construction), and the homology of the finite symmetric groups can be expressed in a precise manner using Dyer-Lashof operations. Analogously, Quillen constructed $\mathrm{BGL}(\mathbf{F}_q)^+ = F\psi^q$ by comparing its cohomology to what he calculated for $\mathrm{BGL}_n(\mathbf{F}_q)$ away from q, for all n.

From a purely calculational point of view, the cases above are very special: the cohomology is detected on abelian subgroups. Hence these results cannot truly reflect the general difficulties of the subject and may not be good examples for relating $H^*(G, k)$ to geometric and algebraic properties of G.

At this point we introduce an ingredient usually ignored by topologists — the classification of finite simple groups. From the topological point of view, only the alternating and groups associated to those of Lie type have been considered (as outlined above). The classification theorem indicates that there are 26 additional groups which are as essential as the two infinite families — the so-called sporadic simple groups. A natural question arises: what rôle do the sporadic simple groups play in this framework?

In this paper I will discuss some recent work on this problem which I have done jointly with R. J. Milgram. By now it is clear that the first step in any such project should be to understand the cohomology first as a calculation and then as a source of algebraic and geometric data. In §1 I will describe the combination of techniques required to calculate the cohomology of groups of this complexity. Then, in §2 I will present some calculations. The particular example which will be dealt with in detail is the Mathieu group M_{12}, a group of order 95,040. The cohomology ring $H^*(M_{12}, \mathbf{F}_2)$ will be fully described. In the following section, a geometric analysis of the results in §2 will be provided. This example will show how well cohomology carries the local structure of the group, as well as plentiful topological information. Finally in §4 I will discuss current work on M_{22}, M_{23}, M^cL and $O'N$, which are even larger groups.

The aim of this paper is to bring the reader up to date on recent progress, hence he will be spared the technical details of the calculations, which are either available ([AMM], [AM]) or will appear soon. All coefficients are taken to be $\mathbf{F}_2$, so they will be suppressed.

A note of caution: it is not the author's viewpoint to encourage mindless calculation, which is seen so often. Rather, the main point is that with some work we can extend our knowledge of group cohomology by understanding the key models on which the whole framework of finite group theory is now based. The sporadic simple groups seen to be the natural objects we should now turn our attention to, if there is any possibility of understanding behaviour in algebraic topology which is modelled on finite group theory.

The results in this paper were presented at a talk given by the author during the 1990 Barcelona Conference on Algebraic Topology. The author is very grateful to the organizers for being invited and in particular to M. Castellet for his kind hospitality.

§1. CALCULATIONAL TECHNIQUES

From the landmark work of Quillen [Q2] it is known that $H^*(G)$ is detected up to nilpotence on its elementary abelian subgroups. Hence a clear understanding of these subgroups is necessary before embarking on a calculation. The relevant data can all be assembled into a finite complex (first introduced by K. Brown) as follows [Q3]. Let

$A_p(G)$ denote the partially ordered set of p-elementary abelian subgroups of G under inclusion. In the usual way $A_p(G)$ can be given the structure of a simplicial complex; the n-simplices correspond to flags of the form $O \neq (\mathbf{Z}/2)^{j_1} \subsetneq \cdots \subsetneq (\mathbf{Z}/2)^{j_n}$. Then G acts on this simplicial set by conjugation, and its realization $|A_p(G)|$ has the structure of a finite G-CW complex, in fact of dimension $\mathrm{rank}_p(G)-1$. If $P \leq G$ is a p-subgroup, then this complex satisfies one key property: $|A_p(G)|^P$ is contractible. We can then apply the following theorem, which is a reformulation of a result due to P. Webb [We]:

THEOREM 1.1

If X is a finite G-CW complex such that $X^P \simeq *$ for each p-subgroup $P \leq G$, then the mod p Leray spectral sequence associated to the map $X \underset{G}{\times} EG \to X/G$ satisfies

$$E_2^{p,q} \cong \begin{cases} 0 & \text{for } p > 0 \\ H^q(G, \mathbf{F}_p) & \text{for } p = 0. \end{cases} \qquad \blacksquare$$

Using the usual combinatorial description of the E_1-term of this spectral sequence yields:

COROLLARY 1.2

$$\left(\bigoplus_{\substack{\sigma_i \in (X/G)^{(i)} \\ i \text{ odd}}} H^*(G_{\sigma_i}) \right) \oplus H^*(G) \cong \bigoplus_{\substack{\sigma_i \in (X/G)^{(i)} \\ i \text{ even}}} H^*(G_{\sigma_i})$$

where X may be taken to be $|A_p(G)|$, and $\mathbf{F}_p$ coefficients are used. $\blacksquare$

The isotropy subgroups for the poset space are intersections of normalizers of p-tori. This formula in general can be of no use either because of its size or the fact that G may normalize a p-torus. However, in case G *is simple*, this reduces the calculation of $H^*(G)$ to calculations for proper, local subgroups.

With this reduction in mind, it seems likely that a cohomology calculation will involve analyzing a local subgroup which contains the 2-Sylow subgroup (recall that we take only $p = 2$ in this paper). To understand how to use the cohomology of such a subgroup, the classical double coset formula of Cartan-Eilenberg is often extremely useful [C-E]. We recall how that goes.

Assume $K \subseteq G$ is a subgroup of odd index. Take a double coset decomposition $G = \coprod_{i=1}^{n} K g_i K$. Using a simple transfer-restriction argument, it is easy to see that $H^*(G) \overset{\mathrm{res}_H^G}{\hookrightarrow} H^*(K)$ is injective. Let $c_x^* : H^*(k) \to H^*(xKx^{-1})$ denote the usual isomorphism induced by conjugation. Recall that $\alpha \in H^*(k)$ is said to be *stable* if $\mathrm{res}^H_{H \cap xHx^{-1}}(\alpha) = \mathrm{res}^{xHx^{-1}}_{H\cap xHx^{-1}} \circ c_x^*(\alpha)$ for $x = g_1, \ldots, g_n$. Then we have, for [G:K] odd.

THEOREM 1.3 (Cartan-Eilenberg)

$\alpha \in H^*(K)$ is in im res_H^G if and only if α is stable. $\blacksquare$

Applying the techniques described above it is in principle possible to reduce the calculation of $H^*(G)$ to understanding the cohomology of its local subgroups, together with the stability conditions. At this point one must dispense with formal reductions and calculate the cohomology of the key local subgroups using the spectral sequences associated to the different extensions they may appear in. Aside from the usual difficulties of understanding the differentials, often some very delicate modular invariant theory must be used to understand the E_2-terms or images of different restriction maps.

A combination of these methods will be used to calculate $H^*(M_{12})$, the main example in §2.

§2. COHOMOLOGY CALCULATIONS FOR M_{12}

M_{12} is the Mathieu group of order 95,040. We will give a complete analysis of its cohomology using the methods outlined in §1.

First, we have that $|A_2(M_{12})|/M_{12}$ is the following 2-dimensional cell-complex:

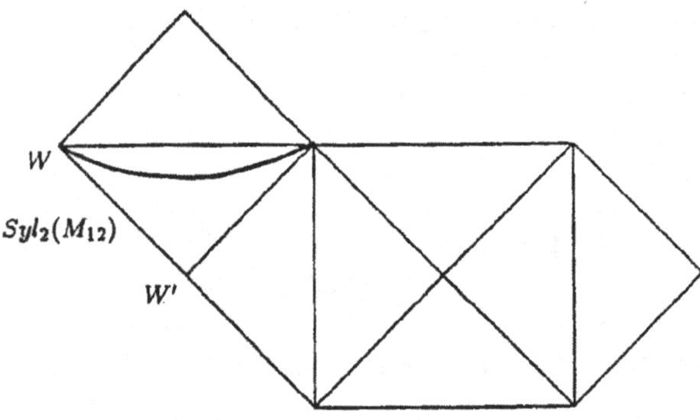

We have only labelled the isotropy on one edge, because 1.2 leads to

Proposition 2.1

Let $H = Syl_2(M_{12})$; then there exist two non-isomorphic subgroups $W, W' \subseteq M_{12}$ of order 192, such that

$$H^*(M_{12}) \oplus H^*(H) \cong H^*(W) \oplus H^*(W'). \qquad \blacksquare$$

Next, we use 1.3 to prove

Proposition 2.2

The image of $\operatorname{res}_H^{M_{12}}$ in $H^*(H)$ is the intersection of the two subalgebras,

$$H^*(M_{12}) = H^*(W) \cap H^*(W'). \qquad \blacksquare$$

In other words, the configuration

$$\begin{array}{rccc} W \cap W' = & H & \hookrightarrow & W' \\ & \downarrow & & \downarrow \\ & W & \hookrightarrow & M_{12} \end{array}$$

completely controls the cohomology of M_{12}.

We now give explicit descriptions of W, W'.

$$W: \qquad 1 \to Q_8 \to W \to \Sigma_4 \to 1$$

a split semidirect product (the holomorph of Q_8).

$$W': \qquad 1 \to (\mathbf{Z}/4 \times \mathbf{Z}/4) \underset{T}{\times} \mathbf{Z}/2 \to W' \to \Sigma_3 \to 1$$

also a split semidirect product.

Although $|H| = 64$, its cohomology is very complicated. It is obtained by using two index two subgroups with accessible cohomology. We have

THEOREM 2.3

$$H^*(Syl_2(M_{12})) \cong \mathbf{F}_2[e_1, s_1, t_1, s_2, t_2, L_3, k_4]/R$$

where R is the set of relations

$$\begin{aligned} &s_1 e_1 = s_2 e_1 = t_1 e_1 = 0,\ s_1^3 = s_1^2 t_1 + s_1 t_2\,, \\ &s_2^2 = s_1 s_2 t_1 + s_2 t_1^2,\ s_1^2 s_2 = t_1 L_3 + s_2 t_1^2 + s_2 t_2\,, \\ &s_1 L_3 = t_1 L_3 + s_1 s_2 t_1 + s_2 t_1^2 + s_2 t_2\,, \\ &s_2 L_3 = s_1 s_2 t_1^2 + s_2 t_1 t_2 + s_2 t_1^3\,, \\ &L_3^2 = t_1 t_2 L_3 + s_1 s_2 t_1^3 + s_2 t_1^2 t_2 + s_2 t_1^4 + e_1^2 k_4 + e_1 t_2 L_3\,. \end{aligned}$$

The Poincaré Series for $H^*(H)$ is given by $p_H(t) = 1/(1-t)^3$. ∎

Similarly, we determine $H^*(W)$, $H^*(W')$ and their intersection in $H^*(H)$. We now describe $H^*(M_{12})$:

DIMENSION	GENERATOR
2	α
3	x, y, z
4	β
5	γ
6	δ
7	ϵ

RELATIONS

$\alpha(x+y+z)=0$	$x^3=\alpha^3x+\alpha\beta x+x\delta$
$xy=\alpha^3+x^2+y^2$	$xz=\alpha^3+y^2$
$x^2y=\alpha^3z+\alpha\beta z+y\delta+\alpha\epsilon$	$yz=\alpha^3+x^2$
$\epsilon x=\beta x^2+\alpha^2x^2$	$\alpha\gamma=\alpha^2y$
$\epsilon y=\alpha^2\delta+\alpha^2y^2+\beta x^2+\beta y^2$	$y\gamma=\alpha y^2$
$\epsilon z=\gamma^2+\alpha^2\delta+\alpha^2x^2+\beta x^2+\beta z^2$	$x\gamma=\alpha^4+\alpha z^2$
$z^4=\gamma\epsilon+x^4+\alpha^4\beta+z^2\delta$	$\epsilon^2=z^3\gamma+\alpha^2\beta\delta+\alpha^5\beta+z\beta\epsilon$
	$+z\delta(\gamma+\alpha z)+\beta^2(\alpha^3+xz+yz)$

The Poincaré Series for $H^*(M_{12})$ is given by

$$p_{M_{12}}(t)=\frac{1+t^2+3t^3+t^4+3t^5+4t^6+2t^7+4t^8+3t^9+t^{10}+3t^{11}+t^{12}+t^{14}}{(1-t^4)(1-t^6)(1-t^7)}. \tag{2.4}$$

§3. AN INTERPRETATION OF THE RESULTS

We now provide a geometric analysis of our calculation of $H^*(M_{12})$. First, note that $\mathbf{F}_2[\beta_4,\delta_6,\epsilon_7]\subseteq H^*(M_{12})$ is a polynomial subalgebra of maximal rank, and in fact the cohomology ring is free and finitely generated over it (this implies that it is Cohen-Macaulay). One can also verify that $Sq^2\beta=\delta$, $Sq^1\delta=\epsilon$, $Sq^4\epsilon=\beta\epsilon$, $Sq^6\epsilon=\delta\epsilon$. On the other hand, let $\mathbf{G_2}$ denote the usual exceptional compact Lie group of automorphisms of the Cayley octaves. A result due to Borel is that

$$H^*(B\mathbf{G_2},\mathbf{F}_2)\cong\mathbf{F}_2[\beta_4,\delta_6,\epsilon_7]$$

with Steenrod operations as above. $\mathbf{G_2}$ is also a 14-dimensional closed manifold. From (2.4) we see that the numerator of $p_{M_{12}}(t)$ looks like the Poincaré Series of a 14-dimensional Poincaré Duality Complex. How does $\mathbf{G_2}$ relate to the group M_{12}?

At this point it is pertinent to mention a technique used by Borel to compute the cohomology of $B\mathbf{G_2}$. Although $\mathbf{G_2}$ has rank 2, it contains a 2-torus of rank 3, denoted by V. Then Borel [Bo] analyzed the spectral sequence for the fibration

$$\begin{array}{ccc}\mathbf{G_2}/V & \hookrightarrow & BV\\ & & \downarrow\\ & & B\mathbf{G_2}\end{array},$$

showing that in fact it collapses, and so

$$p_V(t) = p_{B\mathbf{G_2}}(t) \cdot (P.S.[\mathbf{G_2}/V]).$$

He computes $P.S.[G_2/V] = C_3(t) \cdot C_5(t) \cdot C_6(t)$, where $C_i(t)$ denotes the i-th cyclotomic polynomial.

Now consider, if $V \leq M_{12} \leq \mathbf{G_2}$, it would be reasonable to expect

$$p_{M_{12}}(t) = p_{B\mathbf{G_2}}(t) \cdot (P.S.[\mathbf{G_2}/M_{12}]).$$

Unfortunately, M_{12} is not a subgroup of $\mathbf{G_2}$. However, one does find that $\mathbf{G_2}$ contains a subgroup E of order 1344 which normalizes V, and fits into an extension

$$1 \to V \to E \to \mathrm{GL}_3(\mathbf{F}_2) \to 1 \qquad (3.0)$$

which is *non-split.*

By analogues of Borel's arguments, we have that

$$p_E(t) = p_{B\mathbf{G_2}}(t) \cdot (P.S.[\mathbf{G_2}/E]).$$

Our interest is in M_{12} however, so why does it matter? It matters because of the following result:

THEOREM 3.1

$$H^*(E) \cong H^*(M_{12}) \oplus (H^*(V) \otimes \mathrm{St})^{GL_3(\mathbf{F}_2)},$$

where St *denotes the Steinberg module associated to* $GL_3(\mathbf{F}_2)$. ∎

This result is proved by splitting (3.0) via the Tits Building for $GL_3(\mathbf{F}_2)$, and then reassembling the pieces in the form above. The factor $(H^*(V) \otimes St)^{GL_3(\mathbf{F}_2)}$ is some kind of "error term". What this shows is that the geometry pertaining to $E \subsetneq \mathbf{G_2}$ is transported via cohomology to M_{12}, a rather unexpected occurrence.

The obvious next step is to compare E and M_{12} as groups. They turn out to be close cousins, as manifested by the following result due to Wong [Wo].

THEOREM 3.2

Let G be a finite group with precisely two distinct conjugacy classes of involutions, and such that the centralizer of one of them is isomorphic to $W = Hol(Q_8)$. Then either $G \cong E$ or $G \cong M_{12}$, E as before. ∎

This is a very positive development, as it shows that cohomology measures in a very precise way how close E and M_{12} are from the 2-local point of view. Combined with the previous remarks, we see that group cohomology will inherit geometric properties through a local similarity, with difference measured by a very standard algebraic object. From the geometric point of view, there arises an obvious question: does there exist a 14-dimensional Poincaré Duality complex X with $\pi_1(X) = M_{12}$, and whose mod 2 cohomology has Poincaré Series equal to the numerator of $p_{M_{12}}(t)$? Just as $\mathbf{G_2}$ has important homotopy-theoretic data, so should X.

We now turn to another very unexpected geometric aspect of our calculation. Recall that the cubic tree is defined as the tree whose vertices have valence three — it is clearly infinite. We have

THEOREM 3.3

There exists a group Γ *of automorphism of the cubic tree such that*

(i) there is an extension $1 \to F^{496} \to \Gamma \xrightarrow{\pi} M_{12} \to 1$, *where* F^{496} = *free group on 496 generators, and*

(ii) $\pi^* : H^*(M_{12}, \mathbf{F}_2) \to H^*(\Gamma, \mathbf{F}_2)$ *is an isomorphism.* ∎

Part (i) is a result due to Goldschmidt [G], in fact if we take H, W, W' as in §3, $\Gamma = W \underset{H}{*} W'$. We see then that at the prime $p = 2$, BM_{12} can be modelled using the classifying space of an infinite, virtually-free group. Aside from the relationship to trivalent graphs, it allows us to make the previous geometric considerations for Γ instead of M_{12}. One can then pose the problem of constructing a 14-dim manifold with Γ as its fundamental group and having the required Poincaré Series. The fact that we are now dealing with an amalgamated product may make this considerably easier than for M_{12}.

§4. OTHER SPORADIC SIMPLE GROUPS

The other groups which we are currently considering are M_{22}, M_{23} (Mathieu groups), M^{c}L (the McLaughlin group), J_2, J_3 (the Janko groups) and $O'N$ (the O'Nan group). At this point we have calculated the cohomology of the 2-Sylow subgroups of M_{22} (same for M_{23}) and M^{c}L. For M_{22} we have an odd index subgroup with a three term double coset decomposition, and the stability conditions are clear. We only require some detailed understanding of $\mathbf{F}_2[x_1, x_2, x_3, x_4]^{A_6}$, where $A_6 \subseteq S_{p_4}(\mathbf{F}_2) \cong \Sigma_6$.

For $O'N$, even though the group has order larger than 460 billion, we have calculated its poset space $|A_2(O'N)|$, and used this to show

$$H^*(O'N) \oplus H^*(4^3 \cdot \Sigma_4) \cong H^*(4^3 \cdot GL_3(\mathbf{F}_2)) \oplus H^*(4 \cdot PSL_3(\mathbf{F}_4) \cdot 2)$$

a considerable reduction. Here we are using the Atlas notation for extensions and cyclic groups of order n. Using the Tits Buildings for the Chevalley groups in this formula we have made considerable progress towards a complete calculation (we already have $H^*(PSL_3(\mathbf{F}_4))$ and the cohomology of related extensions [AM]).

For the group $O'N$, the algebra seems to dictate that a 30-dimensional Poincaré Duality complex will play the same rôle as the 14-dimensional one for M_{12}. From the homotopy point of view these are indeed interesting dimensions.

REFERENCES

[AM] A. Adem and R. J. Milgram, "A_5-Invariants and the Cohomology of $L_3(4)$ and Related Extensions", preprint Stanford University 1990.

[AMM] A. Adem, J. Maginnis and R. J. Milgram, "The Geometry and Cohomology of the Mathieu Group M_{12}", J. Algebra, to appear.

[Bo] A. Borel, "La Cohomologie Mod 2 de Certains Espaces Homogènes", Comm. Math. Helv. 27 (1953) 165-197.

[C-E] H. Cartan and S. Eilenberg, *Homological Algebra*, Princeton University Press (1956).

[D-L] E. Dyer and R. Lashof, "Homology of Iterated Loop Spaces", Amer. J. Math. 84 (1962) 35-88.

[G] D. Goldschmidt, "Automorphisms of Trivalent Graphs", Ann. of Math. 111 (1980) 377-406.

[Q1] D. Quillen, "On the Cohomology and K-Theory of the General Linear Groups over a Finite Field", Ann. Math. 96 (1972) 552-586.

[Q2] D. Quillen, "The Spectrum of an Equivariant Cohomology Ring I", Ann. Math. 94 (1971) 549-572.

[Q3] D. Quillen, "Homotopy Properties of the Poset of Non-trivial p-subgroups of a Group", Adv. Math. 28 (1978) 101-128.

[We] P. Webb, "A Local Method in Group Cohomology", Comm. Math. Helv. 62 (1987) 135-167.

[Wo] W. Wong, "A Characterization of the Mathieu Group M_{12}", Math. Z. 84 (1964) 378-388.

Mathematics Department
University of Wisconsin
Madison, WI 53706
U.S.A.

1990 Barcelona Conference
on Algebraic Topology.

RESOLUTIONS AND POINCARÉ DUALITY FOR FINITE GROUPS

D. J. BENSON

Abstract

This talk is a survey of some recent joint work with Jon Carlson on cohomology of finite groups. I shall describe how, for an arbitrary finite group G, one can produce an algebraic analogue of a free G-action on a product of spheres. If k is the field of coefficients, one can use this to build a resolution of k as a kG-module, which consists of a finite Poincaré duality piece and a polynomial piece. This resolution has the same rate of growth as the minimal resolution, but in general is not quite minimal. The deviation from minimality is measured by secondary operations in group cohomology expressible in terms of matric Massey products.

1 Introduction

This talk is intended as a survey of some recent joint work with Jon Carlson [1,2,3] on cohomology of finite groups. Let me begin with an example.

Let G be the quaternion group Q_8 of order eight, and k be the field of two elements. Then it is well known that

$$H^*(G,k) = k[x,y,z]/(x^2+xy+y^2, x^3, y^3)$$

with $\deg(x) = \deg(y) = 1$ and $\deg(z) = 4$, so that we have

n	0	1	2	3	4	5	6	7	8	$\cdots$
$\dim_k H^n(G,k)$	1	2	2	1	1	2	2	1	1	$\cdots$

Thus the cohomology consists of a finite symmetric piece in degrees zero to three, and a polynomial generator in degree four.

There is an easy topological interpretation of this structure, as follows. We can regard the quaternion group as a subgroup of the multiplicative group of unit quaternions. These unit quaternions form a 3-sphere S^3, and so G has a free action on S^3 by left multiplication. If we choose an equivariant CW-decomposition of S^3 and take cellular chains with coefficients in k, we obtain a complex of free kG-modules of length four, with homology one copy of k at the beginning and the end. Thus we may take the Yoneda splice of this complex with itself infinitely often to obtain a projective resolution of k as a kG-module. Taking cohomology, we obtain

$$H^*(G,k) = H^*(S^3/G,k) \otimes k[z],$$

where z is a polynomial generator in degree four. Since S^3/G is a compact manifold, its cohomology satisfies Poincaré duality, which explains the symmetry in the first four degrees.

What I wish to outline here is how this picture can be generalised to all finite groups. Given a finite group G and a field k of characteristic p dividing the order of G (otherwise there is no cohomology), there is a construction of a finite Poincaré duality complex which is a sort of algebraic analogue of a free action of G on a product of spheres. This gives rise to a projective resolution of k as a kG-module which consists of a finite Poincaré duality part and a "polynomial" part. The number of polynomial generators required is equal to the number of spheres; this is the p-rank $r_p(G)$ (the maximal rank of an elementary abelian p-subgroup of G), in accordance with a theorem of Quillen [9,10] on the Krull dimension of the cohomology ring. The resolution we construct has the same polynomial rate of growth as the minimal resolution, but in general it is not minimal. The deviation from minimality is measured by certain secondary operations in group cohomology which are like Massey products. These secondary operations only appear in case the cohomology is not Cohen-Macaulay, so that in case the ring is Cohen-Macaulay we can read off from the Poincaré duality statement a functional equation for the Poincaré series. Thus we have the following theorem from Benson and Carlson [3].

Theorem 1.1 *Suppose that the cohomology ring $H^*(G,k)$ is Cohen-Macaulay. Then the Poincaré series $P_k(t) = \sum_{i\geq 0} t^i \dim_k H^i(G,k)$ satisfies the functional equation*

$$P_k(1/t) = (-t)^{r_p(G)} P_k(t).$$

If $H^*(G,k)$ is not Cohen-Macaulay, we still obtain information in terms of a spectral sequence converging to a finite Poincaré duality ring, and in which the differentials are determined by the secondary operations mentioned above.

2 Resolutions and complexity

We begin with a gentle introduction to the cohomology of finite groups, to set the scene. Let G be a finite group, and k be a field. We think of finitely generated kG-modules as being the same thing as representations of G as matrices with entries in k. Recall that Maschke's theorem says that if the characteristic of k does not divide the order of G then every short exact sequence

$$0 \to M' \to M'' \to M \to 0$$

of kG-modules splits. If the characteristic of k is a prime p dividing the order of G, then there are always counterexamples to Maschke's theorem, and this is one of the fundamental differences between ordinary and modular representation theory. A module M is projective if Maschke's theorem holds for short exact sequences ending in M, and M' is injective if Maschke's theorem holds for short exact sequences beginning with M'.

In fact in this context of finite group algebras over a field, a module is projective if and only if it is injective.

If M is not projective, we can measure the extent to which it "glues" to other modules in the above sense by forming a projective resolution; namely an exact sequence

$$\cdots \to P_2 \to P_1 \to P_0 \to M \to 0$$

in which all the P_i are projective. There is a unique minimal resolution for a given module M, and all other resolutions may be obtained from this one by adding on a split exact sequence of projective modules. The nth kernel in the minimal resolution is denoted $\Omega^n(M)$.

If M' is another module we can take homomorphisms

$$\mathrm{Hom}_{kG}(P_0, M') \to \mathrm{Hom}_{kG}(P_1, M') \to \mathrm{Hom}_{kG}(P_2, M') \to \cdots$$

and define $\mathrm{Ext}^i_{kG}(M, M')$ to be the kernel of the $(i+1)$st map in this sequence modulo the image of the ith map. This is independent of choice of resolution. We define $H^i(G, M)$ to be $\mathrm{Ext}^i_{kG}(k, M)$. This is a finitely generated graded commutative k-algebra in which the multiplication may be described in terms either of cup products or Yoneda compositions.

Let us illustrate with an example. Let $G = GL_3(2)$ be the simple group of order 168 and k be the field of two elements. Then there are four simple kG-modules k, M, N, St of dimensions 1, 3, 3, 8 respectively. The Steinberg module St is projective, while the projective covers P_k, P_M, P_N have the following structures.

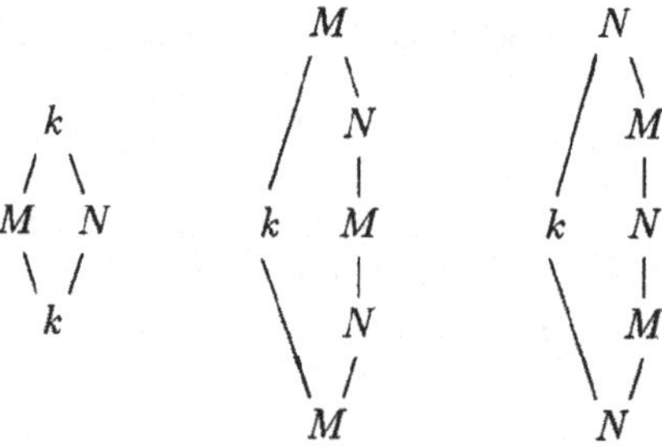

This means P_k has a unique top and bottom composition factor each isomorphic to k, and $\mathrm{Rad}(P_k)/\mathrm{Soc}(P_k) \cong M \oplus N$. Similarly, P_M has a unique top and bottom composition factor each isomorphic to M, and $\mathrm{Rad}(P_M)/\mathrm{Soc}(P_M)$ is a direct sum of k and a uniserial module (i.e., a module with a unique composition series) with composition factors N, M, N.

To obtain a projective resolution of k, we begin with P_k, so that the first kernel is $\Omega(k) = \begin{smallmatrix} M & & N \\ & \diagdown\ \diagup & \\ & k & \end{smallmatrix}$. Thus $\Omega(k)/\mathrm{Rad}\Omega(k) \cong M \oplus N$, so that the projective cover of $\Omega(k)$ is $P_M \oplus P_N$. The next kernel is obtained by cutting and pasting the diagrams for P_M and

P_N as follows.

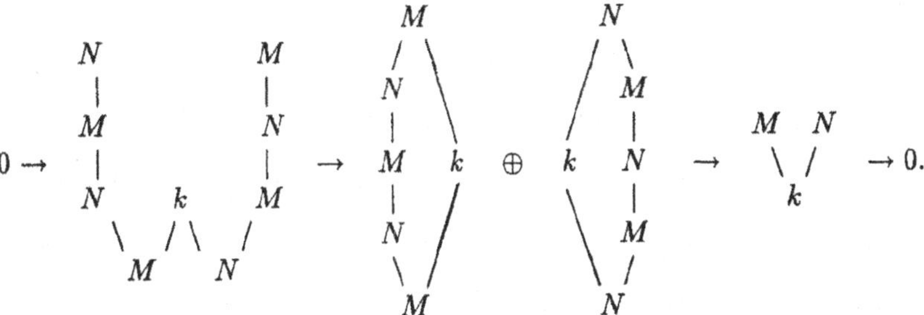

Thus the left-hand module in this sequence is $\Omega^2(k)$. This module modulo its radical is $N \oplus k \oplus M$, and so the projective cover is $P_N \oplus P_k \oplus P_M$ with kernel

$$\Omega^3(k) = \begin{array}{ccccccc} k & & M & & N & & k \\ & \backslash & / \; \backslash & & / \; \backslash & / & \\ & N & & k & & M & \end{array}$$

and so on. At this stage, the reader must presumably have started to wonder about the exact meanings of the diagrams we are drawing. I do not wish to explain that in detail here, but the method is explained in detail in Benson and Carlson [1]. The reader may also get the impression that to calculate an entire projective or injective resolution this way would take infinitely much work. However, after a while the patterns start "repeating." This is formalised, at least for groups of p-rank at most two, in Section 9 of Benson and Carlson [1].

In the above example, the number of summands of the projective modules in the minimal resolution

$$\cdots \to P_k \oplus P_M \oplus P_N \oplus P_k \to P_N \oplus P_k \oplus P_M \to P_M \oplus P_N \to P_k \to k \to 0$$

is growing linearly. Because cohomology of a finite group with coefficients in a finitely generated module is finitely generated (Evens [5]), minimal resolutions always have "polynomial growth" in the following sense. If P_n is the nth module in a minimal resolution of M, we form the Poincaré series

$$f(t) = \sum_{n=0}^{\infty} t^n \dim_k P_n.$$

Then $f(t)$ is the power series expansion of a rational function of t of the form

$$(\text{polynomial with integer coefficients}) / \prod_{i=1}^{r} (1 - t^{k_i}).$$

If $f(t)$ has a pole of order c at $t = 1$ then for some positive real constant μ we have

$$\dim_k P_n \leq \mu n^{c-1},$$

but the exponent $c - 1$ may not be decreased to $c - 1 - \varepsilon$ for any positive value of ε. We say that the integer c is the **complexity** of the module M.

If we look at the above example more closely, we see that the given resolution is the total complex of a double complex of the form

$$\begin{array}{ccccccccc}
\vdots & & \vdots & & \vdots & & \vdots & & \\
\downarrow & & \downarrow & & \downarrow & & \downarrow & & \\
P_k & \leftarrow & P_N & \leftarrow & P_M & \leftarrow & P_k & \leftarrow & \cdots \\
\downarrow & & \downarrow & & \downarrow & & \downarrow & & \\
P_N & \leftarrow & P_M & \leftarrow & P_k & \leftarrow & P_N & \leftarrow & \cdots \\
\downarrow & & \downarrow & & \downarrow & & \downarrow & & \\
P_M & \leftarrow & P_k & \leftarrow & P_N & \leftarrow & P_M & \leftarrow & \cdots \\
\downarrow & & \downarrow & & \downarrow & & \downarrow & & \\
P_k & \leftarrow & P_N & \leftarrow & P_M & \leftarrow & P_k & \leftarrow & \cdots
\end{array}$$

Thus there are vertical and horizontal differentials d_0 and d_1 squaring to zero and anti-commuting, so that the total differential $d_0 + d_1$ squares to zero.

More generally, a c-fold **multiple complex** is like a double complex, except that it has c different directions (vertical, horizontal, ...) and differentials $d_0, \dots, d_{c-1}$ which anticommute and square to zero, so that the total differential, namely the sum of the d_i, squares to zero. The following theorem is proved in Benson and Carlson [2].

Theorem 2.1 *Let M be a finitely generated kG-module of complexity c. Then there is a resolution of M which is the total complex of a c-fold multiple complex in which each row, column, etc. is periodic.*

Note that the growth rate of the total complex of such a c-fold multiple complex is c, so that this is the same as the growth rate of the minimal resolution. In fact, the resolution produced in the proof of the above theorem is a tensor product of periodic complexes, so that the minimal resolution is usually not of the form produced here. In all the examples we have looked at, one can find a c-fold multiple complex whose total complex is the minimal resolution, but one must allow the rows, columns, etc. to be eventually periodic rather than strictly periodic. It is unclear whether one should expect to be able to prove a theorem of this sort.

3 The construction

It is worth sketching the construction used in the proof of the theorem at the end of the last section, because this provides us with the algebraic product of spheres with free G-action mentioned in the introduction.

If M is a finitely generated kG-module, we have a map

$$\mathrm{Ext}^*_{kG}(k,k) \xrightarrow{\otimes M} \mathrm{Ext}^*_{kG}(M,M).$$

given by tensoring with M. Note that while $\mathrm{Ext}^*_{kG}(M,M)$, as a ring under Yoneda composition, is not in general graded commutative (even is M is simple it may have complete matrix rings as quotients), the image of this map is in the graded centre, and by the theorem of Evens [5], $\mathrm{Ext}^*_{kG}(M,M)$ is finitely generated as a module over its image. Thus by Noether's normalisation lemma, one can find a polynomial subring $k[\zeta_1,\dots,\zeta_c]$ of $\mathrm{Ext}^*_{kG}(k,k)$ with the following properties.

(i) The above map is injective on the subring $k[\zeta_1,\dots,\zeta_c]$.

(ii) $\mathrm{Ext}^*_{kG}(M,M)$ is finitely generated as a module over the image of $k[\zeta_1,\dots,\zeta_c]$.

(iii) The elements $\zeta_1,\dots,\zeta_c$ are homogeneous (i.e., live in a single degree, say ζ_i lives in degree m_i).

The number c is the **Krull dimension** of $\mathrm{Ext}^*_{kG}(M,M)$ as a module over $\mathrm{Ext}^*_{kG}(k,k)$, and in fact this is equal to the complexity of M as a kG-module (Carlson [4]).

For each generator $\zeta \neq 0$ (say of degree m) of this polynomial ring, we perform the following construction. Let

$$\cdots \to P_2 \to P_1 \to P_0 \to k \to 0$$

be the minimal resolution of k as a kG-module. Then ζ is represented by a cocycle $P_m \to k$ which is surjective and vanishes on the image of $P_{m+1} \to P_m$. It may therefore be regarded as a surjective map $\hat{\zeta} : \Omega^m(k) \to k$ (which is uniquely determined since we are working with the minimal resolution). We denote the kernel of this map by L_ζ. Thus L_ζ may be regarded as a submodule of P_{m-1} via the inclusion of $\Omega^m(k)$ in P_{m-1}, and we obtain a diagram of the following form, in which the rows are exact.

$$\begin{array}{ccccccccccc}
 & 0 & & 0 & & & & & & & \\
 & \downarrow & & \downarrow & & & & & & & \\
 & L_\zeta & = & L_\zeta & & & & & & & \\
 & \downarrow & & \downarrow & & & & & & & \\
0 \to & \hat{\Omega}^m k & \to & P_{m-1} & \to & P_{m-2} & \to \cdots \to & P_0 & \to & k & \to 0 \\
 & \downarrow \hat{\zeta} & & \downarrow & & \downarrow & & \downarrow & & \| & \\
0 \to & k & \to & P_{m-1}/L_\zeta & \to & P_{m-2} & \to \cdots \to & P_0 & \to & k & \to 0. \\
 & \downarrow & & \downarrow & & & & & & & \\
 & 0 & & 0 & & & & & & &
\end{array}$$

We denote by $\mathbf{C}_\zeta$ the chain complex

$$0 \to P_{m-1}/L_\zeta \to P_{m-2} \to \cdots \to P_0 \to 0$$

formed by truncating the bottom row of this diagram. Thus we have

$$H_i(\mathbf{C}_\zeta) \cong \begin{cases} k & \text{if } i = 0,\ m-1 \\ 0 & \text{otherwise.} \end{cases}$$

We write $\tilde{\zeta}$ for the generator of degree $m-1$, and 1 for the generator of degree zero.

The complex $\mathbf{C}_\zeta$ should be thought of as a sort of algebraic analogue of a sphere with G-action, with ζ being the transgression of the fundamental class of the sphere.

We also write $\mathbf{C}_\zeta^{(\infty)}$ for the chain complex

$$\cdots \to P_1 \to P_0 \to P_{m-1}/L_\zeta \to P_{m-2} \to \cdots \to P_1 \to P_0 \to 0$$

obtained by splicing together infinitely many copies of $\mathbf{C}_\zeta$ in positive degree. It is an exact complex except in degree zero, where the homology is k.

Theorem 3.1 *If $\zeta_1, \ldots, \zeta_c$ are chosen as above, then the complex*

$$\mathbf{C}_{\zeta_1} \otimes \mathbf{C}_{\zeta_2} \otimes \cdots \otimes \mathbf{C}_{\zeta_c} \otimes M$$

is a finite complex of projective modules, whose homology consists of 2^c copies of the module M.

Note that since the tensor product of a projective module with any module is again projective, there is only one module in the above complex which might conceivably not be projective, namely the module

$$P_{m_1-1}/L_{\zeta_1} \otimes \cdots \otimes P_{m_c-1}/L_{\zeta_c} \otimes M.$$

There are two available proofs for the projectivity of this module. One involves the machinery of varieties for modules, and involves showing that the choice of $\zeta_1, \ldots, \zeta_c$ means that the hypersurfaces defined by these elements intersected with the variety of M give the zero variety, and then observing that modules with the zero variety are projective. This proof may be found in Benson and Carlson [2]. The other available proof uses the hypercohomology spectral sequence and Evens' finite generation theorem. This may be found in Benson and Carlson [3].

It follows from the above proposition that the complex

$$\mathbf{C}_{\zeta_1}^{(\infty)} \otimes \cdots \otimes \mathbf{C}_{\zeta_c}^{(\infty)} \otimes M$$

consists of projective modules, and by the Künneth theorem it is exact everywhere except in degree zero, where its homology is M. In other words, it is a projective resolution of M, which is c-fold multiply periodic in the sense of Theorem 2.1. Thus it has the same polynomial growth rate as the minimal resolution.

4 Poincaré duality

Now let us concentrate on the case of the trivial module k. By a theorem of Quillen [9,10], the Krull dimension of $H^*(G,k)$ is $r = r_p(G)$, the maximal rank of an elementary abelian p-subgroup of G. Thus the construction of the last section tells us that we should choose a polynomial subring $k[\zeta_1, \ldots, \zeta_r]$ over which $H^*(G,k)$ is finitely generated as a module. Let $\mathbf{C}$ denote the complex $\mathbf{C}_{\zeta_1} \otimes \cdots \otimes \mathbf{C}_{\zeta_r}$. Then $\mathbf{C}$ is a Poincaré duality complex in the sense that it is chain homotopy equivalent to its dual, suitably shifted in degree. More precisely, the following theorem is proved in Benson and Carlson [3].

Theorem 4.1 *The complex* **C** *is a direct sum of an exact sequence of projective modules* **P**$'$ *and a complex* **C**$'$ *having no such summands, and*

$$\mathbf{C}' \cong \mathbf{Hom}_k(\mathbf{C}', k)[s],$$

where $s = \sum_{i=1}^{r}(\deg \zeta_i - 1)$ *and* $[s]$ *denotes a shift in degree by* s.

Of course, before the removal of P', the complex **C** is not isomorphic to its shifted dual in the above sense, because the modules at one end are much larger than those at the other. But the above theorem is equivalent to saying that **C** is chain homotopy equivalent to its shifted dual. In this form, the statement remains true for more general commutative rings of coefficients.

5 A spectral sequence

The chain complex **C** gives rise to a spectral sequence as follows. Let **P** be a resolution of k as a kG-module, and let C be the chain complex described in the last section. We examine the spectral sequence of the double complex $\mathrm{Hom}_{kG}(\mathbf{P} \otimes \mathbf{C}, M)$, for a kG-module M. If we look at the spectral sequence in which the horizontal differential is performed first, we find that the E_1 page consists of the complex $\mathbf{Hom}_{kG}(\mathbf{C}, M)$ up the left-hand wall, and zero elsewhere. It follows that the spectral sequence converges to $H^*(\mathbf{Hom}_{kG}(\mathbf{C}, M))$. On the other hand, if we perform the vertical differential first, then the E_1 page is $\mathrm{Hom}_{kG}(H_*(\mathbf{C}) \otimes \mathbf{P}, M)$. Thus the E_2 page takes the form

$$\mathrm{Ext}^*_{kG}(H_*(\mathbf{C}), M) \cong H^*(G, M) \otimes \Lambda^*(\tilde{\zeta}_1, \ldots, \tilde{\zeta}_r).$$

Thus the spectral sequence takes the form

$$H^*(G, M) \otimes \Lambda^*(\tilde{\zeta}_1, \ldots, \tilde{\zeta}_r) \Rightarrow H^{p+q}(\mathbf{Hom}_{kG}(\mathbf{C}, M)).$$

Some of the differentials in this spectral sequence are given as follows. If $\alpha \in H^*(G, M)$, then we have

$$d_{n_i}(\alpha\tilde{\zeta}_i) = \alpha\zeta_i.$$

Now if the ζ_i form a regular sequence for $H^*(G, M)$ as a module over $H^*(G, k)$ (for example, this happens in case $M = k$ and $H^*(G, k)$ is a Cohen-Macaulay ring), then these differentials determine all the differentials, and the E_∞ page is $H^*(G, M)/(\zeta_1, \ldots, \zeta_r)$ concentrated along the bottom line. So in this case, we have

$$H^*(G, M)/(\zeta_1, \ldots, \zeta_r) \cong H^*(\mathbf{Hom}_{kG}(\mathbf{C}, M)).$$

So for example if $H^*(G, k)$ is a Cohen-Macaulay ring (see for example Matsumura [7] or Serre [11] for standard properties of Cohen-Macaulay rings), then the Poincaré series of the cohomology ring $P_k(t) = \sum_{i \geq 0} t^i \dim_k H^i(G, k)$ takes the form $P_k(t) =$ (polynomial of degree s with symmetric coefficients)$/\prod_{i=1}^{r}(1 - t^{n_i})$. This is easily seen to imply the functional equation given in Theorem 1.1.

Many cohomology rings are Cohen-Macaulay, but for an example of a cohomology ring which is not Cohen-Macaulay, see for example Evens and Priddy [6].

6 Secondary operations

In general, in the spectral sequence of the last section, not all differentials are described by the formula

$$d_{n_i}(\alpha\tilde{\zeta}_i) = \alpha\zeta_i.$$

For example, if $\alpha \in H^*(G,M)$ with $\alpha\zeta_1 = \alpha\zeta_2 = 0$, then we have not yet described $d_{n_1+n_2-1}(\alpha\tilde{\zeta}_1\tilde{\zeta}_2)$. This is given by a certain secondary operation related to the Massey triple product, and defined as follows. Choose cocycles u, and η_i with $\alpha = [u]$ and $\zeta_i = [\eta_i]$. Then $u\eta_1$ is a coboundary, say $u\eta_1 = du_1$, and similarly $u\eta_2 = du_2$. Also, since $H^*(G,k)$ is graded commutative, we may write

$$\eta_1\eta_2 - (-1)^{n_1n_2}\eta_2\eta_1 = d\eta_{12}.$$

It is easily seen that

$$u_1\eta_2 - (-1)^{n_1n_2}u_2\eta_1 - (-1)^{\deg\alpha}u\eta_{12}$$

is a cocycle, and we denote its cohomology class by $\langle\alpha|\zeta_1,\zeta_2\rangle$. This is well defined modulo the submodule of $H^*(G,M)$ generated by ζ_1 and ζ_2. The differential asked for above is now given by

$$d_{n_1+n_2-1}(\alpha\tilde{\zeta}_1\tilde{\zeta}_2) = \langle\alpha|\zeta_1,\zeta_2\rangle.$$

There are similar definitions of secondary operations $\langle\alpha|\zeta_1,\ldots,\zeta_j\rangle$, defined when the secondary operations involving proper subsets of the ζ's vanish, and defined modulo the images of the previous differentials, with the property that

$$d_{n_1+\cdots+n_j-j+1}(\alpha\tilde{\zeta}_1\ldots\tilde{\zeta}_j) = \langle\alpha|\zeta_1,\ldots,\zeta_j\rangle.$$

When I gave this talk in Sant Feliu, Doug Ravenel was in the audience, and pointed out that in fact the above secondary operations can be expressed in terms of the usual matric Massey products [8]. For example, the first two cases are given by

$$\begin{aligned}\langle\alpha|\zeta_1,\zeta_2\rangle &= \left(\alpha,(\zeta_1,\zeta_2),\begin{pmatrix}\zeta_2\\-\zeta_1\end{pmatrix}\right)\\ \langle\alpha|\zeta_1,\zeta_2,\zeta_3\rangle &= \left(\alpha,(\zeta_1,\zeta_2,\zeta_3),\begin{pmatrix}0&-\zeta_3&\zeta_2\\ \zeta_3&0&-\zeta_1\\ -\zeta_2&\zeta_1&0\end{pmatrix},\begin{pmatrix}\zeta_1\\ \zeta_2\\ \zeta_3\end{pmatrix}\right).\end{aligned}$$

References

[1] D. J. Benson and J. F. Carlson. *Diagrammatic methods for modular representations and cohomology.* Comm. in Algebra 15 (1987), 53–121.

[2] D. J. Benson and J. F. Carlson. *Complexity and multiple complexes.* Math. Zeit. 195 (1987), 221–238.

[3] D. J. Benson and J. F. Carlson. *Projective resolutions and Poincaré duality complexes.* Submitted to Trans. A.M.S.

[4] J. F. Carlson. *Complexity and Krull dimension.* Representations of Algebras, Puebla, Mexico 1980. Springer Lecture Notes in Mathematics 903, 62–67, Springer-Verlag, Berlin/New York 1981.

[5] L. Evens. *The cohomology ring of a finite group.* Trans. Amer. Math. Soc. 101 (1961), 224–239.

[6] L. Evens and S. Priddy. *The cohomology of the semi-dihedral group.* Conf. on Algebraic Topology in honour of Peter Hilton, ed. R. Piccinini and D. Sjerve, Contemp. Math. 37, A.M.S. 1985.

[7] H. Matsumura. *Commutative ring theory.* C.U.P., 1986.

[8] J. P. May. Matric Massey products. J. Algebra 12 (1969), 533–568.

[9] D. Quillen. *The spectrum of an equivariant cohomology ring, I.* Ann. of Math. 94 (1971), 549–572.

[10] D. Quillen. *The spectrum of an equivariant cohomology ring, II.* Ann. of Math. 94 (1971), 573–602.

[11] J.-P. Serre. *Algèbre locale—multiplicités.* Springer Lecture Notes in Mathematics 11, Springer-Verlag, Berlin/New York, 1965.

Mathematical Institute
24–29 St. Giles
Oxford OX1 3LB
Great Britain

1990 Barcelona Conference
on Algebraic Topology.

GROUPS AND SPACES WITH ALL LOCALIZATIONS TRIVIAL

A.J. BERRICK AND CARLES CASACUBERTA

0 Introduction

The *genus* of a finitely generated nilpotent group G is defined as the set of isomorphism classes of finitely generated nilpotent groups K such that the p-localizations K_p, G_p are isomorphic for all primes p [19]. This notion turns out to be particularly relevant in the study of non-cancellation phenomena in group theory and homotopy theory.

In the above definition, the restriction of finite generation is imposed in order to prevent the genera from becoming too large—in fact, with that restriction, genera are always finite sets. Nevertheless, it is perfectly possible to deal with the so-called *extended genus*, in which the groups involved, though still nilpotent, are no longer asked to be finitely generated. This generalization has been found to be useful [9,10,15].

More serious difficulties arise in this context if one attempts to remove the hypothesis of nilpotency. Given any family of idempotent functors $\{E_p\}$ in the category of groups, one for each prime p, extending p-localization of nilpotent groups, one could expect to find groups G such that $E_pG = 1$ for all primes p, that is, belonging to the "genus" of the trivial group. In fact, as shown below, there even exist groups G sharing this property for *every* family $\{E_p\}$ chosen. We call such groups *generically trivial.* In Sections 1 and 2 we exhibit their basic properties and point out several sources of examples.

A group G is called *separable* if for *some* family of idempotent functors $\{E_p\}$ as above, the canonical homomorphism from G to the cartesian product of the groups E_pG is injective. Residually nilpotent groups and many others are separable. In Section 3, we observe that acyclic spaces X with generically trivial fundamental group are relevant because the space $\mathrm{map}_*(X,Y)$ of pointed maps from X to Y is weakly contractible for a very broad class of spaces Y, namely for all those Y such that $\pi_1(Y)$ is separable.

Acyclic spaces X with generically trivial fundamental group deserve to be called *generically trivial* spaces, because, for every family of idempotent functors E_p in the pointed homotopy category of connected CW-complexes extending p-localization of nilpotent CW-complexes, the spaces E_pX are contractible for all p. Familiar examples of generically trivial spaces include all acyclic spaces whose fundamental group is finite.

Note that nilpotent groups (and spaces) form together a class to which p-localization extends in a unique way. The latter provide the obstruction to the existence of localization-completion pullback squares for general groups and spaces.

Finally, Section 4 is devoted to the problem of recognizing generically trivial groups by inspecting their structure. This is in fact rather difficult, and linked to the problem of determining in general the kernel of the universal homomorphism from a given group G to a group in which p'-roots exist and are unique [22].

Acknowledgments. We are both sincerely indebted to the CRM of Barcelona for its hospitality. Some parts of our exposition owe much to discussions with George Peschke.

1 Separable and generically trivial groups

Let P be a set of primes and P' denote its complement. A group G is called *P-local* [16,20] if the map $x \mapsto x^n$ is bijective in G for all positive integers n whose prime divisors lie in P' (written $n \in P'$ for simplicity). For every group G there is a universal homomorphism $l : G \to G_P$ with G_P P-local [13,20,21], which is called *P-localization.* If the set P consists of a single prime p, then we usually write G_p instead of G_P. The properties of the P-localization homomorphism are particularly well understood when G is nilpotent [16].

A group G is called *separable* [22] if the canonical map from G to the cartesian product of its p-localizations

$$\gamma \colon G \longrightarrow \prod_p G_p \tag{1.1}$$

is injective. It is well-known that nilpotent groups are separable [16]. The class of separable groups is in fact much larger. It also contains all groups which are p-local for some prime p, and it is closed under taking subgroups and forming cartesian products; cf. [22, Proposition 6.10]. Thus it is closed under small (inverse) limits. In particular, since every residually nilpotent group embeds in a cartesian product of countably many nilpotent groups, we have

Proposition 1.1 *Residually nilpotent groups are separable.* □

If a group G is not separable, then one cannot expect to recover full information about G from the family of its p-localizations G_p. The worst possible situation occurs, of course, when all these vanish. We introduce new terminology to analyze this case.

Definition 1.2 A group G is called *generically trivial* if $G_p = 1$ for all primes p.

As next shown, it turns out that such groups cannot be detected by *any* idempotent functor extending p-localization of nilpotent groups to all groups. The basic facts about idempotent functors and localization in arbitrary categories are explained in [1,13].

Lemma 1.3 *Assume given a set of primes P and an idempotent functor E in the category of groups such that $E\mathbf{Z} \cong \mathbf{Z}_P$, the integers localized at P. Then, for every group G and every $n \in P'$, the map $x \mapsto x^n$ is bijective in EG.*

Proof. This is in fact a stronger form of a result in [13]. Fix an integer $n \in P'$ and denote by $\rho_n : \mathbf{Z} \to \mathbf{Z}$ the multiplication by n. For a group K, the function $(\rho_n)^* : \mathrm{Hom}(\mathbf{Z}, K) \to \mathrm{Hom}(\mathbf{Z}, K)$ corresponds precisely to the nth power map $x \mapsto x^n$ in K

under the obvious bijection $\mathrm{Hom}(\mathbf{Z}, K) \cong K$. Thus we have to prove that $(\rho_n)^*$ is a bijection when $K = EG$ for some G; in other words, that ρ_n is an E-equivalence. By assumption, there is a commutative diagram

$$\begin{array}{ccccc} \mathbf{Z} & \xrightarrow{\eta} & E\mathbf{Z} & \xrightarrow{\cong} & \mathbf{Z}_P \\ {\scriptstyle \rho_n}\downarrow & & {\scriptstyle E\rho_n}\downarrow & & {\scriptstyle \psi}\downarrow \\ \mathbf{Z} & \xrightarrow{\eta} & E\mathbf{Z} & \xrightarrow{\cong} & \mathbf{Z}_P, \end{array} \tag{1.2}$$

in which η denotes the natural transformation associated to E. Here η cannot be identically zero, because there is a non-trivial homomorphism from $\mathbf{Z}$ to an E-local group—namely, $\mathbf{Z}_P$. Thus, if $x \in \mathbf{Z}_P$ is the image of $1 \in \mathbf{Z}$ under the top composition in (1.2), then $x \neq 0$ and $\psi(x) = nx$. It follows that ψ is multiplication by n and hence an isomorphism. This implies that $E\rho_n$ is also an isomorphism, i.e. that ρ_n is an E-equivalence, as desired. □

Theorem 1.4 (a) *A group G is generically trivial if and only if $E_pG = 1$ for every prime p and every idempotent functor E_p in the category of groups satisfying $E_p\mathbf{Z} \cong \mathbf{Z}_p$.*

(b) *A group G is separable if and only if the canonical homomorphism $G \to \prod_p E_pG$ is injective for some family $\{E_p\}$ of idempotent functors in the category of groups, one for each prime p, satisfying $E_p\mathbf{Z} \cong \mathbf{Z}_p$.*

Proof. One implication is trivial in both part (a) and part (b). To prove the converse in (a), note that, for every prime p and every choice of E_p, the homomorphism $G \to 1$ is an E_p-equivalence when $G_p = 1$, because every homomorphism $\varphi\colon G \to K$ with K E_p-local factorizes through G_p by Lemma 1.3. To prove the converse in (b), use again Lemma 1.3 to obtain, for every family $\{E_p\}$, a factorization

$$G \xrightarrow{\gamma} \prod_p G_p \to \prod_p E_pG. \qquad \square$$

Corollary 1.5 *Generically trivial groups are perfect.*

Proof. Choose E_p to be Bousfield's $H\mathbf{Z}_p$-localization [8]. It follows from Theorem 1.4 that, if G is generically trivial, then $H_1(G; \mathbf{Z}_p) = 0$ for all primes p, and hence $H_1(G; \mathbf{Z}) = 0$. Actually, there is another argument available: It suffices to apply Corollary 2.1 below to the projection of a generically trivial group G onto its abelianization $G \twoheadrightarrow G/[G, G]$, using the fact that a generically trivial abelian group is necessarily trivial. □

In general, a perfect group need not be generically trivial. There are indeed perfect groups which are locally free [3, Lemma 3.1] and hence separable [22, §9]. Further, in Section 4 below, we present an example of a countable perfect group whose localizations contain the localizations of all finitely generated nilpotent groups. However, as next shown, finite perfect groups are generically trivial.

Proposition 1.6 *For a finite group G, the following assertions are equivalent:*

(a) *G is generically trivial;*

(b) *G is perfect;*

(c) *for every prime p, G is generated by p'-torsion elements.*

PROOF. The implications (c)⇒(a)⇒(b) hold for all groups G. If G is finite, then, for each set of primes P, $l: G \to G_P$ is an epimorphism onto a P-group, and $\operatorname{Ker} l$ is generated by the set of P'-torsion elements of G; cf. [22, §7]. This shows that (a)⇒(c). To prove that (b)⇒(a), observe that, given a prime p, G_p is perfect because it is a homomorphic image of G, and also nilpotent because it is a finite p-group. This forces $G_p = 1$. □

The implication (a)⇒(c) in Proposition 1.6 still holds if we assume G locally finite, by Lemma 2.5 below. Its failure to be true for arbitrary groups is discussed in Section 4, cf. Theorem 4.7.

One of the most characteristic features of generically trivial groups G is that homomorphisms $\varphi: G \to K$ are trivial for a broad class of groups K. It follows from Corollary 1.5 that this happens whenever K is residually nilpotent. More generally,

Proposition 1.7 *A group G is generically trivial if and only if every homomorphism $\varphi: G \to K$ with K separable is trivial.*

PROOF. If G is generically trivial, then the composition

$$G \xrightarrow{\varphi} K \xrightarrow{\gamma} \prod_p K_p \overset{\text{proj}}{\twoheadrightarrow} K_p$$

is trivial for all p. Hence, $\gamma(\varphi(x)) = 1$ for every $x \in G$, and, if γ is a monomorphism, then $\varphi(x) = 1$ for every $x \in G$. To prove the converse, take $K = G_p$ for each prime p. □

2 Other properties and examples

The class of generically trivial groups is closed under several constructions which we list below in the form of corollaries of Proposition 1.7. Direct proofs of these statements can also be given using basic properties of the P-localization functor [20,21,22].

Corollary 2.1 *Every homomorphic image of a generically trivial group is generically trivial.* □

Corollary 2.2 *If $N \rightarrowtail G \twoheadrightarrow Q$ is a group extension in which N and Q are generically trivial, then G is also generically trivial.* □

Corollary 2.3 *The (restricted) direct product of a family of generically trivial groups is generically trivial.* □

Corollary 2.4 *The free product of a family of generically trivial groups is generically trivial.* □

Thus, the class of generically trivial groups is closed under small colimits. This is not a surprise, in view of the next general fact: Since the P-localization functor has a right adjoint—namely, the inclusion of the subcategory of P-local groups in the category of groups—it preserves colimits [18, V.5], that is

Lemma 2.5 *Let F be a diagram of groups, and denote by F_P the diagram of P-local groups induced by functoriality. Then* $(\operatorname{colim} F)_P \cong (\operatorname{colim} F_P)_P$. □

(Note that, to construct colimits in the category of P-local groups, one computes the corresponding colimit in the category of groups and takes its P-localization.)

We next describe other important sources of examples of generically trivial groups. As already observed, groups satisfying condition (c) in Proposition 1.6 are always generically trivial. Among such groups are all strongly torsion generated groups (a group is *strongly torsion generated* [7] if, for every $n \geq 2$, there is an element $x \in G$ of order n whose conjugates generate G). Examples include the subgroup $E(R)$ generated by the elementary matrices within the general linear group $GL(R)$, for an arbitrary associative ring R with 1; see [6].

A simple group G containing elements of each finite order is strongly torsion generated. Interesting examples of this kind are

- the infinite alternating group $\mathcal{A}_\infty$;
- Philip Hall's countable universal locally finite group [14];
- all non-trivial algebraically closed groups [17].

This last example shows, by [17, IV.8.1], that

Theorem 2.6 *Every infinite group G can be embedded in a generically trivial group of the same cardinality as G.* $\square$

By [6], every abelian group is the centre of a generically trivial group. This prompts the question: which groups are normal in generically trivial groups?

Although generically trivial groups are perfect, no further restriction can be made on their integral homology in general, because

Theorem 2.7 *For any sequence $A_2, A_3, \ldots, A_n, \ldots$ of abelian groups, there exists a generically trivial group G such that $H_n(G; \mathbf{Z}) \cong A_n$ for all $n \geq 2$.*

PROOF. By [7, Theorem 1], one can always find a strongly torsion generated group with this property. $\square$

3 Some implications in homotopy theory

Our setting in this section is the pointed homotopy category $\mathbf{Ho}_*$ of connected CW-complexes. Our main tool will be the functor $(\ \ )_P$ defined in [11,12,13] for a set of primes P. It is an idempotent functor in $\mathbf{Ho}_*$ which is left adjoint to the inclusion of the subcategory of spaces X for which the nth power map $\rho_n \colon \Omega X \to \Omega X$, $\omega \mapsto \omega^n$, is a homotopy equivalence for every $n \in P'$. The map $l \colon X \to X_P$ turns out to be indeed P-localization if X is nilpotent, and, for every space X, the induced homomorphism $l_* \cdot \pi_1(X) \to \pi_1(X_P)$ P-localizes in the sense of Section 1. As explained in [13], the universality of Bousfield's $H_*(\ \ ; \mathbf{Z}_P)$-localization [8] in $\mathbf{Ho}_*$ implies that

$$l_* \colon H_*(X; \mathbf{Z}_P) \cong H_*(X_P; \mathbf{Z}_P) \tag{3.1}$$

for every space X. In fact, the map $l_* \colon H_*(X; A) \to H_*(X_P; A)$ is an isomorphism for a broader class of (twisted) coefficient modules A, namely those which are P-local as $\mathbf{Z}[\pi_1(X_P)]$-modules; see [11].

Theorem 3.1 *The following assertions are equivalent.*

(a) *X is acyclic and $\pi_1(X)$ is a generically trivial group.*

(b) *For every space Y such that $\pi_1(Y)$ is separable, the space* $\mathrm{map}_*(X,Y)$ *of pointed maps from X to Y is weakly contractible.*

(c) *E_pX is contractible for every prime p and every idempotent functor E_p in* $\mathbf{Ho}_*$ *satisfying $E_pS^1 \simeq (S^1)_p$.*

(d) *X_p is contractible for all primes p.*

For the proof we need to remark the following fact.

Lemma 3.2 *If the space X is acyclic and* $\mathrm{Hom}(\pi_1(X), \pi_1(Y))$ *consists of a single element, then* $\mathrm{map}_*(X,Y)$ *is weakly contractible.*

PROOF. Every map $f\colon X \to Y$ can be extended to the cone of X by obstruction theory, because $f_*\colon \pi_1(X) \to \pi_1(Y)$ is trivial and the cohomology groups of X with untwisted coefficients are zero. Thus $[X,Y]$ consists of a single element, and hence $\mathrm{map}_*(X,Y)$ is path-connected. The higher homotopy groups $\pi_k(\mathrm{map}_*(X,Y)) \cong [\Sigma^k X, Y]$, $k \geq 1$, vanish because the suspension of an acyclic space is contractible. □

PROOF OF THEOREM 3.1. The implication (a)⇒(b) follows from Proposition 1.7 and Lemma 3.2. Now assume given a prime p and an idempotent functor E_p as in (c). Then, by the same argument used in the proof of Lemma 1.3, the standard map $\rho_n\colon S^1 \to S^1$ of degree n is an E_p-equivalence if $(n,p) = 1$, and thus, for every space X, the map $(\rho_n)^*\colon [S^1, E_pX] \to [S^1, E_pX]$ is a bijection. This tells us that $\pi_1(E_pX)$ is a p-local group and hence separable. If we assume that (b) holds, then in particular $\eta\colon X \to E_pX$ is nullhomotopic, and the universal property of η forces E_pX to be contractible. This shows that (b)⇒(c). The implication (c)⇒(d) is trivial. Finally, if X_p is contractible then $\pi_1(X)_p = 1$ and, by (3.1), $H_k(X; \mathbf{Z}_p) = 0$ for all $k \geq 1$, where $\mathbf{Z}_p$ denotes the integers localized at p. If this happens for all primes p, then X is acyclic and $\pi_1(X)$ is generically trivial. Thus, (d)⇒(a). □

We call *generically trivial* those spaces satisfying the equivalent conditions of Theorem 3.1. Such spaces are not rare. For example,

Proposition 3.3 *Every acyclic space X with finite fundamental group is generically trivial.*

PROOF. If X is acyclic, then $\pi_1(X)$ is a perfect group. But finite perfect groups are generically trivial by Proposition 1.6. □

Also classifying spaces of generically trivial acyclic groups G are generically trivial spaces. Such groups G exist by Theorem 2.7. One explicit example is Philip Hall's countable universal locally finite group; see [5]. Another example is the general linear group on the cone of a ring [5,7]. A third example is the universal finitely presented strongly torsion generated acyclic group constructed in [7].

Since many groups are separable (cf. Section 1), Theorem 3.1 provides a good number of examples of mapping spaces which are weakly contractible.

4 Structure of generically trivial groups

In this final section we address the question of how to characterize generically trivial groups in terms of their structure.

Given a group G and a set of primes P, the kernel of $l\colon G \to G_P$ is hard to compute in general. If G is nilpotent, then $\operatorname{Ker} l$ is precisely the set of P'-torsion elements of G [16]. For other groups G, it can be considerably bigger, as we next explain.

An element $x \in G$ is said to be *of type* $T_{P'}$ [20,21] if there exist $a, b \in G$ and an integer $n \in P'$ such that

$$x = ab^{-1}, \qquad a^n = b^n.$$

Note that the P-localization homomorphism kills all elements of type $T_{P'}$. We may recursively define a sequence of normal subgroups of G

$$1 = T_0 \leq T_1 \leq T_2 \leq \dots \tag{4.1}$$

by letting T_{i+1}/T_i be the subgroup of G/T_i generated by all its elements of type $T_{P'}$. Set $T_{P'}(G) = \bigcup_i T_i$. This subgroup of G was first analyzed by Ribenboim [20]. It has, among others, the following elementary properties.

Proposition 4.1 (a) *If G is P-local, then $T_{P'}(G) = 1$.*

(b) *Every homomorphism $\varphi\colon G \to K$ satisfies $\varphi(T_{P'}(G)) \subseteq T_{P'}(K)$.*

(c) *For every group G, P'-roots are unique in $G/T_{P'}(G)$.* □

It follows from (a) and (b) that $T_{P'}(G)$ is always contained in the kernel of $l\colon G \to G_P$. However, $\operatorname{Ker} l$ can be bigger still. To understand this, we introduce the following notion.

Definition 4.2 Given a set of primes P, a monomorphism $\iota\colon G \to K$ is called *P-faithful* if $\iota_P\colon G_P \to K_P$ is also a monomorphism.

By [16, I.3.1] and Lemma 2.5, every embedding into a locally nilpotent group is P-faithful for all sets P. On the other hand, an embedding of the cyclic group of order 3 in the symmetric group Σ_3 is not 3-faithful, since $(\Sigma_3)_3 = 1$.

The authors wish to acknowledge remarks of Derek Robinson helpful to the following example.

Example 4.3 Let R denote the ring $\mathbf{Z} \oplus (\oplus_p \mathbf{F}_p)$ and consider the group $K = M(\mathbf{Q}, R)$ of all upper unitriangular R-matrices with finitely many non-zero off-diagonal entries indexed by the rational numbers. By [4], K is acyclic and locally nilpotent. Now, by Hirsch [23, p. 139], every finitely generated nilpotent group embeds in the direct product of a torsion-free finitely generated nilpotent group and finitely many finite p-groups. By Hall [23, p. 159], the first factor admits a faithful representation by integral unitriangular matrices (of finite size). Also, via its regular representation, a finite p-group of order n embeds in the group of $n \times n$ unitriangular matrices over $\mathbf{F}_p$. Thus every finitely generated nilpotent group embeds in K. By the above remark, such an embedding must be P-faithful. Thus K is a countable, acyclic group such that, for all sets of primes P, K_P contains a copy of G_P for every finitely generated nilpotent group G.

A group is called *P′-radicable* [23] if each of its elements has at least one nth root for every $n \in P'$.

Lemma 4.4 *For every group G there exists a P-faithful embedding $\iota\colon G \hookrightarrow K$ into a group K which is P'-radicable.*

PROOF. This follows from [21, Proposition 5.2]. In fact, ι can be chosen so that ι_P is an isomorphism. □

We denote by $ET_{P'}(G)$ the subset of G of those elements x belonging to $T_{P'}(K)$ for some P-faithful embedding $G \hookrightarrow K$. This is in fact a normal subgroup of G and, moreover, we have

Proposition 4.5 *Assume given a group G and a set of primes P. Then the kernel of $l\colon G \to G_P$ is precisely $ET_{P'}(G)$.*

PROOF. The inclusion $ET_{P'}(G) \subseteq \operatorname{Ker} l$ is clear. To check the converse, choose a P-faithful embedding $G \hookrightarrow K$ with K P'-radicable, as given by Lemma 4.4. Then, by part (c) of Proposition 4.1, the group $K/T_{P'}(K)$ is P-local. Since every homomorphism $\varphi\colon K \to L$ with L P-local satisfies $\varphi(T_{P'}(K)) = 1$, the projection $K \twoheadrightarrow K/T_{P'}(K)$ is a P-equivalence and hence a P-localization. It follows that the kernel of $l\colon G \to G_P$ is $G \cap T_{P'}(K)$, which is contained in $ET_{P'}(G)$. □

Example 4.6 Let the infinite cyclic group $C = \langle \xi \rangle$ act on the abelian group $A = \mathbf{Z}[1/q]$ by $\xi \cdot a = (1/q)a$, for a certain positive integer q. Let S be the semidirect product $A \rtimes C$ with respect to this action. One can directly check that the nth power map is injective in S for all positive integers n, and hence $T_{P'}(S) = 1$ for every set of primes P. Now denote $a = (0, \xi)$, $b = (1, 1)$ in S, and let $G = \langle S, c \mid c^r = b^m a^n \rangle$, where $b^m a^n$ is not a kth power in S for any $k > 1$ dividing r. Then $T_{P'}(G)$ is still trivial. However, $ET_{P'}(G)$ need not be trivial in general. For example, choose $q = 2$, $r = 2$, $m = 1$, $n = -2$, so that $G = \langle a, b, c \mid a^{-1}ba = b^2,\ b = c^2a^2 \rangle$. (This example is due to B. H. Neumann [2].) Let P be any set of primes such that $2, 3 \notin P$, and $G \hookrightarrow K$ be a P-faithful embedding in which b has a cube root d. Write $T_i = T_i(K)$ as in (4.1). Then one readily checks that $(a^{-1}da)^3 = d^6$, and hence $a^{-1}dad^{-2} \in T_1$; further $(da^{-1})^2c^{-2} \in T_1$, which implies $d(ca)^{-1} \in T_2$, and so $(ca)^{-3}b \in T_2$. This element belongs to $ET_{P'}(G)$ and is not trivial.

We can now give the following characterization of generically trivial groups.

Theorem 4.7 *A group G is generically trivial if and only if, for every prime p, there exists a p-faithful embedding $G \hookrightarrow K$ such that $G \subseteq T_{p'}(K)$.* □

References

[1] J. F. ADAMS, *Localisation and Completion*, Lecture Notes University of Chicago (1975).

[2] G. BAUMSLAG, Wreath products and p-groups, *Math. Proc. Cambridge Philos. Soc.* **55** (1959), 224–231.

[3] G. Baumslag, E. Dyer and A. Heller, The topology of discrete groups, *J. Pure Appl. Algebra* **16** (1980), 1–47.

[4] A. J. Berrick, Two functors from abelian groups to perfect groups, *J. Pure Appl. Algebra* **44** (1987), 35–43.

[5] A. J. Berrick, Universal groups, binate groups and acyclicity, *in:* Group Theory, Proceedings of the 1987 Singapore Conference, Walter de Gruyter, Berlin-New York, 1989, 253–266.

[6] A. J. Berrick, Torsion generators for all abelian groups, *J. Algebra*, to appear.

[7] A. J. Berrick and C. F. Miller III, Strongly torsion generated groups, preprint (1990).

[8] A. K. Bousfield, The localization of spaces with respect to homology, *Topology* **14** (1975), 133–150.

[9] C. Casacuberta and P. Hilton, On the extended genus of finitely generated abelian groups, *Bull. Soc. Math. Belg. Sér. A* **41** (1989), 51–72.

[10] C. Casacuberta and P. Hilton, A note on extensions of nilpotent groups, preprint (1990).

[11] C. Casacuberta and G. Peschke, Localizing with respect to self maps of the circle, preprint (1990).

[12] C. Casacuberta, G. Peschke and M. Pfenniger, Sur la localisation dans les catégories avec une application à la théorie de l'homotopie, *C. R. Acad. Sci. Paris Sér. I Math.* **310** (1990), 207–210.

[13] C. Casacuberta, G. Peschke and M. Pfenniger, On orthogonal pairs in categories and localization, preprint (1990).

[14] P. Hall, Some constructions for locally finite groups, *J. London Math. Soc.* **34** (1959), 305–319.

[15] P. Hilton, On the extended genus, *Acta Math. Sinica (N. S.)* **4** (1988), no. 4, 372–382.

[16] P. Hilton, G. Mislin and J. Roitberg, *Localization of Nilpotent Groups and Spaces*, North-Holland Math. Studies 15 (1975).

[17] R. C. Lyndon and P. E. Schupp, *Combinatorial Group Theory*, Ergeb. Math. Grenzgeb. 89, Springer-Verlag, 1977.

[18] S. MacLane, *Categories for the Working Mathematician*, Graduate Texts in Math. 5, Springer-Verlag, 1975.

[19] G. Mislin, Nilpotent groups with finite commutator subgroup, Lecture Notes in Math. 418, Springer-Verlag, 1974, 103–120.

[20] P. RIBENBOIM, Torsion et localisation de groupes arbitraires, Lecture Notes in Math. 740, Springer-Verlag, 1978, 444–456.

[21] P. RIBENBOIM, Equations in groups, with special emphasis on localization and torsion I, *Atti Accad. Naz. Lincei Mem. Cl. Sci. Fis. Mat. Natur. Sez. Ia* (8) **19** (1987), no. 2, 23–60.

[22] P. RIBENBOIM, Equations in groups, with special emphasis on localization and torsion II, *Portugal. Math.* **44** (1987), fasc. 4, 417–445.

[23] D. J. S. ROBINSON, *Finiteness Conditions and Generalized Soluble Groups, Part 2*, Ergeb. Math. Grenzgeb. 63, Springer-Verlag, 1972.

A.J. Berrick: Department of Mathematics
National University of Singapore
10 Kent Ridge Crescent
Singapore 0511

C. Casacuberta: Departament d'Àlgebra i Geometria
Universitat de Barcelona
Gran Via de les Corts Catalanes, 585
E-08007 Barcelona
Spain

1990 Barcelona Conference
on Algebraic Topology.

MORE EXAMPLES OF NON-CANCELLATION IN HOMOTOPY

IMRE BOKOR AND IRENE LLERENA

It was shown in [1] that for $\mathcal{C}_n^1$, the class of all spaces of the homotopy type of a CW-complex with precisely three cells, one in each of the dimensions 0, $2n$ and $4n$, there is a factorisation theorem with respect to the one-point union (or $\vee$-product) of spaces. The theorem states that as long as the Hopf invariant of the attaching map of the $4n$-cell is not 0, the homotopy type of a finite wedge of such spaces determines the homotopy type of each of the spaces in the wedge.

This was generalised in [2] to $\mathcal{C}_n^K$, the class of all spaces of the homotopy type of a CW-complex with precisely one cell in dimensions 0 and $4n$ and K cells in dimension $2n$. The generalised theorem states that as long as the Hilton-Hopf quadratic form of the attaching map of the $4n$-cell is non-singular, the homotopy type of a finite wedge of such spaces determines the homotopy type of each of the spaces in the wedge.

This note is devoted to showing that the results in [2] are the best possible. The counter-examples we present arose from dicussions during the Barcelona Conference.

We use the notation from [1] and begin by recalling that (up to homotopy) the spaces in $\mathcal{C}_n^K$ are mapping cones of maps

$$f : S^{4n-1} \longrightarrow \bigvee_{k=1}^{K} S^{2n}$$

and that homotopy classes of maps between such spaces correspond to homotopy classes of homotopy commutative diagrams

$$\begin{array}{ccc} S^{4n-1} & \xrightarrow{f} & \bigvee S^{2n} \\ {\scriptstyle\psi}\big\downarrow & & \big\downarrow{\scriptstyle\varphi} \\ S^{4n-1} & \xrightarrow[g]{} & \bigvee S^{2n} \end{array}$$

For the investigation of $\vee$-cancellation phenomena involving N such spaces, the relevant diagrams are of the form

$$\begin{array}{ccc} \bigvee_{h=1}^{N} S^{4n-1} & \xrightarrow{f} & \bigvee_{j=1}^{NK} S^{2n} \\ \psi \downarrow & & \downarrow \varphi \\ \bigvee_{h=1}^{N} S^{4n-1} & \xrightarrow{f} & \bigvee_{j=1}^{NK} S^{2n} \end{array}$$

with N copies of S^{4n-1} and NK copies of S^{2n} in the respective $\vee$-products. These diagrams can be characterised algebraically by

$$\underline{H}(g)(\underline{A}(\psi) \otimes 1_K) = \underline{A}(\varphi)\underline{H}(f)(1_K \otimes \underline{A}(\varphi)^t)$$

and

$$\Sigma g \underline{A}(\psi) = \underline{A}(\varphi)\Sigma f,$$

where $\underline{H}(f)$ is the Hilton-Hopf quadratic form of f, $\underline{A}(\varphi)$ is the matrix of φ and $\otimes$ denotes the Kronecker product of matrices. Moreover, we treat $\Sigma(f)$ as a column K-vector over T, the torsion subgroup of $\pi_{4n-1}(S^{2n})$. (cf. [1]).

Recall further that homotopy (resp. p-)equivalences correspond to diagrams where both $det(\underline{A}(\varphi))$ and $det(\underline{A}(\psi)) = \pm 1$ (resp. both coprime to p). We restrict attention to $n \neq 1, 2, 4$ merely to avoid the minor technical inconviences occasioned by the existence of maps Hopf invariant 1.

Take $y \in T$ of order m. Take $1 < k < m - 1$ with k co-prime to m. (Note that this is possible if and only if $m \neq 1, 2, 3, 4, 6$) Choose $f_i, g_i : S^{4n-1} \longrightarrow S^{2n} \vee S^{2n}$ given by

$$\underline{H}(f_1) = \begin{bmatrix} 2 & 0 \\ 0 & 0 \end{bmatrix} \quad \Sigma(f_1) = \begin{bmatrix} 0 \\ y \end{bmatrix} \quad \underline{H}(f_2) = \begin{bmatrix} 0 & 0 \\ 0 & 0 \end{bmatrix} \quad \Sigma(f_2) = \begin{bmatrix} ky \\ 0 \end{bmatrix}$$

$$\underline{H}(g_1) = \begin{bmatrix} 2 & 0 \\ 0 & 0 \end{bmatrix} \quad \Sigma(g_1) = \begin{bmatrix} 0 \\ ky \end{bmatrix} \quad \underline{H}(g_2) = \begin{bmatrix} 0 & 0 \\ 0 & 0 \end{bmatrix} \quad \Sigma(g_2) = \begin{bmatrix} y \\ 0 \end{bmatrix}$$

We claim that:

(i) $C_{f_1} \vee C_{f_2}$ and $C_{g_1} \bigvee C_{g_2}$ are homotopy equivalent.
(ii) C_{f_2} and C_{g_2} are homotopy equivalent.
(iii) $C_{f_1} \bigvee C_{f_1}$ and $C_{g_1} \bigvee C_{g_1}$ are homotopy equivalent.
(iv) C_{f_1} and C_{g_1} are not homotopy equivalent

Since k and m are coprime, there are integers u and v with $uk - vm = 1$. Choose homotopy self-equivalences ψ of $S^{4n-1} \bigvee S^{4n-1}$ and φ of $S^{2n} \bigvee S^{2n} \bigvee S^{2n} \bigvee S^{2n}$ given by

$$\underline{A}(\psi) = \begin{bmatrix} 1 & 0 \\ v & 1 \end{bmatrix} \text{ and } \underline{A}(\varphi) = \begin{bmatrix} 1 & 0 & 0 & 0 \\ 0 & k & m & 0 \\ 0 & v & u & 0 \\ 0 & 0 & 0 & 1 \end{bmatrix}$$

Then

$$\underline{H}(\varphi \circ (f_1 \vee f_2)) = \underline{A}(\varphi)\underline{H}(f_1 \vee f_2)(\underline{1}_2 \otimes \underline{A}(\varphi)^t)$$
$$= \begin{bmatrix} 2 & 0 & 0 & 0 & 0 & 0 & 0 & 0 \\ 0 & 0 & 0 & 0 & 0 & 0 & 0 & 0 \\ 0 & 0 & 0 & 0 & 0 & 0 & 0 & 0 \\ 0 & 0 & 0 & 0 & 0 & 0 & 0 & 0 \end{bmatrix}$$
$$= \underline{H}(g_1 \vee g_2)(\underline{A}(\varphi) \otimes \underline{1}_4)$$
$$= \underline{H}((g_1 \vee g_2) \circ \psi)$$

and

$$\Sigma(\varphi \circ (f_1 \vee f_2)) = \underline{A}(\varphi)\Sigma(f_1 \vee f_2)$$
$$= \begin{bmatrix} 1 & 0 & 0 & 0 \\ 0 & k & m & 0 \\ 0 & v & u & 0 \\ 0 & 0 & 0 & 1 \end{bmatrix} \begin{bmatrix} 0 & 0 \\ y & 0 \\ 0 & ky \\ 0 & 0 \end{bmatrix}$$
$$= \begin{bmatrix} 0 & 0 \\ ky & 0 \\ vy & uky \\ 0 & 0 \end{bmatrix}$$

whereas

$$\Sigma((g_1 \vee g_2) \circ \psi) = \begin{bmatrix} 0 & 0 \\ ky & 0 \\ 0 & y \\ 0 & 0 \end{bmatrix} \begin{bmatrix} 1 & 0 \\ v & 1 \end{bmatrix}$$
$$= \begin{bmatrix} 0 & 0 \\ ky & 0 \\ vy & y \\ 0 & 0 \end{bmatrix}$$

Hence $\Sigma(\varphi \circ (f_1 \vee f_2)) \simeq \Sigma((g_1 \vee g_2) \circ \psi)$ since $uk \equiv 1 \pmod m$, showing that $C_{f_1} \bigvee C_{f_2}$ and $C_{g_1} \bigvee C_{g_2}$ are homotopy equivalent, thereby completing the verification of (i).

To verify (ii), it suffices to compare $\Sigma(f_2)$ and $\Sigma(g_2)$ since $\underline{H}(f_2) = \underline{H}(g_2) = \begin{bmatrix} 0 & 0 \\ 0 & 0 \end{bmatrix}$. Let ζ be the self homotopy equivalence of $S^{2n} \bigvee S^{2n}$ given by

$$\underline{A}(\zeta) = \begin{bmatrix} k & v \\ m & u \end{bmatrix}.$$

Then

$$\Sigma(\zeta \circ g_2) = \begin{bmatrix} k & v \\ m & u \end{bmatrix} \begin{bmatrix} y \\ 0 \end{bmatrix} = \begin{bmatrix} ky \\ 0 \end{bmatrix} = \Sigma(f_2)$$

since m is the order of y. Thus ζ and the identity map on S^{4n-1} induce a homotopy equivalence between C_{f_2} and C_{g_2}.

To verify (iii), let θ be the homotopy self-equivalence of $S^{2n} \bigvee S^{2n} \bigvee S^{2n} \bigvee S^{2n}$ given by

$$\underline{A}(\theta) = \begin{bmatrix} 1 & 0 & 0 & 0 \\ 0 & k & v & 0 \\ 0 & m & u & 0 \\ 0 & 0 & 0 & 1 \end{bmatrix}.$$

A simple direct calculation shows that $\underline{H}(\theta \circ (f_1 \bigvee f_1)) = \underline{H}(g_1 \bigvee g_1)$. Moreover

$$\Sigma\big(\theta \circ (f_1 \vee f_1)\big) = \begin{bmatrix} 1 & 0 & 0 & 0 \\ 0 & k & v & 0 \\ 0 & m & u & 0 \\ 0 & 0 & 0 & 1 \end{bmatrix} \begin{bmatrix} 0 & 0 \\ y & y \\ 0 & 0 \\ 0 & 0 \end{bmatrix} = \begin{bmatrix} 0 & 0 \\ ky & ky \\ 0 & 0 \\ 0 & 0 \end{bmatrix} = \Sigma(g_1 \vee g_1),$$

so that θ together with the identity map of $S^{4n-1} \bigvee S^{4n-1}$ define a homotopy equivalence between $C_{f_1} \bigvee C_{f_1}$ and $C_{g_1} \bigvee C_{g_1}$.

Finally, suppose we have a homotopy commutative diagram

$$\begin{array}{ccc} S^{4n-1} & \xrightarrow{f_1} & S^{2n} \vee S^{2n} \\ \lambda \downarrow & & \downarrow \xi \\ S^{4n-1} & \xrightarrow{g_1} & S^{2n} \vee S^{2n} \end{array}$$

with

$$deg(\lambda) = \ell \text{ and } \underline{A}(\xi) = \begin{bmatrix} a & b \\ c & d \end{bmatrix}.$$

Then

$$\begin{bmatrix} a & b \\ c & d \end{bmatrix} \begin{bmatrix} 2 & 0 \\ 0 & 0 \end{bmatrix} \begin{bmatrix} a & c \\ b & d \end{bmatrix} = \begin{bmatrix} 2\ell & 0 \\ 0 & 0 \end{bmatrix},$$

so that $c = 0$ and $det\underline{A}(\xi) = ad$. But we also have the equality

$$\begin{bmatrix} a & b \\ c & d \end{bmatrix} \begin{bmatrix} 0 \\ y \end{bmatrix} = \begin{bmatrix} 0 \\ k\ell y \end{bmatrix},$$

so that $d \equiv k\ell \pmod m$. Since by choice $1 < k < m-1$, either $\ell \neq \pm 1$ — in which case λ is not a homotopy equivalence — or $d \neq \pm 1$ — in which case $det(\underline{A}(\xi)) = ad \neq \pm 1$ and ξ is not a homotopy equivalence. Hence C_{f_1} and C_{g_1} cannot be homotopy equivalent, completing the verification of (iv).

Note, however, that C_{f_1} and C_{g_1} are in fact of the same genus. To verify this, it is only necessary to localise at the primes dividing m, the order of y. But choosing $a = 1, b = 0, d = k$ in ξ and $\lambda = id_{S^{4n-1}}$ above defines an equivalence at each such prime, since k is coprime to m. This situation is typical as shown by the main result in **[2]**.

References

1. I. BOKOR, On Genus and Cancellation in Homotopy, *Israel Journal of Mathematics (in press)*.
2. I. LLERENA, Wedge Cancellation of Certain Mapping Cones, *Compositio Mathematica (to appear)*.

I. Bokor: Mathematik
Berninaplatz 3
CH–8057 – Zürich
Switzerland

I. Llerena: Departament d'Àlgebra i Geometria
Universitat de Barcelona
Gran Via de les Corts Catalanes, 585
E–08007 Barcelona
Spain

1990 Barcelona Conference
on Algebraic Topology.

ON SUB-$\mathcal{A}_p^*$-ALGEBRAS OF H^*V

CARLOS BROTO AND SAÏD ZARATI

Introduction

Let p be a prime number, $\mathcal{A}_p^*$ the mod p Steenrod algebra and $\mathcal{U}$ (resp. $\mathcal{K}$) the category of unstable $\mathcal{A}_p^*$-modules (resp. unstable $\mathcal{A}_p^*$-algebras). Following L. Schwartz [S] we consider the full subcategory $\mathcal{N}il_m$, $m \geq 0$, of $\mathcal{U}$ whose objects are the m-nilpotent unstable $\mathcal{A}_p^*$-modules. The category $\mathcal{N}il_m$ is a localizing subcategory of $\mathcal{U}$, so for any object M of $\mathcal{U}$ we have a $\mathcal{N}il_m$-localization $\mu_m: M \to \mathcal{N}_m^{-1}(M)$. We verify that $M \cong \varprojlim_m \mathcal{N}_m^{-1}(M)$ and that $\mathcal{N}_m^{-1}(K)$ is still an object of $\mathcal{K}$ if K is so. The aim of this note is to compute the $\mathcal{N}il_m$-localization of sub-$\mathcal{A}_p^*$-algebras of $H^*(V;\mathbf{F}_p)$, V an elementary abelian p-group, which are noetherian with Krull dimension the rank of V and weakly integrally closed. It is known that the $\mathcal{N}il_0$-localization of such an algebra, K, is a subalgebra of invariants of $H^*(V;\mathbf{F}_p)$ by the action of a subgroup, G, of $GL(V)$ ([**AW**] [**BZ**]). Our result, which will be more precise in paragraph 2, states that for any $m \geq 0$ the $\mathcal{N}il_m$-localization of K is the algebra of invariants $\left(R_m(K)\right)^G$, same group as before, where $R_m(K)$ is the smallest $\mathcal{N}il_m$-closed sub-$\mathcal{A}_p^*$-algebra of $H^*(V;\mathbf{F}_p)$ that contains K and whose $\mathcal{N}il_0$-localization is precisely $H^*(V;\mathbf{F}_p)$. Recent results of H.-W. Henn [**H**] show that this K is $\mathcal{N}il_n$-closed for some n, and so, therefore that $K = \left(R_n(K)\right)^G$.

The paper is organized as follows. In the first paragraph we introduce the categories of m-nilpotent, m-reduced and $\mathcal{N}il_m$-closed unstable $\mathcal{A}_2^*$-modules and we give some of their properties. In the second paragraph we state and prove our main result for $p=2$. The modifications that one needs to get the odd prime case are discussed in the third paragraph.

§1 $\mathcal{N}il_m$-localization

In this paragraph we generalize the concept of $\mathcal{N}il$-localization introduced in [**LZ1**] (see also [**BZ**], [**GLZ**] and [**HLS**]).

Definitions and notations. Let $\mathcal{A}_2^*$ be the mod 2 Steenrod algebra and $\mathcal{U}$ the category of unstable $\mathcal{A}_2^*$-modules and $\mathcal{A}_2^*$-linear maps of degree zero. We denote by $J(n)$, $n \geq 0$, the Brown-Gitler module (see [**C**], [**M**], [**LZ1**]) which might be defined by $\mathrm{Hom}_{\mathcal{U}}\left(M, J(n)\right) \cong \mathrm{Hom}_{\mathbf{F}_2}(M^n, \mathbf{F}_2)$ for any unstable $\mathcal{A}_2^*$-module $M = \{M^n, n \geq 0\}$. In particular, $J(n)$ in an injective object of the category $\mathcal{U}$ ($\mathcal{U}$-injective for short).

Let V be an elementary abelian 2-group ($V \cong (\mathbf{Z}/2)^d, d \geq 0$); we denote by H^*V the mod 2 cohomology of V, then $H^*V \otimes J(n)$ is $\mathcal{U}$-injective (see [**C**] for $n=0$ and [**LZ1**] for $n \geq 1$) and this gives essentially all injective objects of the category $\mathcal{U}$, [**LS**].

The category of m-nilpotent unstable $\mathcal{A}_2^*$-modules. After L. Schwartz we introduce the concept of m-nilpotent unstable $\mathcal{A}_2^*$-modules, which is a natural generalization of the concept of nilpotent $\mathcal{A}_2^*$-modules (0-nilpotent is equivalent to nilpotent).

1.1 Definition. *An unstable $\mathcal{A}_2^*$-module N is called m-nilpotent, $m \geq 0$, if and only if* $\mathrm{Hom}_{\mathcal{U}}(N, H^*V \otimes J(n)) = 0$ *for all elementary abelian 2-group V and all n: $0 \leq n \leq m$. We denote $\mathcal{N}il_m$ the full subcategory of all m-nilpotent objects of $\mathcal{U}$.*

1.2 Remark: Let $T_V: \mathcal{U} \to \mathcal{U}$ denote Lannes' T-functor associated to the elementary abelian 2-group V, [L]. An unstable $\mathcal{A}_2^*$-module N is m-nilpotent if and only if $T_V N$ is m-connected for any V ($T_V^n N = 0$, $0 \leq n \leq m$). As a consequence $T_V N$ is also m-nilpotent and N is m-connected.

1.3 Examples: (1) $\Sigma^{m+1}M$, the $(m+1)$-fold suspension of M, is m-nilpotent for every M because T_V commutes with suspensions.

(2) Again by properties of T_V, if M is m-nilpotent and N is n-nilpotent then $M \otimes N$ is $(m+n+1)$-nilpotent for $m, n \geq -1$ with the convention $\mathcal{N}il_{-1} = \mathcal{U}$.

(3) Let $0 \to M' \to M \to M'' \to 0$ be an exact sequence in $\mathcal{U}$, then M is m-nilpotent if and only if both M' and M'' are m-nilpotent. This follows by definition of m-nilpotent module and injectivity of $H^*V \otimes J(n)$, $n \geq 0$.

(4) Any object M of $\mathcal{U}$ such that $M^k \neq 0$ only if $m < k < n$ is m-nilpotent.

The examples above show that $\mathcal{N}il_m$ is a Serre category as defined in [**Sw**]. So we are given the usual properties associated to the following definition:

1.4 Definition. (1) *A morphism of $\mathcal{U}$, $f: M \to N$ is a $\mathcal{N}il_m$-isomorphism if both* $\ker f$ *and* $\operatorname{coker} f$ *are m-nilpotent.*

(2) *An object M of $\mathcal{U}$ is called m-reduced if* $\mathrm{Hom}_{\mathcal{U}}(N, M) = 0$ *for any object N of $\mathcal{N}il_m$.*

(3) *An object L of $\mathcal{U}$ is called $\mathcal{N}il_m$-closed if for any given $\mathcal{N}il_m$-isomorphism, $f: M \to N$, the induced map: f^*:* $\mathrm{Hom}_{\mathcal{U}}(N, L) \to \mathrm{Hom}_{\mathcal{U}}(M, L)$ *is an isomorphism.*

(4) *Let M be an object of $\mathcal{U}$, a map $\mu_m: M \to \mathcal{N}_m^{-1}(M)$ is called a $\mathcal{N}il_m$-localization of M if μ_m is a $\mathcal{N}il_m$-isomorphism and $\mathcal{N}_m^{-1}(M)$ is a $\mathcal{N}il_m$-closed object of $\mathcal{U}$.*

1.5 Remark: Provided it exists, the $\mathcal{N}il_m$-localization of an object M is unique up to isomorphism and initial among maps from M to $\mathcal{N}il_m$-closed objects. Hence the notation.

1.6 Theorem. *$\mathcal{N}il_m$ is a localizing subcategory of $\mathcal{U}$. That is, for every object M of $\mathcal{U}$ there exists its $\mathcal{N}il_m$-localization.*

Proof: Since $\mathcal{U}$ has injective envelopes, according to [**Sw**, 1.8] we only need to show that every object M of $\mathcal{U}$ has a largest m-nilpotent submodule. This is given by $\{ x \in M \mid \mathcal{A}_2^* \cdot x \text{ is } m\text{-nilpotent} \}$, where $\mathcal{A}_2^* \cdot x$ is the sub-$\mathcal{A}_2^*$-module of M generated by x. ∎

1.7 Proposition. *For every object M of $\mathcal{U}$, $M \xrightarrow{\cong} \varprojlim_m \mathcal{N}_m^{-1}(M)$.*

Proof: Since $\mathcal{N}_m^{-1}(M)$ is also $\mathcal{N}il_{(m+1)}$-closed, we obtain the inverse system: $\cdots \to \mathcal{N}_{(m+1)}^{-1}(M) \to \mathcal{N}_m^{-1}(M) \to \cdots$. Now, $\mu_m : M \to \mathcal{N}_m^{-1}(M)$ can be decomposed as $0 \to \ker \mu_m \to M \to \operatorname{Im} \mu_m \to 0$ together with $0 \to \operatorname{Im} \mu_m \to \mathcal{N}_m^{-1}(M) \to \operatorname{coker} \mu_m \to 0$, where $\ker \mu_m$ is precisely the maximal m-nilpotent submodule of M and $\operatorname{coker} \mu_m$ is m-nilpotent, too. So, in particular, $\ker \mu_m$ and $\operatorname{coker} \mu_m$ are m-connected and then $\varprojlim_m^i \ker \mu_m = \varprojlim_m^i \operatorname{coker} \mu_m = 0$ for $i = 0, 1$. So, therefore: $M \cong \varprojlim_m \operatorname{Im} \mu_m \cong \varprojlim_m \mathcal{N}_m^{-1}(M)$. ■

Characterization of m-nilpotent unstable $\mathcal{A}_2^*$-modules. As in **[LZ1]** we denote by Ω (resp. $\widetilde{\Sigma}$) the left adjoint (resp. the right adjoint) of the suspension functor, Σ. It is clear that Ω iterated k times: $\Omega^k : \mathcal{U} \to \mathcal{U}$ (resp. $\widetilde{\Sigma}^k : \mathcal{U} \to \mathcal{U}$) is the left adjoint functor (resp. right adjoint functor) of the k-fold suspension functor, Σ^k.

We denote by $\Phi : \mathcal{U} \to \mathcal{U}$ the "doubler" functor defined by

$$(\Phi M)^n = \begin{cases} M^{\frac{n}{2}} \text{ if } n \equiv 0 \pmod 2 \\ 0 \text{ if } n \equiv 1 \pmod 2 \end{cases} \quad \text{and} \quad Sq^i(\Phi x) = \begin{cases} \Phi(Sq^{\frac{i}{2}} x) \text{ if } i \equiv 0 \pmod 2 \\ 0 \text{ if } i \equiv 1 \pmod 2 \end{cases}$$

The functor Φ has a right adjoint functor $\tilde{\Phi}$.

To finish, we denote by $Sq_0 : \Phi M \to M$, $M \in \mathcal{U}$, the $\mathcal{A}_2^*$-linear map defined by $Sq_0 x = Sq^n x$ for any $x \in M^n$, and by $\widetilde{Sq_0} : M \to \tilde{\Phi} M$ its adjoint map. We have the following exact sequences in $\mathcal{U}$ (see **[LZ1]**, **[LZ2]**).

$$\text{(E1)} \qquad 0 \to \Sigma\Omega_1 M \to \Phi M \xrightarrow{Sq_0} M \to \Sigma\Omega M \to 0,$$

where Ω_1 is the first (and unique) derived functor of Ω, and the Mahowald exact sequence:

$$\text{(E2)} \qquad 0 \to \Sigma\widetilde{\Sigma} M \to M \xrightarrow{\widetilde{Sq_0}} \tilde{\Phi} M \to \Sigma(R^1\widetilde{\Sigma})(M) \to 0$$

where $R^1\widetilde{\Sigma}$ is the first derived functor of $\widetilde{\Sigma}$.

We have the following properties of m-nilpotent unstable $\mathcal{A}_2^*$-modules.

1.8 Lemma. *Let N be an m-nilpotent unstable $\mathcal{A}_2^*$-modules. We have:*

(1) *$\widetilde{\Sigma}^k N$ is $(m-k)$-nilpotent for any k: $0 \le k \le m$.*

(2) *$\tilde{\Phi}^k N$ is nilpotent for any $k \ge 0$.*

Proof: (1) From (E2) we have that $\Sigma\widetilde{\Sigma} N$ is a sub-$\mathcal{A}_2^*$-module of N and therefore m-nilpotent. Then, it follows from the formula $\widetilde{\Sigma}(H^*V \otimes J(n)) \cong H^*V \otimes J(n-1)$, for any elementary abelian 2-group V and $n \ge 0$, (**[LZ1]**), that $\widetilde{\Sigma} N$ is $(m-1)$-nilpotent.

(2) It suffices to remark that for any $M \in \mathcal{U}$ the cokernel of $\widetilde{Sq_0} : M \to \tilde{\Phi} M$ is nilpotent (see (E2)) then $\tilde{\Phi} M$ is nilpotent if M is nilpotent. ■

1.9 Proposition. [Compare with [S]] *An unstable $\mathcal{A}_2^*$-module N is m-nilpotent if and only if $\Omega^k N$ is nilpotent for any k: $0 \leq k \leq m$.*

Proof: Suppose N m-nilpotent and k: $0 \leq k \leq m$. The natural surjection $N \to \Sigma^k\Omega^k N$ shows that $\Sigma^k\Omega^k N$ is m-nilpotent. The point (1) of lemma 1.8 proves that $\Omega^k N$ is $(m-k)$-nilpotent, so, in particular, it is nilpotent. To prove the converse we use induction on m and the Mahowald exact sequence (E2) for the Brown-Gitler modules $J(n)$, $n \geq 0$. ∎

1.10 Remark: Let Sq_k, $k \in \mathbf{Z}$, the operation defind by $Sq_k x = Sq^{|x|-k}x$ if x is an element of degree $|x|$ of an $\mathcal{A}_2^*$-module M. We verify that $\Sigma^k\Omega^k M$, $k \geq 0$ and $M \in \mathcal{U}$, is the quotient of M by its sub-$\mathcal{A}_2^*$-module generated by $\operatorname{Im} Sq_0 + \ldots + \operatorname{Im} Sq_{k-1}$. It is proved in [LS] that N is nilpotent if and only if for any $x \in N$, there exists $r \geq 0$ such that $Sq_0^r x = 0$ (Sq_0 r times). The proposition 1.9 is now equivalent to the following statement. An unstable $\mathcal{A}_2^*$-module is m-nilpotent if and only if for any $x \in N$ and for any k: $0 \leq k \leq m$, there exists $r_k \geq 0$ such that $Sq_k^{r_k} x = 0$ (Sq_k r_k-times) (see [S]).

Characterization of m-reduced unstable $\mathcal{A}_2^*$-modules.

1.11 Proposition. *Let M be an unstable $\mathcal{A}_2^*$-module. The following are equivalent.*

(1) *M is m-reduced.*

(2) *M embeds in $\prod_\alpha H^*V_\alpha \otimes J(n_\alpha)$ where V_α is an elementary abelian 2-group and $0 \leq n_\alpha \leq m$.*

(3) $\widetilde{\Sigma}^{m+1} M = 0$.

Proof: (1) ⇒ (2) It is enough to verify that the map $\prod u\colon M \to \prod H^*V \otimes J(n)$ is injective, where the product is taken over the set of all maps $u\colon M \to H^*V \otimes J(n)$, for all classes of elementary abelian 2-groups and $0 \leq n \leq m$.

(2) ⇒ (3) This follows easily from the formula $\widetilde{\Sigma}(H^*V \otimes J(n)) \cong H^*V \otimes J(n-1)$ ([**LZ1**]) since $\widetilde{\Sigma}$ is left exact and commutes with products.

(3) ⇒ (1) The proof is by induction on m. The case $m = 0$ is proved in [**LZ1**]. Let N be m-nilpotent and $f\colon N \to M$ $\mathcal{A}_2^*$-linear. $\operatorname{Im} f$ is also m-nilpotent. By induction, $\widetilde{\Sigma}M$ is $(m-1)$-reduced and by 1.8 (1) $\widetilde{\Sigma}\operatorname{Im} f$ is $(m-1)$-nilpotent then, since $\widetilde{\Sigma}$ is left exact and using the inductive hypothesis we get that $\widetilde{\Sigma}(\operatorname{Im} f)$ is trivial. That is, $(\operatorname{Im} f)$ is reduced, but it is also nilpotent, hence it is trivial and this means $f = 0$. ∎

1.12 Remark: For an unstable $\mathcal{A}_2^*$-module, M, we have the following description of its sub-$\mathcal{A}_2^*$-module $\Sigma^k\widetilde{\Sigma}^k M$:

$$\Sigma^k\widetilde{\Sigma}^k M = \left\{x \in M \mid Sq_0 x = \ldots = Sq_{k-1}x = 0\right\}, \quad k \geq 1.$$

1.13 Example: If $M \in \mathcal{U}$ is reduced then $\Sigma^k M$ is m-reduced for all $m \geq k$, but not $(k-1)$-reduced because it is $(k-1)$-nilpotent.

1.14 Lemma. *If M is i-reduced then $\mathcal{N}_s^{-1}(M)$ is i-reduced for any $s \geq 0$.*

Proof: This is obvious for $i \geq s$. Assume $i < s$. Then M has no s-nilpotent sub-$\mathcal{A}_2^*$-module and $M \to \mathcal{N}_s^{-1}(M)$ is injective; its cokernel, C, is s-nilpotent. Applying $\mathcal{N}_i^{-1}(-)$ to the obtained exact sequence, we get:

$$\begin{array}{ccccc} M & \longrightarrow & \mathcal{N}_s^{-1}(M) & \longrightarrow & C \\ \downarrow & & \downarrow & & \downarrow \\ \mathcal{N}_i^{-1}(M) & \xrightarrow{\cong} & \mathcal{N}_i^{-1}(\mathcal{N}_s^{-1}(M)) & \longrightarrow & 0 \end{array}$$

It follows that $\ker\big(\mathcal{N}_s^{-1}(M) \to \mathcal{N}_i^{-1}(\mathcal{N}_s^{-1}(M))\big) \subset C$ hence, it is s-nilpotent, but $\mathcal{N}_s^{-1}(M)$ is s-reduced so this kernel is trivial:

$$\mathcal{N}_s^{-1}(M) \hookrightarrow \mathcal{N}_i^{-1}(\mathcal{N}_s^{-1}(M)) = \mathcal{N}_i^{-1}(M).$$

This implies that $\mathcal{N}_s^{-1}(M)$ is i-reduced. ■

Characterization of $\mathcal{N}il_m$-closed unstable $\mathcal{A}_2^*$-modules.

1.15 Proposition. *Let M be an unstable $\mathcal{A}_2^*$-module. The following properties are equivalent*

(1) *M is $\mathcal{N}il_m$-closed, $m \geq 0$.*
(2) *$\mathrm{Ext}^i_{\mathcal{U}}(N, M) = 0$ for any $N \in \mathcal{N}il_m$ and $i = 0, 1$.*
(3) *There exists an injective resolution of M starting:*

$$0 \to M \to \prod_\alpha H^*V_\alpha \otimes J(n_\alpha) \to \prod_\beta H^*V_\beta \otimes J(n_\beta)$$

where V_α and V_β are elementary abelian 2-groups and $0 \leq n_\alpha,\ n_\beta \leq m$.
(4) *$\mathrm{Ext}^i_{\mathcal{U}}(\Sigma^{m+1}L, M) = 0$ for any $L \in \mathcal{U}$ and $i = 0, 1$.*
(5) *$\widetilde{\Sigma}^{m+1}M = R^1\widetilde{\Sigma}^{m+1}(M) = 0$.*

Proof: (1) $\Longleftrightarrow$ (2). (2) $\Rightarrow$ (1) clearly. To show that (1) $\Rightarrow$ (2) we only need to apply $\mathrm{Hom}_{\mathcal{U}}(-, M)$ to an exact sequence $\ker\epsilon \xrightarrow{i} L \xrightarrow{\epsilon} N$ where L is a free object of $\mathcal{U}$ for any m-nilpotent$N$.

(2) $\Longleftrightarrow$ (3). Use proposition 1.11.

(2) $\Rightarrow$ (4) is evident, (4) $\Rightarrow$ (5) follows from $\big(R^i\widetilde{\Sigma}^{m+1}(M)\big)^n \cong \mathrm{Ext}^i_{\mathcal{U}}\big(\Sigma^{m+1}F(n), M\big)$, $i = 0, 1$, where $F(n)$ is the free unstable $\mathcal{A}_2^*$-module generated by one generator of degree n, and (5) $\Rightarrow$ (3) is again a consequence of proposition 1.11. ■

1.16 Examples: (1) Assume that M has dimension m, that is $M^k = 0$ if $k > m$ then M is $\mathcal{N}il_m$-closed. But not $(m-1)$-reduced if $M^m \neq 0$. In fact, we can write an injective resolution of M that starts as: $M \to \prod_\alpha J(n_\alpha) \to \prod_\beta J(n_\beta)$ with $0 \leq n_\alpha \leq m$

and $0 \le n_\beta \le m-1$. If $M^m \ne 0$, $\Sigma^m \mathbf{F}_2 \subset M$ hence M is not $(m-1)$-reduced. In particular $J(n) \otimes J(m)$ is $\mathcal{N}il_{(n+m)}$-closed but not $(n+m-1)$-reduced.

(2) Let V be an elementary abelian 2-group and M a $\mathcal{N}il_m$-closed object of $\mathcal{U}$, then $H^*V \otimes M$ is still $\mathcal{N}il_m$-closed (use point (3) of proposition 1.15).

(3) Again by the point (3) of proposition 1.15 and the preceding examples we have that if M is $\mathcal{N}il_m$-closed and N is $\mathcal{N}il_n$-closed, then $M \otimes N$ is $\mathcal{N}il_{(m+n)}$-closed.

As a consequence we get a refinement of example (3):

1.17 Proposition. *Assume that M is (exactly) i-reduced and $\mathcal{N}il_m$-closed, and N is (exactly) j-reduced and $\mathcal{N}il_n$-closed. Then, $M \otimes N$ is (exactly) $(i+j)$-reduced and $\mathcal{N}il_{\max\{(i+n),(j+m)\}}$-closed.*

Proof: We already know $M \otimes N$ is (exactly) $(i+j)$-reduced. Suppose $(n+i) \ge (m+j)$. Let $N \to \mathcal{N}_j^{-1}(N) \to C$ be the $\mathcal{N}il_j$-localization of N. Then, for $k > j$, $\tilde{\Sigma}^k C \cong (R^1\tilde{\Sigma}^k)(N)$, so C is (exactly) n-reduced. Now we consider the exact sequence $0 \to M \otimes N \to M \otimes \mathcal{N}_j^{-1}(N) \to M \otimes C \to 0$. $M \otimes \mathcal{N}_j^{-1}(N)$ is $\mathcal{N}il_{(m+j)}$-closed, so if $k > m+j$, $\tilde{\Sigma}^k(M \otimes C) \cong R^1\tilde{\Sigma}^k(M \otimes N)$ but $M \otimes C$ is (exactly) $(i+n)$-reduced; so therefore $M \otimes N$ is (exactly) $\mathcal{N}il_{(i+n)}$-closed. ∎

1.18 Proposition. *Every projective limit of $\mathcal{N}il_m$-closed objects is $\mathcal{N}il_m$-closed.*

Proof: Let $F(k)$ be the free unstable $\mathcal{A}_2^*$-module on one generator ι_k of degree k. For any $m \ge 0$ and any $n \ge 1$ we denote by $\ker_{m,n}$ the kernel of the surjective $\mathcal{A}_2^*$-linear map $F(n+m) \to \Sigma^{m+1}F(n-1)$, $\iota_{m+n} \mapsto \sigma^{m+1}\iota_{n-1}$. Let M be an object of $\mathcal{U}$, by applying $\mathrm{Hom}_{\mathcal{U}}(-, M)$ to the exact sequence

$$0 \to \ker_{m,n} \to F(m+n) \to \Sigma^{m+1}F(n-1) \to 0.$$

and since we have isomorphisms $\mathrm{Hom}_{\mathcal{U}}\left(\Sigma^{m+1}F(n-1), M\right) \cong \left(\Sigma^{m+1}\tilde{\Sigma}^{m+1}M\right)^{n+m}$ and $\mathrm{Ext}^1_{\mathcal{U}}\left(\Sigma^{m+1}F(n-1), M\right) \cong \left(\Sigma^{m+1}R^1\tilde{\Sigma}^{m+1}M\right)^{n+m}$, and $F(m+n)$ is $\mathcal{A}_2^*$-free, we get that $\mathrm{Hom}_{\mathcal{U}}\left(F(n+m), M\right) \xrightarrow{\cong} \mathrm{Hom}_{\mathcal{U}}\left(\ker_{m,n}, M\right)$ if and only if M is $\mathcal{N}il_m$-closed.

The proposition 1.18 is now a consequence of this last isomorphism. ∎

Let $m \ge 1$ be an integer. We denote by 2^{α_m} the biggest power of 2 in m, that is: $m = 2^{\alpha_m} + l$, $0 \le l \le 2^{\alpha_m} - 1$. We have the following properties of $\mathcal{N}il_m$-closed objects.

1.19 Proposition. (1) *Let M be a $\mathcal{N}il_m$-closed unstable $\mathcal{A}_2^*$-module. Then*

$$\tilde{\Phi}M \quad \textit{is} \quad \begin{cases} \mathcal{N}il_{\frac{m}{2}}\textit{-closed if } m \equiv 0 \pmod 2. \\ \mathcal{N}il_{\frac{m-1}{2}}\textit{-closed if } m \equiv 1 \pmod 2. \end{cases}$$

In particular, $\tilde{\Phi}^{\alpha_m+1}M$ is $\mathcal{N}il_0$-closed.

(2) *Let M be a $\mathcal{N}il_0$-closed unstable $\mathcal{A}_2^*$-module. Then $\Phi^m M$ is $\mathcal{N}il_{2(m-1)}$-closed, $m \ge 1$.*

Proof: (1) We use essentially the point (3) of the proposition 1.15 and the following formula: $\tilde{\Phi}(H^*V \otimes J(n)) \cong H^*V \otimes J(\frac{n}{2})$, $\frac{n}{2} = 0$ if $n \equiv 1 \pmod 2$ (see [**LZ1**]).

(2) The proof is by induction on m. First, assume $m = 1$. If M is reduced the exact sequence (E1) becomes: $(*)\ 0 \to \Phi M \to M \to \Sigma\Omega M \to 0$. To prove that ΦM is $\mathcal{N}il_1$-closed it suffices (because M is $\mathcal{N}il_0$-closed) to show that $\Sigma\Omega M$ is 1-reduced or equivalently ΩM reduced. This is a result of [Z] where it is proved that M is $\mathcal{N}il_0$-closed if and only if M and ΩM are reduced.

Suppose $\Phi^{m-1}M$ is $\mathcal{N}il_{2(m-2)}$-closed, $m \geq 2$. Because Φ is exact, we get from $(*)$ the following exact sequence:

$$0 \to \Phi^m M \to \Phi^{m-1}M \to \Phi^{m-1}(\Sigma\Omega M) = \Sigma^{2^{m-1}}\Phi^{m-1}\Omega M \to 0$$

This shows that $\Phi^m M$ is $\mathcal{N}il_{2(m-1)}$-closed because $\Sigma^{2^{m-1}}\Phi^{m-1}\Omega M$ is 2^{m-1}-reduced and $\Phi^{m-1}M$ is $\mathcal{N}il_{2(m-2)}$-closed. ∎

1.20 Lemma. *Let $M \in \mathcal{U}$ then, for any $k \geq 0$, the natural map $\Phi^k\tilde{\Phi}^k M \to M$ is a monomorphism.*

Proof: If suffices to prove the lemma for $k = 1$ (the case $k = 0$ is trivial). By induction on n we prove the lemma for Brown-Gitter modules $J(n)$. For this we use the following Mahowald exact sequence (see (E2)): $0 \to \Sigma J(n-1) \to J(n) \to J(\frac{n}{2}) \to 0$ ($\frac{n}{2}$ is zero if n is odd). This implies that the lemma is true for $H^*V \otimes J(n)$, where V is an elementary abelian 2-group; for this we use the formula of [**LZ1**]: $\tilde{\Phi}(H^*V \otimes J(n)) = H^*V \otimes J(\frac{n}{2})$. For the general case we only need to embed M in $\prod_\alpha H^*V_\alpha \otimes J(n_\alpha)$, $n_\alpha \geq 0$ and V_α elementary abelian 2-groups. The result follows because Φ and $\tilde{\Phi}$ are left exact. ∎

1.21 Proposition. *Assume that M is a $\mathcal{N}il_m$-closed unstable $\mathcal{A}_2^*$-module, then $\Phi^{\alpha_m+1}(\mathcal{N}_0^{-1}(M))$ is contained in M.*

Proof: By lemma 1.20 we have $\Phi^{\alpha_m+1}\tilde{\Phi}^{\alpha_m+1}M$ is contained in M. It suffices to observe that $\mathcal{N}_0^{-1}(\tilde{\Phi}^{\alpha_m+1}) \cong \tilde{\Phi}^{\alpha_m+1}M$ (see proposition 1.19) and that, in general, $\mathcal{N}_0^{-1}(L) \cong \mathcal{N}_0^{-1}(\tilde{\Phi}L)$ for any $L \in \mathcal{U}$. ∎

$\mathcal{N}il_m$-localization of algebras. Let $\mathcal{K}$ denote the category of unstable algebras over the Steenrod algebra and $\mathcal{A}_2^*$-linear algebra maps.

1.22 Definition. *An object K of $\mathcal{K}$ is called m-nilpotent (resp. m-reduced, $\mathcal{N}il_m$-closed) if the underlying unstable $\mathcal{A}_2^*$-module is m-nilpotent (resp. m-reduced, $\mathcal{N}il_m$-closed)*

Assume that M and N are objects of $\mathcal{U}$, $\mu_r\colon M \to \mathcal{N}_r^{-1}(M)$ the $\mathcal{N}il_r$-localization of M and $\mu_s\colon N \to \mathcal{N}_s^{-1}(N)$ the $\mathcal{N}il_s$-localization of N. Then the diagram

$$\begin{array}{ccc} M \otimes N & \xrightarrow{\mu_r \otimes 1} & \mathcal{N}_r^{-1}(M) \otimes N \\ {\scriptstyle 1\otimes\mu_s}\downarrow & & \downarrow{\scriptstyle 1\otimes\mu_s} \\ M \otimes \mathcal{N}_s^{-1}(M) & \xrightarrow{\mu_r \otimes 1} & \mathcal{N}_r^{-1}(M) \otimes \mathcal{N}_s^{-1}(N) \end{array}$$

has $\mathcal{N}il_s$-isomorphism vertical arrows and $\mathcal{N}il_r$-isomorphism horizontal arrows. In particular, $M \otimes N \to \mathcal{N}_r^{-1}(M) \otimes \mathcal{N}_s^{-1}(N)$ is $\mathcal{N}il_{\min(r,s)}$-isomorphism. So we have a natural commutative diagram:

$$\begin{array}{ccc} M \otimes N & \longrightarrow & \mathcal{N}_r^{-1}(M) \otimes \mathcal{N}_s^{-1}(N) \\ & \searrow \quad \swarrow & \\ & \mathcal{N}^{-1}_{\min(r,s)}(M \otimes N) & \end{array}$$

in which all arrows are $\mathcal{N}il_{\min(r,s)}$-isomorphisms. We are interested in the case $r = s$:

1.23 Proposition. (1) *For every M, N there is a natural commutative diagram*

$$\begin{array}{ccc} M \otimes N & \xrightarrow{\mu_s \otimes \mu_s} & \mathcal{N}_s^{-1}(M) \otimes \mathcal{N}_s^{-1}(N) \\ & \searrow \quad \swarrow & \\ & \mathcal{N}_s^{-1}(M \otimes N) & \end{array}$$

in which every arrow is a $\mathcal{N}il_s$-isomorphism.

(2) *If M and N are reduced, then $\mathcal{N}_s^{-1}(M) \otimes \mathcal{N}_s^{-1}(N) \xrightarrow{\cong} \mathcal{N}_s^{-1}(M \otimes N)$*

Proof: It only remains to show that if M and N are reduced then $\mathcal{N}_s^{-1}(M) \otimes \mathcal{N}_s^{-1}(N)$ is $\mathcal{N}il_s$-closed. This follows from 1.17 ■

1.24 Proposition. *Let K be an object of $\mathcal{K}$, then $\mathcal{N}_s^{-1}(K)$ has a natural algebra structure such that*

$$\mu_s: K \longrightarrow \mathcal{N}_s^{-1}(K)$$

is a morphism of $\mathcal{K}$.

Proof: Let $\phi: K \otimes K \to K$ be the multiplication. We have a natural commutative diagram:

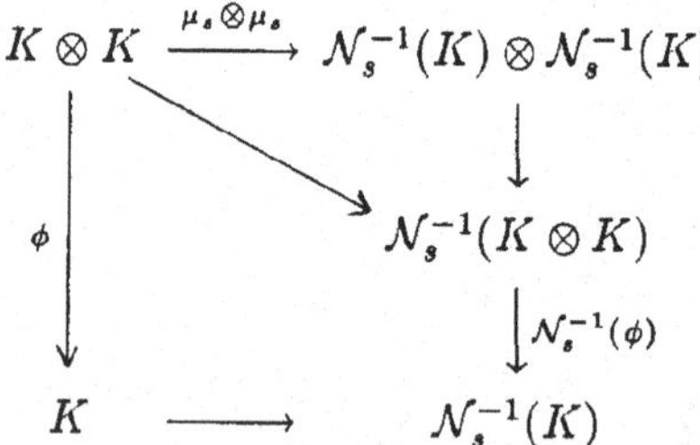

defining so a multiplication in $\mathcal{N}_s^{-1}(K)$ that endows $\mathcal{N}_s^{-1}(K)$ with an $\mathcal{A}_2^*$-algebra structure. It remains to show the unstability formula $Sq^{|x|}x = x^2$. For an $\mathcal{A}_2^*$-algebra K this is equivalent to the commutativity of the diagram:

$$\begin{array}{ccc} (K \otimes K)^{\Sigma_2} & \xrightarrow{\delta} & \Phi K \\ m \downarrow & \swarrow Sq_0 & \\ K & & \end{array}$$

where δ is the surjection in the exact sequence $0 \to \Lambda^2 K \to (K \otimes K)^{\Sigma_2} \xrightarrow{\delta} \Phi K \to 0$ (see [**GLZ**]). It is not hard to get the corresponding diagram for $\mathcal{N}_s^{-1}(K)$ from the $\mathcal{N}il_s$-localization of the one given for K. ■

§2 $\mathcal{N}il_m$-localization of sub-$\mathcal{A}_2^*$-algebras of H^*V

In this paragraph we generalize the result of Adams-Wilkerson [**AW**] (see also [**BZ**]) concerning the computation of the $\mathcal{N}il_0$-localization of some noetherian unstable $\mathcal{A}_2^*$-algebras. More precisely we compute the $\mathcal{N}il_m$-localization of such algebras for $m \geq 0$. (See [**HLS2**] for the case of H^*BG, G a compact Lie group.)

Let V be an elementary abelian 2-group ($V \cong (\mathbf{Z}/2)^d, d \geq 0$), K be a sub-$\mathcal{A}_2^*$-algebra of H^*V and $m \geq 1$. We write $m = 2^{\alpha_m} + l$, $0 \leq l \leq 2^{\alpha_m} - 1$.

2.1 Proposition. *Let K be a sub-$\mathcal{A}_2^*$-algebra of H^*V. There exists a smallest sub-$\mathcal{A}_2^*$-algebra $R_m(K)$, $m \geq 0$, in H^*V with the following properties:*

(1) *K is contained in $R_m(K)$.*
(2) *$R_m(K)$ is $\mathcal{N}il_m$-closed.*
(3) *$\mathcal{N}_0^{-1}(R_m(K)) = H^*V$.*

Proof: Let $R_m(K) = \mathcal{N}_m^{-1}(\langle K, \Phi^{\alpha_m+1}H^*V\rangle)$ the $\mathcal{N}il_m$-localization of the sub-$\mathcal{A}_2^*$-algebra of H^*V generated by $K + \Phi^{\alpha_m+1}H^*V$. Evedently we have (1) and (2). The point (3) is clear because of the following: (*) $\Phi^{\alpha_m+1}H^*V \hookrightarrow R_m(K) \hookrightarrow H^*V$ and (**) $\mathcal{N}_0^{-1}(\Phi H^*V) \cong H^*V$.

To prove that $R_m(K)$ is the smallest satisfying (1) (2) and (3) we consider B a sub-$\mathcal{A}_2^*$-algebra of H^*V which verifies (1), (2) and (3). By proposition 1.21, we have $\Phi^{\alpha_m+1}H^*V$ is contained in B. The point (1) implies taht K is in B so the algebra $\langle K, \Phi^{\alpha_m+1}H^*V\rangle$ generated by $K + \Phi^{\alpha_m+1}H^*V$ is contained in B and, since B is $\mathcal{N}il_m$-closed, $R_m(K)$ is contained in B. ∎

2.2 Examples of algebras satisfying (2) and (3) of proposition 2.1: Observe that an algebra R satisfying (2) and (3) of proposition 2.1, has an injective resolution that starts

$$0 \to R \to H^*V \to \prod_{\beta} H^*V_\beta \otimes J(n_\beta) \qquad 0 \leq n_\beta \leq m$$

(1) $\Phi^{\alpha+1}H^*\mathbf{Z}/2$ is the kernel of the map $H^*\mathbf{Z}/2 \to H^*\mathbf{Z}/2 \otimes J(2^\alpha)$ obtained from the diagonal $H^*\mathbf{Z}/2 \to H^*\mathbf{Z}/2 \otimes H^*\mathbf{Z}/2$ and the only non trivial map $H^*\mathbf{Z}/2 \to J(2^\alpha)$. Then $\Phi^{\alpha+1}H^*\mathbf{Z}/2$ is $\mathcal{N}il_{2^\alpha}$-closed and $\mathcal{N}_0^{-1}(\Phi^{\alpha+1}H^*\mathbf{Z}/2) = H^*\mathbf{Z}/2$.
(2) Define $\tau_m H^*\mathbf{Z}/2$ as the kernel of the product of the non trivial maps $H^*\mathbf{Z}/2 \to \prod_{j=1}^m J(j)$, then $\tau_m H^*\mathbf{Z}/2$ is $\mathcal{N}il_m$-closed and $\mathcal{N}_0^{-1}(\tau_m H^*\mathbf{Z}/2) = H^*\mathbf{Z}/2$. Observe that $\tau_m H^*\mathbf{Z}/2 = H^*(\mathbf{R}P^\infty/\mathbf{R}P^m)$.
(3) Proposition 1.17 and example (1) above show that if $0 \leq r_1 \leq r_2 \leq \cdots \leq r_n$ then $\mathbf{F}_2[u_1^{2^{r_1}}, \ldots, u_n^{2^{r_n}}] = \Phi^{r_1}\mathbf{F}_2[u_1] \otimes \cdots \otimes \Phi^{r_n}\mathbf{F}_2[u_n]$ is $\mathcal{N}il_{(2^{r_n-1})}$-closed if $r_n \neq 0$ and $\mathcal{N}il_0$-closed if $r_n = 0$. By a result due to C. Wilkerson (see [**K**]) if our algebra is noetherian and integrally closed then it is of this form.

Assume that K is an integral domain. We say that K is weakly integrally closed [**BZ**] if given an element x in the field of fractions of K such that: $x^n + a_1x^{n-1} + \ldots + a_n = 0$, $a_i \in K$, there is an integer $l \geq 0$ such that $x^{2^l} \in K$. Then Adams-Wilkerson embedding states that if K is noetherian, integral domain and weakly integrally closed object of

$\mathcal{K}$, then the $\mathcal{N}il_0$-localization of K, $\mathcal{N}_0^{-1}(K)$, is an algebra of invariants: $\mathcal{N}_0^{-1}(K) = (H^*V)^G$, for certain elementary abelian 2-group V and $G \leq GL(V)$.

Our next result states that the $\mathcal{N}il_m$-localization is obtained again as a ring of invariants: the invariant subalgebra of $R_m(K)$ by the action of the same group, G.

2.3 Theorem. *Let K be a connected unstable $\mathcal{A}_2^*$-algebra such that:*

(1) K is an integral domain.
(2) K is noetherian.
(3) K is weakly integrally closed.

Then, there exists an elementary abelian 2-group V and a subgroup G of $GL(V)$ such that, for any $m \geq 0$, the $\mathcal{N}il_m$-localization of K is isomorphic to the ring of invariants $(R_m(K))^G$:

$$\mathcal{N}_m^{-1}(K) \cong (R_m(K))^G.$$

Recent results of H.-W. Henn ([H]) show that an unstable $\mathcal{A}_2^*$-algebra as in the theorem is $\mathcal{N}il_m$-closed for some $m \geq 0$. As a consequence we have:

2.4 Corollary. *Let K be a connected unstable $\mathcal{A}_2^*$-algebra with the properties (1), (2) and (3) of theorem 2.3. Then, there exists V an elementary abelian 2-group, G a subgroup of $GL(V)$ and $m = m(K) \in \mathbf{N}$ such that K is isomorphic to the ring of invariant $\left(R_m(K)\right)^G$.*

2.5 Remark: The action of the group G can be extended to the integral closure of $R_s(K)$ in its field of fractions: $\overline{R_s(K)}$, which is again a sub-$\mathcal{A}_2^*$-algebra of H^*V (see [W]) and therefore of the type in example 2.2(3). It is easy to show that the algebra of invariants $\overline{R_s(K)}^G$ is now the integral closure of $\mathcal{N}_s^{-1}(K)$ in its field of fractions. As a consequence, if K in the corollary above is integrally closed we also have $K = \overline{R_m(K)}^G$.

Proof of theorem: Let K be a connected unstable $\mathcal{A}_2^*$-algebra. Properties (1) and (2) ensure that there exists an elementary abelian 2-group V with dimension the krull-dimension of K and an embedding $K \subset H^*V$, [**AW**]. Moreover if K satisfies (3) then there exists a subgroup $G \leq GL(V)$ such that the $\mathcal{N}il_0$-localization of K is the algebra of invariants: H^*V^G ([**AW**] [**BZ**]).

Suppose $m \geq 1$, then the $\mathcal{N}il_m$-localization will be a sub-algebra of H^*V^G:

$$K \hookrightarrow \mathcal{N}_m^{-1}(K) \hookrightarrow H^*V^G$$

and it is clear that the $\mathcal{N}il_0$-localization of $\mathcal{N}_m^{-1}(K)$ is H^*V^G, then by proposition 1.21 $\Phi^{\alpha_m+1}H^*V^G \subset \mathcal{N}_m^{-1}(K)$ and therefore

$$(1) \qquad \langle K, \Phi^{\alpha_m+1}H^*V^G\rangle \subset \mathcal{N}_m^{-1}(K)$$

Recall from Proposition 2.1 that $R_m(K)$ is defined as

$$R_m(K) = \mathcal{N}_m^{-1}(\langle K, \Phi^{\alpha_m+1}H^*V\rangle)$$

Since $\mathcal{N}il_m$-localization commutes with action of finite groups and the group G acts on $\langle K, \Phi^{\alpha_m+1}H^*V\rangle$, then G acts on $R_m(K)$ and the $\mathcal{N}il_m$-localization of $\langle K, \Phi^{\alpha_m+1}H^*V\rangle^G$ is

$$(2) \qquad \langle K, \Phi^{\alpha_m+1}H^*V\rangle^G \hookrightarrow R_m(K)^G$$

Finally $\mathcal{N}_m^{-1}(K) \subset \mathcal{N}_m^{-1}\langle K, \Phi^{\alpha_m+1}H^*V^G\rangle \subset R_m(K)^G$ so we have the following diagram

$$\begin{array}{ccc} \langle K, \Phi^{\alpha_m+1}H^*V^G\rangle & \xrightarrow{i_1} & \mathcal{N}_m^{-1}(K) \\ {\scriptstyle j_1}\downarrow & & \downarrow{\scriptstyle j_2} \\ \langle K, \Phi^{\alpha_m+1}H^*V\rangle^G & \xrightarrow[i_2]{} & R_m(K)^G \end{array}$$

The inclusions $K \subset \langle K, \Phi^{\alpha_m+1}H^*V^G\rangle \subset \mathcal{N}_m^{-1}(K)$ shows that i_1 is $\mathcal{N}il_m$-isomorphism. i_2 is so.

In order to prove that j_2 is a $\mathcal{N}il_m$-isomorphism it suffices to prove that j_1 is $\mathcal{N}il_m$-isomorphism. Once we have this, it is clear that j_2 is actually an isomorphism since both, $\mathcal{N}_m^{-1}(K)$ and $R_m(K)^G$ are $\mathcal{N}il_m$-closed.

To finish the proof of the theorem it only remains to show:

2.6 Lemma. *Let G be a subgroup of $GL(V)$ and K a sub-$\mathcal{A}_2^*$-algebra of H^*V which is invariant under the action of G on H^*V; that is, $K \subset H^*V^G$. Then the inclusion*

$$\langle K, \Phi^\alpha H^*V^G\rangle \subset \langle K, \Phi^\alpha H^*V\rangle^G$$

is a $\mathcal{N}il_{2^\alpha-1}$-isomorphism.

Proof: We first observe that this inclusion is a $\mathcal{N}il_{2^\alpha-1}$-isomorphism if and only if the inclusion in degree i:

$$T_W^i\left(\langle K, \Phi^\alpha H^*V^G\rangle\right) \subset T_W^i\left(\langle K, \Phi^\alpha H^*V\rangle^G\right)$$

is an equality for any elementary abelian 2-group, W, and $0 \le i \le 2^\alpha - 1$. Next, the facts: (1) $T_W\langle M\rangle \cong \langle T_W M\rangle$ for any sub-$\mathcal{A}_2^*$-module M of an algebra H, and (2) $(\Phi^\alpha H^*V)^i = 0$ if $1 \le i \le 2^\alpha - 1$, reduces the problem to prove that the inclusion:

$$\langle T_W K, T_W^0 H^*V^G\rangle \subset \langle T_W K, T_W^0 H^*V\rangle^G$$

is actually an equality.

For this recall ([**LZ2**]) that the dual of $T_W^0 H^*V$ is a $\mathbf{F}_2$-vector space with basis $\mathcal{L}(W,V)$, the set of $\mathbf{F}_2$-linear maps from W to V. We denote by $\{e_f | f \in \mathcal{L}(W,V)\}$ a basis of $T_W^0 H^*V$ dual to $\mathcal{L}(W,V)$ and verify that $e_f^2 = e_f$, $e_f e_h = 0$ if $f \ne h$ and $1 = \sum_{f\in\mathcal{L}(W,V)} e_f$. Moreover, $GL(V)$ acts on $T_W^0 H^*V$ by permuting the elements of this basis: $g(e_f) = e_{fg^{-1}}$ for $f \in \mathcal{L}(W,V)$, $g \in GL(V)$.

In order to simplify our notation we will write $[a]$ for the sum of all elements in the G-orbit of the element a.

Now, an element $x \in \langle T_W K, T_W^0 H^* V\rangle$ is written as $x = \sum_{f\in\mathcal{L}(W,V)} x_f e_f$, $x_f \in T_W K$. Assume that this element is G-invariant: $x \in \langle T_W K, T_W^0 H^* V\rangle^G$, that is, $g(x) = x$, for all $g \in G$:

$$\sum_{f\in\mathcal{L}(W,V)} x_f e_f = g\Big(\sum_{f\in\mathcal{L}(W,V)} x_f e_f\Big) = \sum_{f\in\mathcal{L}(W,V)} x_f e_{fg^{-1}}$$

Multiply this equality by $e_{fg^{-1}}$: $x_{fg^{-1}} e_{fg^{-1}} = x_f e_{fg^{-1}} = g(x_f e_f)$. Repeat it for all $g \in G$. It follows that all the elements in the orbit of $x_f e_f$ appear in x, that is, x can be written as:

$$x = [x_{f_1} e_{f_1}] + \cdots + [x_{f_k} e_{f_k}]\,.$$

Finally, it is easy to show that the orbits of $x_f e_f$ and e_f have the same number of elements and then that $[x_f e_f] = x_f [e_f]$. So we have

$$x = x_{f_1}[e_{f_1}] + x_{f_2}[e_{f_2}] + \cdots + x_{f_k}[e_{f_k}]$$

This means $x \in \langle T_W K, T_W^0 H^* V^G\rangle$. ∎

§3 The odd prime case

Now, $\mathcal{U}$ (resp. $\mathcal{K}$) denote the category of unstable $\mathcal{A}_p^*$-modules and $\mathcal{A}_p^*$-linear maps of degree zero (resp. unstable $\mathcal{A}_p^*$-algebras and $\mathcal{A}_p^*$-linear algebra maps of degree zero) for an odd prime p.

In this paragraph we will discuss the modifications that are needed in order to obtain the analogue of theorem 2.3 at odd primes. For this we introduce the full subcategory $\mathcal{C}' = \mathcal{U}', \mathcal{K}'$ of $\mathcal{C} = \mathcal{U}, \mathcal{K}$ whose objects are concentrated in even degree. We denote by $\mathcal{O}: \mathcal{C}' \to \mathcal{C}$ the forgetful functor and by $\tilde{\mathcal{O}}: \mathcal{C} \to \mathcal{C}'$ its left adjoint functor (see [**LZ1**], [**GSZ**]).

The notion of m-nilpotent, m-reduced and $\mathcal{N}il_m$-closed are defined in $\mathcal{U}$ (resp. $\mathcal{K}$) in the same way as in the case $p = 2$ (see paragraph 2) and most of their properties are still true for $p > 2$. The only result which needs modification is that given by proposition 1.21. More precisely, let $\Phi: \mathcal{U} \to \mathcal{U}$ the functor defined by:

$$(\Phi M)^n = \begin{cases} M^{n/2p} \text{ if } n \equiv 0\ (2p) \\ M^{\frac{n-2}{2p}+1} \text{ if } n \equiv 2\ (2p) \\ 0 \quad \text{otherwise} \end{cases} \qquad \text{and}$$

$$\begin{cases} \mathcal{P}^i(\Phi x) = \Phi(\mathcal{P}^{i/p} x) \quad \text{if } |x| \equiv 0\ (2p) \\ \mathcal{P}^i(\Phi x) = \Phi(\mathcal{P}^{i/p} x) + \Phi(\beta \mathcal{P}^{\frac{i-1}{p}} x) \text{ if } |x| \equiv 0\ (2p) \\ \beta(\Phi x) = 0 \qquad \forall x \in M \end{cases}$$

where $|x|$ is the degree of x and $\mathcal{P}^{j/p} = 0$ if $j \not\equiv 0 \pmod p$. We denote by $\tilde{\Phi}: \mathcal{U} \to \mathcal{U}$ its left adjoint functor (see [**LZ1**]).

Let V be an elementary abelian p-group of rank ≥ 2 then the map $\lambda: \Phi H^*V \to H^*V$, $\Phi x \longmapsto \beta^\varepsilon P^i x$ if $|x| = 2i+\varepsilon$; $\varepsilon = 0,1$, is not injective. This shows that the proposition 1.21 is not true for $p > 2$ (H^*V is $\mathcal{N}il_0$-closed so, in particular, is $\mathcal{N}il_1$-closed). For $m \geq 1$ we denote by p^{α_m} the biggest power of p in m: $m = p^{\alpha_m} + l$, $0 \leq l \leq p^{\alpha_m} - 1$. We have the following analogue of proposition 1.21, for $p > 2$.

3.1 Proposition. *Assume that M is a $\mathcal{N}il_m$-closed unstable $\mathcal{A}_p^*$-module. Then $\Phi^{\alpha_m+1}\left(\tilde{O}\mathcal{N}_o^{-1}(M)\right)$ is contained in M.*

To prove proposition 3.1 we need the following lemma. (Compare with 1.20).

3.2 Lemma. *Let $M \in \mathcal{U}$ then, for any $k \geq 0$, the natural map $\Phi^k\tilde{O}\tilde{\Phi}^k M \to M$ is a monomorphism.*

Proof: The map $\Phi^k\tilde{O}\tilde{\Phi}^k M \to M$ is the composition $\Phi^k\tilde{O}\tilde{\Phi}^k M \to \Phi^k\tilde{\Phi}^k M \to M$ where the first map is induced by the inclusion $\tilde{O}\tilde{\Phi}^k M \hookrightarrow \tilde{\Phi}^k M$ and the second is the natural map. It suffices to prove the lemma for $k = 1$ ($k = 0$ is trivial). For this we follow the idea of the proof of Lemma 1.20. ■

Proof of proposition 3.1: Now one proves this proposition as in the case $p = 2$. ■

We are ready to state the main theorem about the $\mathcal{N}il_m$-localization of sub-$\mathcal{A}_p^*$-algebras of H^*V. Let K be a sub-$\mathcal{A}_p^*$-algebra of H^*V then, the algebra $R_m(K) = \mathcal{N}_m^{-1}(\langle K, \Phi^{\alpha_m+1}\tilde{O}H^*V\rangle)$, (recall that $\tilde{O}H^*V$ is the polynomial part of H^*V see [**LZ1**]), is the smallest sub-$\mathcal{A}_p^*$-algebra of H^*V with the following properties:

(1) K is contained in $R_m(K)$.
(2) $R_m(K)$ is $\mathcal{N}il_m$-closed.
(3) $\mathcal{N}_0^{-1}(R_m(K)) = H^*V$.

(same proof as for $p = 2$, see proposition 2.1).

3.3 Theorem. *Let K be a connected and reduced unstable $\mathcal{A}_p^*$-algebra such that:*

(1) *$\tilde{O}K$ is an integral domain.*
(2) *$\tilde{O}K$ is noetherian.*
(3) *$\tilde{O}K$ is weakly integrally closed.*

Then, there exists an elementary abelian p-group V and a subgroup G of $GL(V)$ such that, for any $m \geq 0$, the $\mathcal{N}il_m$-localization of K is isomorphic to the ring of invariant $(R_m(K))^G$.

Proof: Let K be a connected unstable $\mathcal{A}_p^*$-algebra. Properties (1) and (2) ensure that there exists an elementary abelian p-group V and an embedding $K \subset H^*V$, [**BZ**]. Moreover if K satisfies (3) then there exists a subgroup $G \leq GL(V)$ such that the $\mathcal{N}il_0$-localization of K is the algebra of invariants H^*V^G, [**BZ**].

Having in mind the new construction of $R_m(K)$ and that $T_W^0 H^*V \cong T_W^0(\tilde{O}H^*V)$ the proof goes like in the case $p = 2$ for $m \geq 1$. ■

References

[AW] J.F. Adams and C.W. Wilkerson, Finite H-spaces and algebras over the Steenrod algebra, *Ann. of Math.* **111** (1980), 95–143.

[BZ] C. Broto and S. Zarati, Nil-localization of unstable algebras over the Steenrod algebra, *Math. Zeit.* **199** (1988), 525–537.

[C] G. Carlsson, G.B. Segal burnside ring conjecture for $(\mathbf{Z}/2)^k$, *Topology* **22** (1983), 83–102.

[GLZ] J.H. Gunawardena, J. Lannes and S. Zarati, Cohomologie des groupes symétriques et application de Quillen, in "*Homotopy Theory*," L.M.S., Camb. Univ. Press, 1990.

[GSZ] P. Goerss, L. Smith and S. Zarati, Sur les $\mathcal{A}_p^*$-algèbres instables, in "*Algebraic Topology, Barcelona 1986*," Lecture Notes in Math. vol. 1298, Springer, 1987.

[H] H.W. Henn, Some finiteness results in the category of unstable modules over the Steenrod algebra and application to stable splittings; Preprint.

[HLS] H.W. Henn, J. Lannes and L. Schwartz, The categories of unstable modules and unstable algebras modulo nilpotent objects; Preprint.

[HLS2] ————, Nil-localization of cohomology of BG; In preparation.

[K] R.M. Kane, "*The homology of Hopf Spaces*," North Holland Math. Library, 1988.

[L] J. Lannes, Sur la cohomologie modulo p des p-groupes abeliens élémentaries, *L. M. S, Camb. Univ. Press*; Proc. Durham Symposium on Homotopy Theory 1985, (1987), 97–116.

[LS] J. Lannes and L. Schwartz, Sur la structure des $\mathcal{A}_p^*$-modules instables injectifs, *Topology* **28** (1989), 153–169.

[LZ1] J. Lannes and S. Zarati, Sur les $\mathcal{U}$-injectifs, *Ann. Scient. Ec. Norm. Sup.* **19** (1986), 303–333.

[LZ2] ————, Foncteurs dérivés de la déstabilisation, *Math. Z.* **194** (1987), 25–59.

[M] H.R. Miller, The Sullivan conjecture on maps from classifying spaces, *Annals of Math* **120** (1984), 39–87.

[S] L. Schwartz, La filtration nilpotente de la catégorie $\mathcal{U}$ et la cohomologie des espaces de lacets, in "*Algebraic Topology Rational Homotopy, Louvain 1986*," Lecture Notes in Math 1318, Springer, 1988.

[Sw] R.G. Swan, "*Algebraic K-Theory*," L.N.M. 76, 1968.

[W] C.W. Wilkerson, Integral closure of unstable Steenrod algebra actions, *JPAA* **13** (1978), 49–55.

[Z] S. ZARATI, Derived functors of the Destabilization functor and the Adams spectral sequence, in "*Algebraic Topology, Luminy 1988;*" to appear in Astérisque.

C. Broto: Universitat Autònoma de Barcelona,
Departament de Matemàtiques,
E-08193 Bellaterra (Barcelona)
Spain.

S. Zarati: Université de Tunis,
Faculté des Sciences - Mathematics
TN-1060 Tunis,
Tunisie
and
Centre de Recerca Matemàtica
E-08193 Bellaterra (Barcelona)
Spain.

1990 Barcelona Conference
on Algebraic Topology.

THE CLASSIFICATION OF 3-MANIFOLDS WITH SPINES RELATED TO FIBONACCI GROUPS (*)

ALBERTO CAVICCHIOLI AND FULVIA SPAGGIARI

Abstract
We study the topological structure of closed connected orientable 3–manifolds which admit spines corresponding to the standard presentation of Fibonacci groups.

1. Introduction.

A spine of a closed 3–manifold M is a 2–dimensional subpolyhedron such that $M\setminus$ (open 3–cell) collapses onto it. It is well–known that any closed 3–manifold admits spines with a single vertex, corresponding to suitable presentations of its fundamental group. On the other hand, let $P =< x_1, x_2, ..., x_n/r_1, r_2, ..., r_n >$ be a group presentation and let $K(P)$ be the canonical 2–complex associated to P. By definition, $K(P)$ has one vertex v and n 1–cells (resp. 2–cells) corresponding to generators (resp. relators) of P. We always label each 1-cell of $K(P)$ by the corresponding generator x_i of P and denote the 2–cells of $K(P)$ by $c_1, c_2, ..., c_n$. Recall that the fundamental group $\Pi_1(K(P), v)$ is isomorphic to the group presented by P. Through the paper we will identify a group presentation with its associated group.

In [**14**] L. Neuwirth described an algorithm for deciding when $K(P)$ is a spine of a closed orientable 3–manifold. He used three permutations directly deduced from P (also compare section 2). Then other authors investigated some group presentations of special forms by using the Neuwirth algorithm (see for example [4], [**16**], [**17**], [**22**]).

Here we are interested in the standard presentation of the Fibonacci group $F(r, n)$ so defined:

$$F(r,n) =< x_1, x_2, ..., x_n / x_i x_{i+1} \dots x_{i+r-1} x_{i+r}^{-1} = 1 \quad (i = 1, 2, \dots, n; \text{ indices mod n}) >$$

for any two positive integers r, n.

(*) Work performed under the auspicies of the G.N.S.A.G.A. of the C.N.R. and financially supported by Ministero della Ricerca Scientifica e Tecnologica of Italy within the project "Geometria Reale e Complessa".

AMS Subject classification (1980): 57 M 05 , 57 M 12 , 57 N 10. Key words and phrases: 3–manifold, spine, Fibonacci group, Seifert fibered space, branched covering, Heegaard diagram, knot, link.

Indeed, it was proved in [4] that $F(n-1,n)$, $n \geq 3$, corresponds to a spine of the Heegaard genus $n-1$ Seifert fibered 3–manifold

$$(0\ o\ 0 \mid -1 \underbrace{(2,1)(2,1)\dots(2,1)}_{n\ \text{times}}).$$

Thus we are motivated to study the same problem for Fibonacci groups $F(r,n)$, $r \neq n-1$. More precisely, let $K(r,n)$ be the canonical 2–complex associated to $F(r,n)$. Then we wish to know if $K(r,n)$ is a spine of a closed 3–manifold $M(r,n)$ and, in this case, we will study the topological structure of $M(r,n)$. In order to do this, we also use some algebraic results about Fibonacci groups for which we refer to [**1**], [**10**], [**11**], [**12**], [**24**], [**25**].

A 3–manifold M is said to be irreducible if any embedded 2–sphere in M bounds a 3–ball. We say that M is prime if for any decomposition $M = M_1 \# M_2$ as a connected sum, one of M_1, M_2 is homeomorphic to S^3 (3–sphere). Clearly an irreducible manifold is prime. The converse is not true but the only 3–manifolds which are prime and not irreducible are the two S^2–bundles over S^1. All considered manifolds will be closed, connected and orientable. For basic results about 3–manifold topology see [**2**], [**9**], [**13**], [**15**], [**19**].

2. The Neuwirth theorem.

As above, let us given a group presentation P and its canonical 2–complex $K(P)$. Let V, N_i be regular neighbourhoods of v, x_i in $K(P)$ respectively, and let us denote the points of $x_i \cap \partial V$ by $e_i, \bar{e}_i$. Let e_i^h, $\bar{e}_i^h$ be the points of ∂N_i which lie in regular neighbourhoods of e_i, $\bar{e}_i$ in ∂V respectively ($h = 1,2,\dots,\alpha(i)$; $i = 1,2,\dots,n$). Further, we can suppose that e_i^h and $\bar{e}_i^h$ are joined by an arc in $\partial N_i \backslash V$. Then we set $E_i = \{e_i^h \mid h = 1,2,\dots,\alpha(i)\}$, $\overline{E_i} = \{\bar{e}_i^h \mid h = 1,2,\dots,\alpha(i)\}$ and $E = \cup_{i=1}^{n}(E_i \cup \overline{E_i})$. A simple curve near each ∂c_j intersecs ∂V in a set of simple arcs γ_r with endpoints in E. Interchanging the endpoints of those arcs γ_r's defines an involution $\mathbf{A}(P)$ of E. Let $\mathbf{B}(P)$ be the involutory permutation of E in which e_i^h and $\bar{e}_i^h$ are corresponding points. An arbitrary numbering of the elements of E around each vertex e_i (resp. $\bar{e}_i$) determines a permutation $\mathbf{C}(P)$ of E whose orbit sets are E_i, $\overline{E_i}$. With the above notation, we have the following result proved in [**14**].

THEOREM 1. *$K(P)$ is a spine of a closed prime orientable 3–manifold $M(P)$ if and only if the permutation group generated by $\mathbf{A}(P)\mathbf{C}(P)$ and $\mathbf{B}(P)\mathbf{C}(P)$ (resp. $\mathbf{A}(P)$ and $\mathbf{C}(P)$) is transitive and the relation*

$$|\mathbf{A}(P)| - |\mathbf{C}(P)| + 2 = |\mathbf{A}(P)\mathbf{C}(P)|$$

holds.
Here $|\theta|$ denotes the number of cycles of a permutation $\theta : E \longrightarrow E$.

By selecting all possible cyclic orderings of the elements of E_i together their opposites in $\overline{E_i}$, we may determine which permutations $\mathbf{C}(P)$ satisfy theorem 1. In this manner we obtain a catalogue with possible repetitions (also using a computer) of all orientable prime closed 3–manifolds having $K(P)$ as spine. As pointed out in [**3**] and [**4**], the permutations $\mathbf{A}(P)$ and $\mathbf{C}(P)$ allow us to draw a Heegaard diagram $H(P)$ which represents $M(P)$. Recall that a Heegaard splitting of a closed orientable 3–manifold M is a closed connected orientable surface T_g of genus g imbedded in M and dividing M into two homeomorphic handlebodies. The Heegaard genus of M is the smallest integer g such that M has a Heegaard splitting of genus g. The Heegaard diagram of M consists of the splitting surface upon which are drawn a certain number of non–intersecting simple closed curves. These curves describe the manner in which the homeomorphic handlebodies are attached to obtain the manifold M (see [**9**]). Now the unbarred cycles of $\mathbf{C}(P)$ represent the holes of the splitting surface and the set of arcs joining $\mathbf{A}(P)$–correspondent points induces the system of non–intersecting simple closed curves (for details see [**3**], [**4**]). In [**4**] it is also described a very simple construction to obtain a 4–coloured graph $G(P)$ representing $M(P)$ directly deduced from the Heegaard diagram $H(P)$. Here the manifold representation by coloured graphs is meant in the usual sense, i. e. it is given by taking the 1–skeleton of the cellular subdivision dual to a suitable triangulation (minimal with respect to the vertices) of a 3–manifold and by labelling the dual of each 2–simplex by the vertex it does not contain (see for example [**4**], [**8**] and their references). Finally we recall that, for any two 4–coloured graphs of the same manifold, a finite sequence of moves (named adding or cancelling dipoles) exists which transform one graph to the other. Thus we will often say reduced the graph $G^*(P)$ obtained from $G(P)$ by cancelling all possible dipoles.

3. Lens spaces.

In this section we consider the Fibonacci groups $F(1,n) \cong Z$ and $F(r,1) \cong Z_{r-1}$ for any two positive integers n, r , $r \geq 2$. We have the following

Proposition 2.

1) *$S^1 \times S^2$ is the unique closed orientable prime 3–manifold which admits $K(1,n)$ as spine , $n \geq 1$.*
2) *For any two positive integers r, p , $r \geq 2$, $(r-1, p) = 1$, $K(r,1)$ is a spine of the lens space $L(r-1, p)$.*

Proof: 1) Let us consider the standard presentation

$$F(1,n) = < x_i / x_i x_{i+1}^{-1} = 1 \ \ (i = 1, 2, \ldots, n \ \text{, indices mod } n) > .$$

As usual, we denote the oriented 1–cells of $K(1,n)$ by $x_1, x_2, \ldots, x_n$ and the 2–cells of $K(1,n)$ by $c_1, c_2, \ldots, c_n$. Then ∂c_i is attached to the 1–skeleton $K^{(1)}(1,n) = x_1 \vee x_2 \vee \cdots \vee x_n$ by a map f_i representing the word $x_i x_{i+1}^{-1} = 1$ (see fig. 1). We choose a numbering of the elements of $E = E(1,n)$ so that an appropriate closed simple curve parallel to and near ∂c_i intersects (i,i), $\overline{(i,i)}$, $\overline{(i,i+1)}$, $(i,i+1)$, pairs mod n, in this order. Thus the set $E(1,n)$ consists of the following elements

$$E(1,n) = \left\{(i,i), \overline{(i,i)}, \overline{(i,i+1)}, (i,i+1) / i = 1, 2, \ldots, n, \text{pairs mod } n\right\}.$$

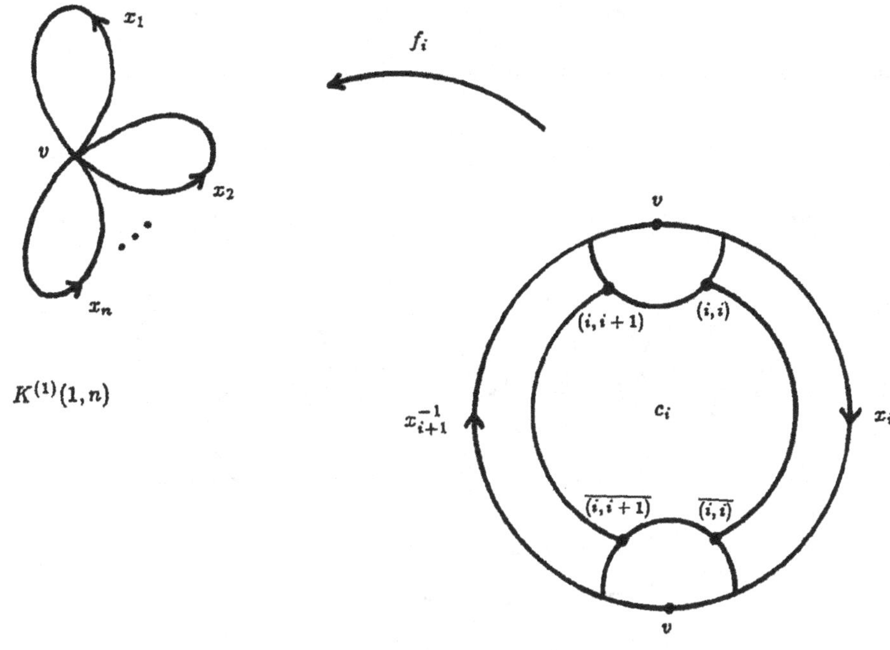

fig. 1

Then we have

$$\mathbf{A}(1,n)=\prod_{i=1}^{n}\big((i,i)(i,i+1)\big)\big(\overline{(i,i)}\ \overline{(i,i+1)}\big)$$

$$\mathbf{B}(1,n)=\prod_{i=1}^{n}\big((i,i)\overline{(i,i)}\big)\big((i,i+1)\overline{(i,i+1)}\big)$$

and $\mathbf{C}(1,n)$ cyclically acts on the orbit sets

$$E_i=\{(i-1,i),(i,i)\}\ ,\ \overline{E_i}=\{\overline{(i-1,i)},\overline{(i,i)}\},\quad i=1,2,\dots,n,$$

pairs mod n.

Therefore there is exactly one possibility for $\mathbf{C}(1,n)$, i. e.

$$\mathbf{C}(1,n)=\prod_{i=1}^{n}\big((i-1,i)(i,i)\big)\big(\overline{(i-1,i)}\ \overline{(i,i)}\big).$$

Since $|\mathbf{A}(1,n)|=|\mathbf{C}(1,n)|=2n$ and

$$\mathbf{A}(1,n)\mathbf{C}(1,n)=\big((1,1)(2,2)\dots(n,n)\big)\big(\overline{(1,1)}\ \overline{(2,2)}\dots\overline{(n,n)}\big),$$

the relation $|\mathbf{A}|-|\mathbf{C}|+2=|\mathbf{AC}|$ is verified. Further, it is easily checked that $\mathbf{A}(1,n)$ and $\mathbf{C}(1,n)$ (resp. $\mathbf{A}(1,n)\mathbf{C}(1,n)$ and $\mathbf{B}(1,n)\mathbf{C}(1,n)$) generate a transitive group. Thus theorem 1 implies that there exists a unique closed 3–manifold $M(1,n)$ with spine $K(1,n)$. In fig. 2 we show the Heegaard diagram $H(1,n)$ of $M(1,n)$ determined by the permutations $\mathbf{A}(1,n)$ and $\mathbf{C}(1,n)$.

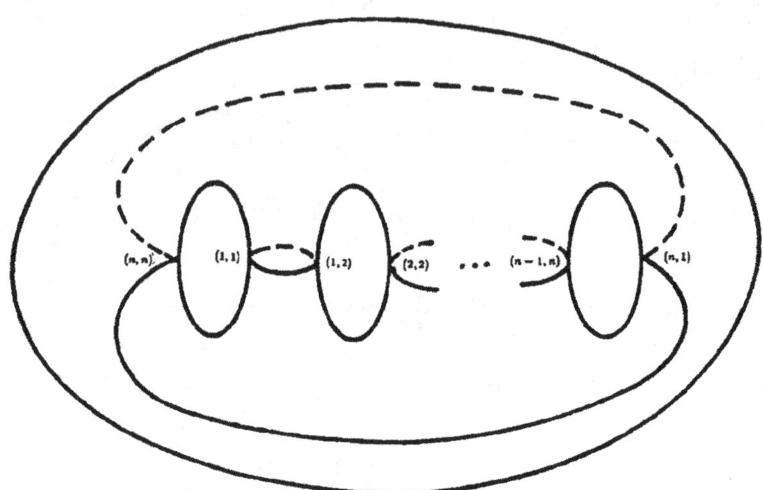

The Heegaard $H(1,n)$ of $S^1 \times S^2$

fig. 2

Now $H(1,n)$ is easily proved to be equivalent to the standard genus one diagram of $S^1 \times S^2$ by using the Singer moves (see [**21**]).
2) The standard genus one Heegaard diagram (see [**9**]) of $L(r-1,p)$ corresponds to a spine whose fundamental group has the presentation

$$\Pi_1\left(L(r-1,p)\right) \cong < x_1 / x_1^{r-1} = 1 > \cong Z_{r-1}.$$

We can slightly isotope this diagram to obtain another one which corresponds to $F(r,1) =< x_1/x_1^r x_1^{-1} = 1 >$.
Indeed, we use the following permutations:

$$\mathbf{A}(r,1) = (1\ r+1)(\bar{r}\ \overline{r+1}) \prod_{i=1}^{r-1} (\bar{i}\ i+p)$$

$$\mathbf{B}(r,1) = \prod_{i=1}^{r+1} (i\ \bar{i})$$

$$\mathbf{C}(r,1) = (1\ 2 \ldots r+1)(\overline{r+1} \ldots \bar{2}\ \bar{1})$$

$$(i \bmod r-1).$$

4. Cyclic branched coverings.

In this section we consider the Fibonacci groups $F(2,n)$, $n \geq 2$. It is well-known that $F(2,2) \cong \{1\}$, $F(2,3) \cong Q_8$ (quaternion group), $F(2,4) \cong Z_5$, $F(2,5) \cong Z_{11}$, $F(2,7) \cong Z_{29}$ and $F(2,n)$ is infinite for either $n = 6$ or $n \geq 8$. In [4] it was proved that the Seifert manifold $(0\ o\ 0| -1\ (2,1)(2,1)(2,1))$ is the unique closed orientable 3-manifold which admits $K(2,3)$ as spine. However, we have verified by direct computations that $K(2,5)$ and $K(2,7)$ can not be spines of a closed 3-manifold. Thus it seems natural to think that there are no 3-manifolds with spine $K(2,2m+1)$, $m \geq 2$. On the contrary, for $n = 2m$, we have theorem 3 below. In order to prove it we need some notations and results about the Kirby–Rolfsen calculus on links with coefficients (see [**19**]).

Let $L = L_1 \cup L_2 \cup \ldots \cup L_n$ be a tame oriented link in S^3. We consider disjoint closed tubular neighbourhoods N_i of L_i in S^3 and specify a preferred framing for N_i in which the longitude λ_i is oriented in the same way as L_i and the meridian μ_i has linking number $+1$ with L_i. The result of a Dehn surgery on S^3 along L with surgery coefficient $r_1, r_2, \ldots, r_n$ is the closed orientable 3-manifold

$$M = \left(S^3 \setminus \cup_{i=1}^n \overset{o}{N}_i\right) \cup_h \left(\cup_{i=1}^n N_i\right)$$

where h is the union of homeomorphisms $h_i : \partial N_i \longrightarrow \partial N_i$ such that $h_{i*}([\mu_i]) = a_i[\lambda_i] + b_i[\mu_i]$, a_i, b_i being coprime integers such that $r_i = b_i/a_i$. If $a_i = 0$, then $b_i = \pm 1$, hence we set $r_i = \infty$. This case corresponds to the trivial surgery in which N_i is replaced using the identity map on the boundary. Thus the surgery manifold M is unchanged by erasing all components of L which have coefficient ∞. There exists a modification of surgery instructions which yields the same 3-manifold, up to homeomorphism. Indeed, if a component L_i of L is unknotted, then we twist the solid torus $S^3 \setminus \overset{o}{N}_i$ so that the meridian μ_i of N_i is carried to a curve representing the homology class $\tau[\lambda_i] + [\mu_i]$. Here the integer τ represents the number of twists, and it is positive (resp. negative) whenever the twist is in the right–hand (resp. left–hand) sense. Following [**19**], the formulas for the new surgery coefficients are the following:

$$\text{twisted component} \qquad r_i' = \frac{1}{\tau + 1/r_i}$$

$$\text{other components} \qquad r_j' = r_j + \tau\left[lk(L_i, L_j)\right]^2.$$

These formulas are consistent with the conventions $\pm 1/0 = \infty$ and $\pm 1/\infty = 0$.

Now we are going to prove the following

Theorem 3.

1) *S^3 is the unique closed orientable 3-manifold which admits $K(2,2)$ as spine.*
2) *The unique closed orientable prime 3-manifold $M(2,2m)$ with spine $K(2,2m)$, $m \geq 2$, is the m-fold branched cyclic covering of S^3 branched over the figure-eight knot. In particular, $M(2,4)$ is the lens space $L(5,2)$.*

Moreover, $M(2,2m)$, $m \geq 3$, *has Heegaard genus two, hence it is also a double branched cyclic covering of* S^3 *branched over a link of bridge number three.*

Corollary 4. *For any integer* $m \geq 2$, *we have*

$$F(2,2m) \cong \Pi_1(M(2,2m)) \cong [mZ \ltimes (Z * Z)] *_{Z\oplus Z} Z$$

where $\ltimes$ *and* $*$ *denote the semidirect product and the free product respectively. The order* q_n *of the abelianized group* $AbF(2,n)$ *of* $F(2,n)$ *grows according to the Fibonacci sequence*

$$q_{2m} = q_{2m-1} + q_{2m-2}$$

$$q_{2m+1} = q_{2m} + q_{2m-1} + 2.$$

In particular, we have

$$AbF(2,2m) \cong H_1(M(2,2m)) \cong \begin{cases} Z_{q_{2m}} & \textit{if } 2m \equiv 0 (mod\ 4) \\ Z_\xi \oplus Z_\xi & \textit{if } 2m \equiv 2(mod\ 4), \\ & \textit{hence } \xi^2 = q_{2m} \end{cases}$$

Proofs:

1) is trivial. For 2) we set $n = 2m$ and consider the standard presentation

$$F(2,n) =< x_i / x_i x_{i+1} x_{i+2}^{-1} = 1 \ (i = 1,2,\ldots,n, \text{ indices mod} \, n) > .$$

As usual $K(2,n)$ has one vertex v and n 1-cells (resp. 2-cells) labelled by $x_1, x_2, \ldots, x_n$ (resp. $c_1, c_2, \ldots, c_n$). Then ∂c_i is attached to $K^{(1)}(2,n) = x_1 \vee x_2 \vee \cdots \vee x_n$ by a map f_i representing the word $x_i x_{i+1} x_{i+2}^{-1} = 1$ (see fig. 3). The choosen numbering of the elements of $E = E(2,n)$ implies that an appropriate closed simple curve parallel to and near ∂c_i intersects $(i,i), \overline{(i,i)}, (i,i+1), \overline{(i,i+1)}, \overline{(i,i+2)}, (i,i+2)$ in this order (pairs mod n). Thus the set $E(2,n)$ consists of $6n$ elements as below:

$$E(2,n) = \{(i,i), \overline{(i,i)}, (i,i+1), \overline{(i,i+1)}, \overline{(i,i+2)}, (i,i+2) / \\ i = 1,2,\ldots,n, \text{ pairs mod } n\}.$$

Then we have

$$\mathbf{A}(2,n) = \prod_{i=1}^{n} ((i,i)(i,i+2)) (\overline{(i,i)}(i,i+1)) (\overline{(i,i+1)}\ \overline{(i,i+2)})$$

$$\mathbf{B}(2,n) = \prod_{i=1}^{n} ((i,i)\overline{(i,i)}) ((i,i+1)\overline{(i,i+1)}) ((i,i+2)\overline{(i,i+2)})$$

(pairs mod n).

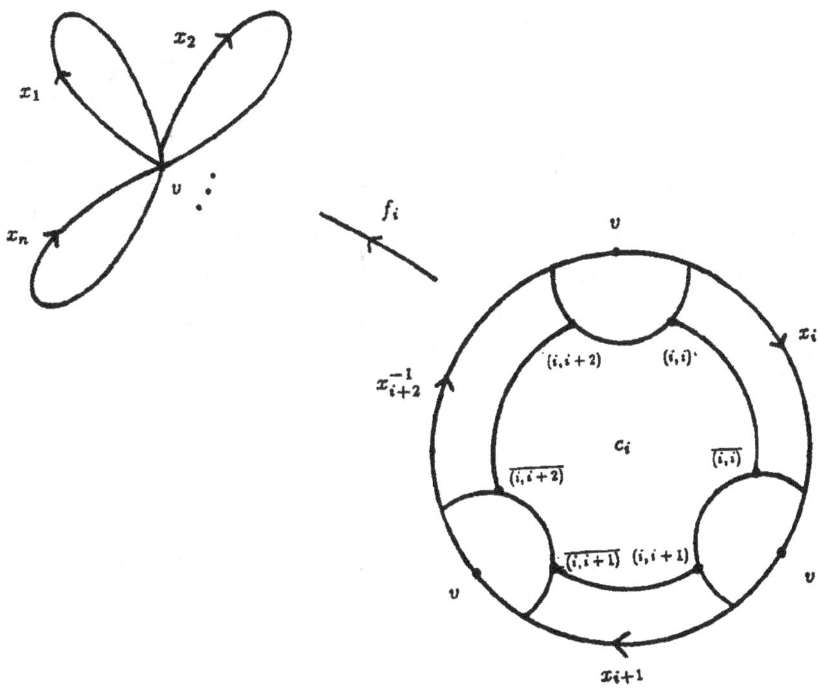

fig. 3

The permutations **C** cyclically act on the orbit sets $E_i = \{(i-2,i),(i-1,i),(i,i)\}$, $\overline{E_i} = \{\overline{(i-2,i)},\overline{(i-1,i)},\overline{(i,i)}\}$, $i = 1,2,\ldots,n,$ pairs mod n.
We consider for example the following permutation

$$\begin{aligned}\mathbf{C}(2,n) = &\prod_{i=1}^{m}\big((2i-2,2i)(2i,2i)(2i-1,2i)\big)\\ &\prod_{i=1}^{m}\big(\overline{(2i-1,2i)}\;\overline{(2i,2i)}\;\overline{(2i-2,2i)}\big)\\ &\prod_{i=1}^{m}\big((2i-1,2i+1)(2i,2i+1)(2i+1,2i+1)\big)\\ &\prod_{i=1}^{m}\big(\overline{(2i+1,2i+1)}\;\overline{(2i,2i+1)}\;\overline{(2i-1,2i+1)}\big)\end{aligned}$$

(pairs mod n).

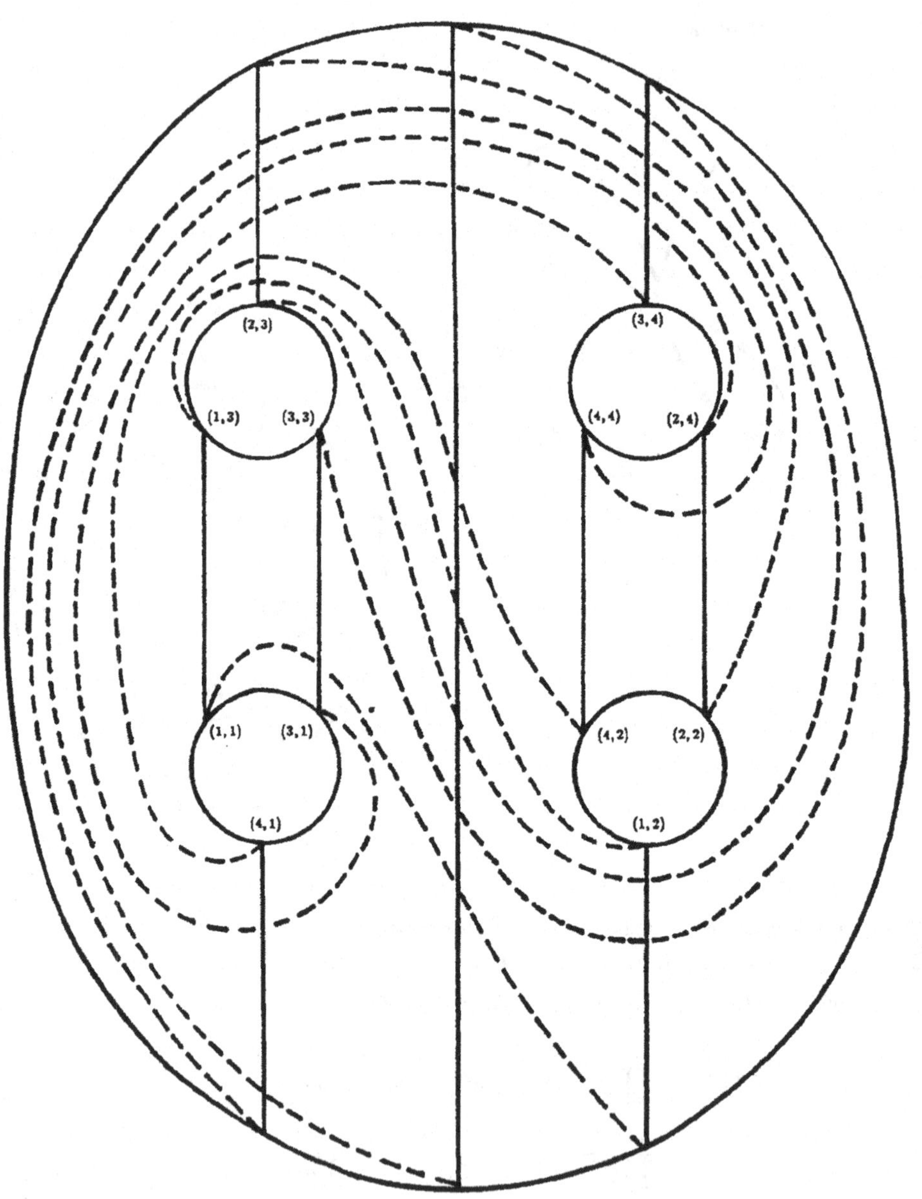

The Heegaard diagram $H(2,4)$ of the manifold $M(2,4)$

fig. 4

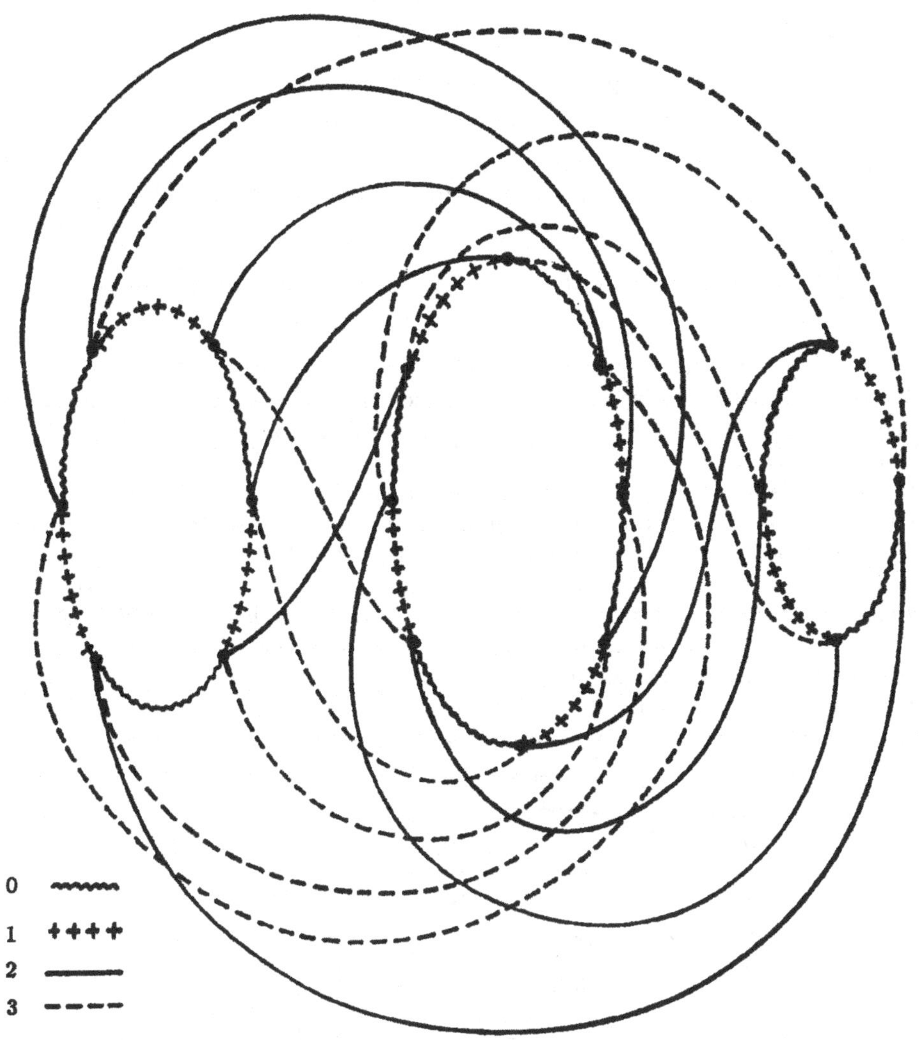

A genus two 4–coloured graph representing $M(2,4)$

fig. 5

The product $\mathbf{A}(2,n)\mathbf{C}(2,n)$ has exactly two cycles of length m and n cycles of length 5, hence it follows that $|\mathbf{A}| - |\mathbf{C}| + 2 = 3n - 2n + 2 = n + 2 = |\mathbf{AC}|$.

Indeed, we have

$$\mathbf{A}(2,n)\mathbf{C}(2,n) = ((2,2)\,(4,4)\dots(2i,2i)\dots(n,n))$$
$$((1,3)\,(n-1,1)\dots(n-2i-1,n-2i+1)\dots(3,5))$$
$$\prod_{i=0}^{m-1}((2i+1,2i+1)\,(2i+2,2i+3)\,\overline{(2i,2i+2)}\;\overline{(2i-1,2i+1)}\;\overline{(2i,2i)})$$
$$\prod_{i=0}^{m-1}(\overline{(2i+1,2i+1)}\,(2i,2i+2)\,(2i-1,2i)\,\overline{(2i-2,2i-1)}\;\overline{(2i-1,2i)}).$$

Further, it is easily checked that $\mathbf{A}(2,n)$ and $\mathbf{C}(2,n)$ (resp. $\mathbf{A}(2,n)\mathbf{C}(2,n)$ and $\mathbf{B}(2,n)$ $\mathbf{C}(2,n)$) generate a transitive group, hence by theorem 1 there exists a closed orientable prime 3–manifold $M(2,n)$, $n = 2m$, which admits $K(2,n)$ as spine. For $m = 2$, the permutations $\mathbf{C}$ cyclically act on the orbit sets E_i, $\overline{E_i}$ where $i = 1,2,3,4$. Since the cyclic orderings induced by $\mathbf{C}$ on E_i, $\overline{E_i}$ must be opposite, we have exactly $2^4 = 16$ possible cases for $\mathbf{C}$. By a direct computation, we have verified that there are only two permutations $\mathbf{C}$, satisfying the formula $|\mathbf{A}| - |\mathbf{C}| + 2 = |\mathbf{AC}| = 6$, which are $\mathbf{C}(2,4)$ and its inverse. Thus the manifold $M(2,4)$ is unique. In fig. 4 we show the Heegaard diagram $H(2,4)$, representing $M(2,4)$, induced by the permutations $\mathbf{A}(2,4)$ and $\mathbf{C}(2,4)$. Let $G(2,4)$ be the 4–coloured graph representing $M(2,4)$, obtained from $H(2,4)$ by the construction described in [4]. Cancelling dipoles in $G(2,4)$ yields a genus two 4–coloured graph (see fig. 5) also representing $M(2,4)$ as double branched cyclic covering of S^3 branched over the knot illustrated in fig. 6 (compare also [**8**]).

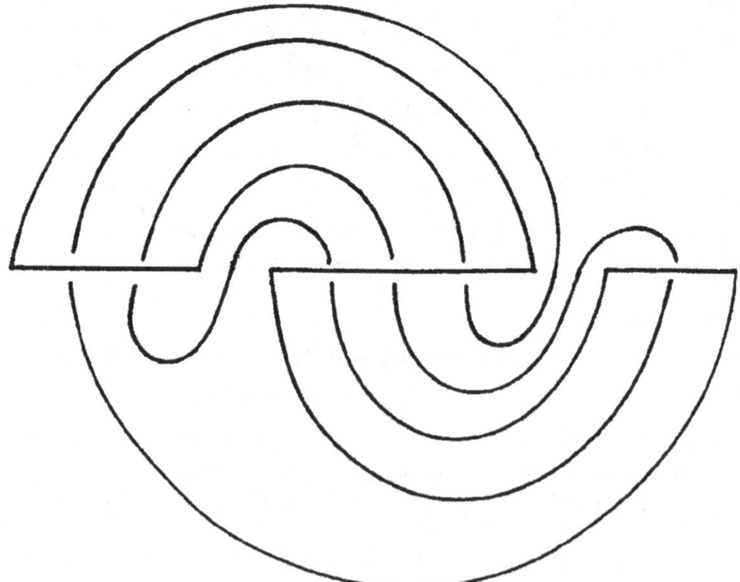

$M(2,4)$ is the double cyclic covering of S^3 branched over this knot

fig. 6

By isotopy this knot is easily proved to be equivalent to the one shown in fig. 7, hence $M(2,4)$ is the lens space $L(5,2)$ (see [**13**], p. 35-A, fig. 18 and [**19**], p. 303).

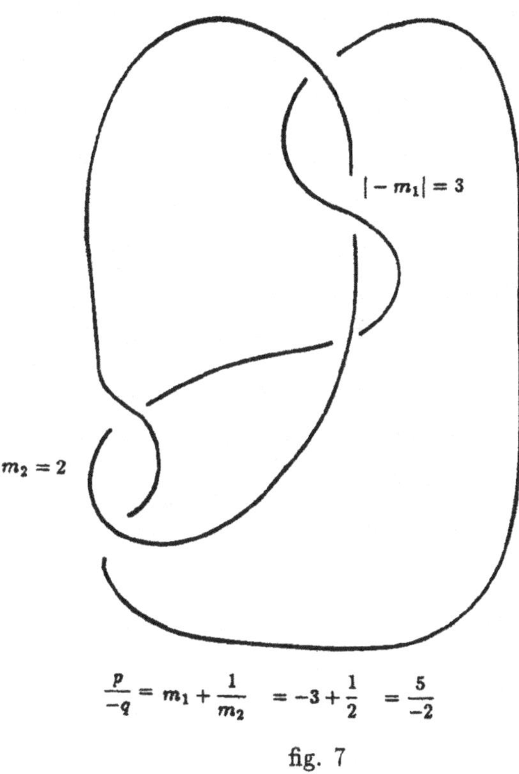

fig. 7

In order to prove that $M(2,2m)$ is unique for $m > 2$ and that it is homeomorphic to the m–fold branched cyclic covering of S^3 branched over the figure-eight knot (see fig. 10), we use the Kirby-Rolfsen calculus on links with coefficients (see [**19**]). Let us consider the link L_4 illustrated in fig. 8a.

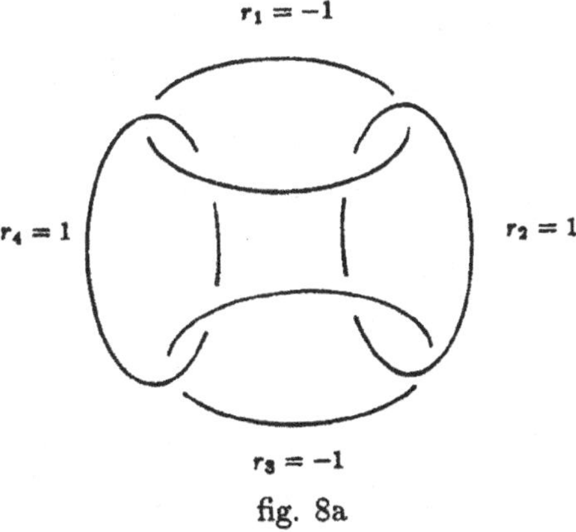

fig. 8a

In [23] M.Takahashi proved that the surgery manifold M_4 has fundamental group $F(2,4)$. Here we prove that M_4 is just homeomorphic to $M(2,4) \cong L(5,2)$ by using the Kirby-Rolfsen calculus on L_4. Twist about the topmost component of L_4 once in the right-hand sense ($\tau = 1$)

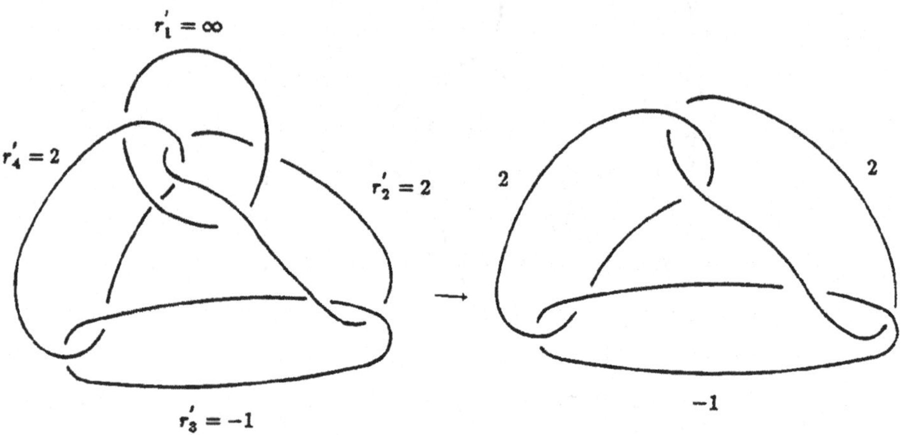

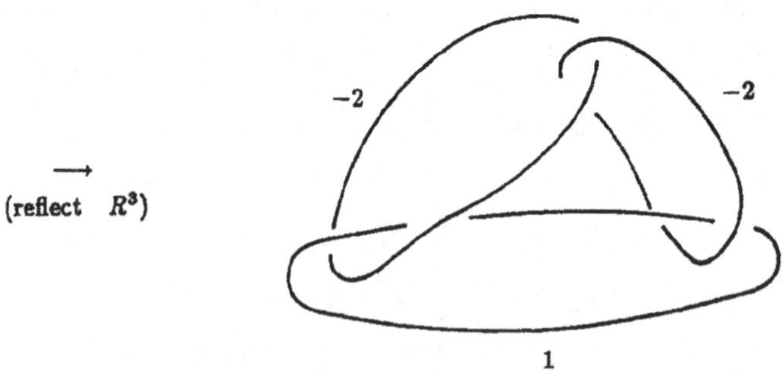

fig. 8b

Similarly twist about the lower most component and rub it out ($\tau = -1$)

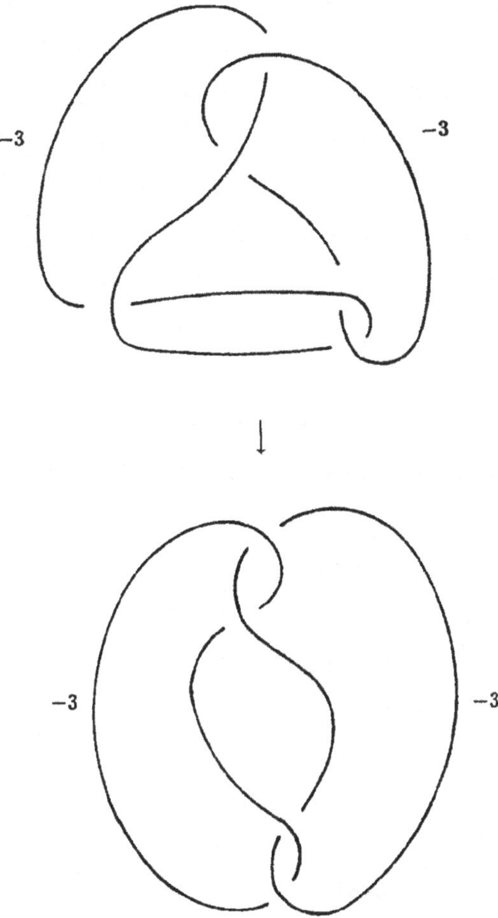

fig. 8c

A right-hand twist about the left component yields ($\tau = 1$)

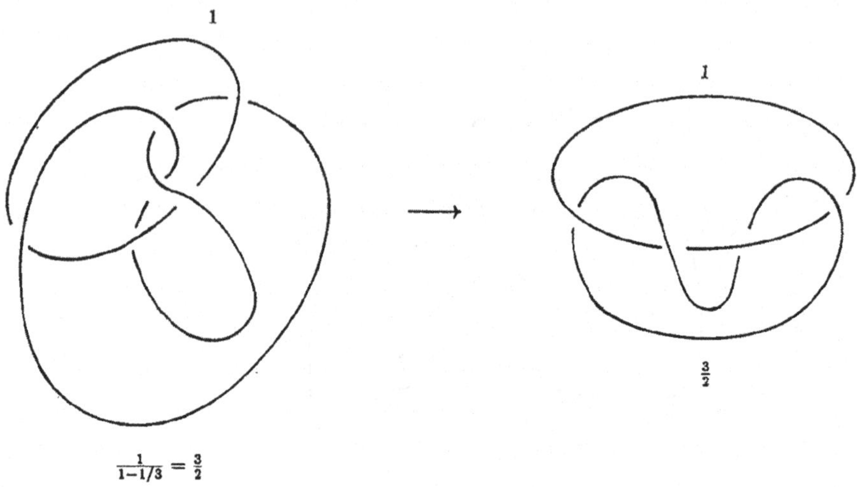

fig. 8d

A left-handed twist about the top most component yields ($\tau = -1$)

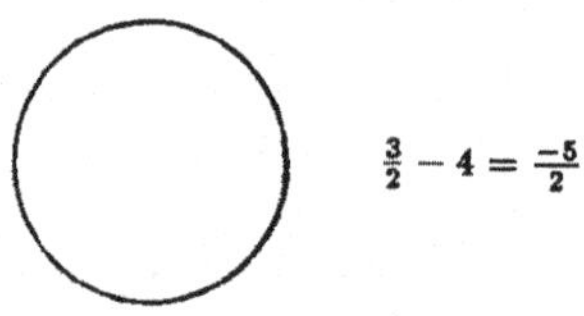

fig. 8e

since the linking number between the two components is two. Hence M_4 is homeomorphic to $L(-5,2) \cong L(5,3) \cong L(5,2)$ as required.

Now we consider the link L_{2m} illustrated in fig. 9. In **[23]** p. 181, M.Takahashi proved that the surgery manifold M_{2m} has fundamental group $\Pi_1(M_{2m}) \cong F(2,2m)$. Here we show that M_{2m} is the m–fold branched cyclic covering $\tilde{X}_m$ of S^3 branched over the figure-eight knot (see **[19]**, pp. 58–299 and fig. 10). Indeed, twist ($\tau = -1$) about each component of L_{2m} with surgery coefficient $+1$ and rub it out; it is easily seen that M_{2m} can be obtained by surgery on the link L'_m shown in fig. 11. Thus M_{2m} is homeomorphic to $\tilde{X}_m$ as pointed out in **[19]**, p. 299. In order to prove that $M(2,2m)$ is homeomorphic to $\tilde{X}_m$, we recall the construction of $\tilde{X}_m$ according to **[2]**, p. 51 ,**[19]**, p. 131.

Let X denote the complement of a knot K in S^3. The finite m–fold cyclic covering X_m of X is the space corresponding to the kernel of the composite homomorphism

$$\Pi_1(X) \longrightarrow H_1(X) \cong Z \longrightarrow Z_m.$$

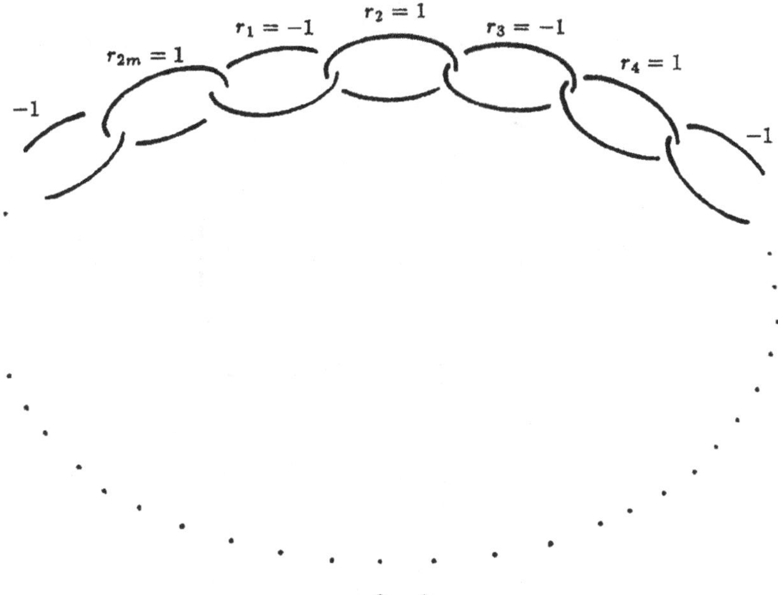

fig. 9

Let μ be a meridian of a suitable regular neighbourhood of K. It is well-known that the m-th power μ^m of μ is a simple closed curve on the torus ∂X_m (see [2], p. 114). By attaching a solid torus $S^1 \times D^2$ to X_m, $h : S^1 \times S^1 \longrightarrow \partial X_m$, such that the meridian of $S^1 \times D^2$ is mapped onto μ^m, we obtain the m-fold cyclic branched covering $\tilde{X}_m = X_m \cup_h (S^1 \times D^2)$. If K is the figure-eight knot, then $\tilde{X}_m$ $(m \geq 3)$ is an irreducible aspherical closed 3-manifold since $\Pi_1(\tilde{X}_m) \cong F(2,2m)$ is infinite (see [24] and [26], p. 59). Moreover, $\Pi_1(\tilde{X}_m) \cong \Pi_1(X_m) *_{Z \oplus Z} Z$ is a non trivial free product with amalgamation, where $\Pi_1(X_m) \cong mZ \quad [\Pi_1(X), \Pi_1(X)]$ (see [2], p. 114). Here $[\Pi_1(X), \Pi_1(X)]$ represents the commutator subgroup of $\Pi_1(X)$, X being the complement of the figure-eight knot in S^3. Hence $[\Pi_1(X), \Pi_1(X)]$ is free of rank 2 (see [2], p. 35). By lemma 1.1.6 of [26], p. 59, it follows that $\tilde{X}_m$ is <u>sufficiently large</u> (i. e. it contains some incompressible surface), and whence $\tilde{X}_m$ is also a <u>Haken manifold</u> (see [5], p. 130). Now the orientable manifold $M(2,2m)$ is irreducible since it is prime (use Theorem 1) and different from $S^1 \times S^2$. Indeed, the first homology group of $M(2,2m)$ is finite by Corollary 2 of [11]. Since $\Pi_1(M(2,2m)) \simeq F(2,2m) \simeq \Pi_1(\tilde{X}_m)$ is a non-trivial free product with amalgamation, it follows from [26], Lemma 1.1.6, that $M(2,2m)$ is sufficiently large too. Since closed Haken manifolds with isomorphic fundamental groups are homeomorphic (see [20], p. 5), we have $\tilde{X}_m \cong M_{2m} \cong M(2,2m)$ as required. This proves the unicity and the first part of corollary 4; the second one is easily verifiable. Finally we observe that the permutations $\mathbf{A}(2,2m)$ and $\mathbf{C}(2,2m)$, $m \geq 3$, allow us to obtain a Heegaard diagram $H(2,2m)$ of $M(2,2m)$ and a genus two reduced 4-coloured graph $G^*(2,2m)$ representing $M(2,2m)$ as double cyclic covering of S^3 branched over a link of bridge number three (see also [8]). In figg. 12 – 13 – 14 we show for example the above-mentioned constructions for $M(2,6)$.

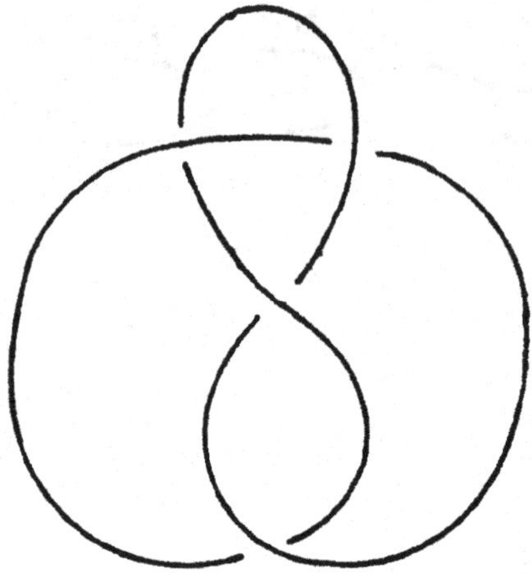

The figure–eight knot

fig. 10

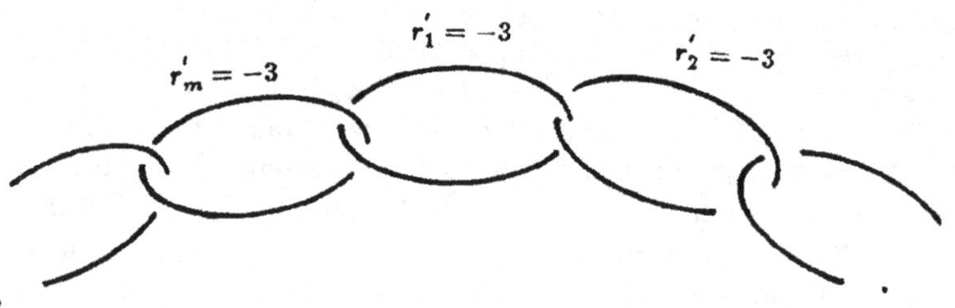

fig. 11

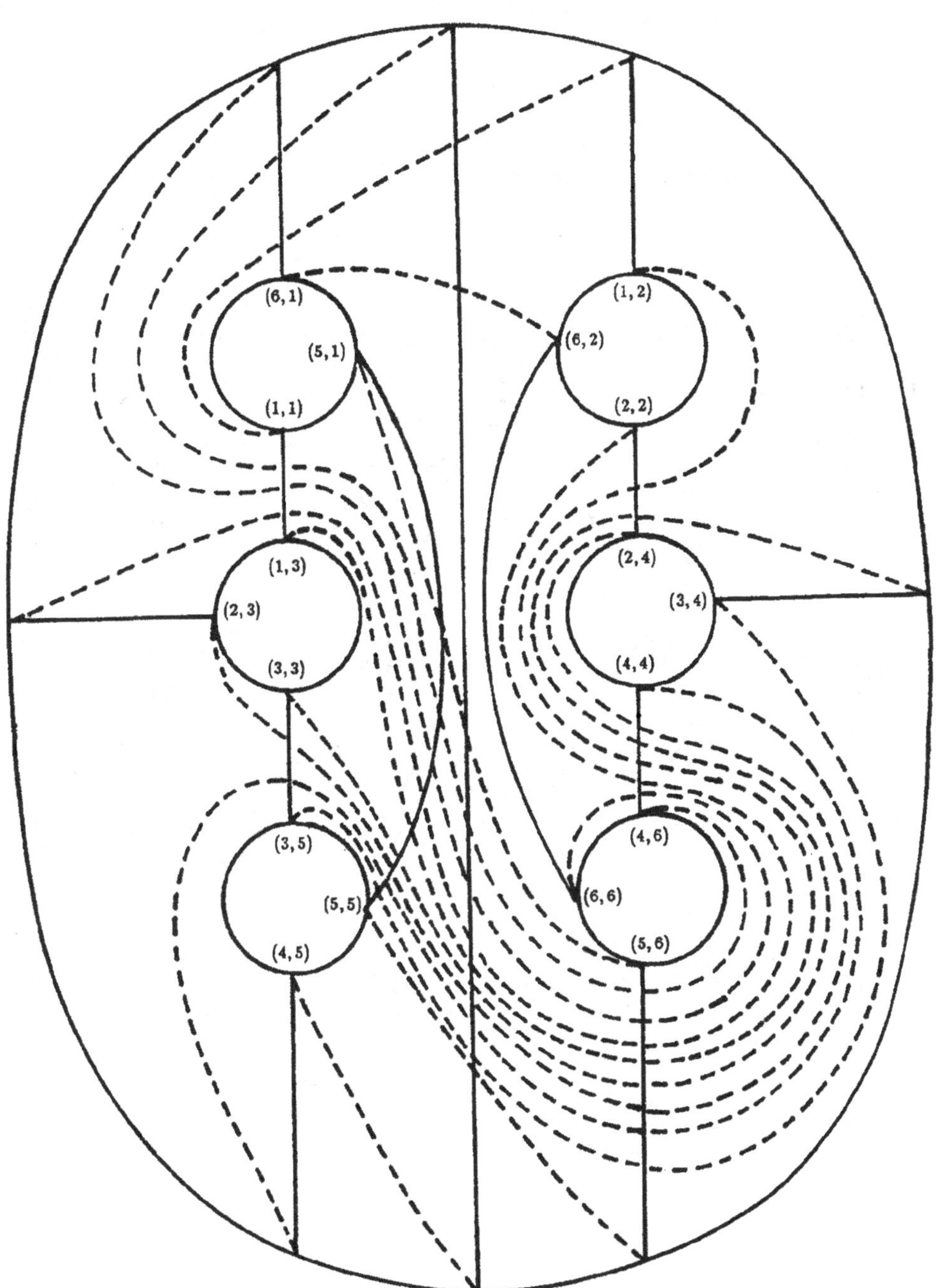

The Heegaard diagram $H(2,6)$ of $M(2,6)$

fig. 12

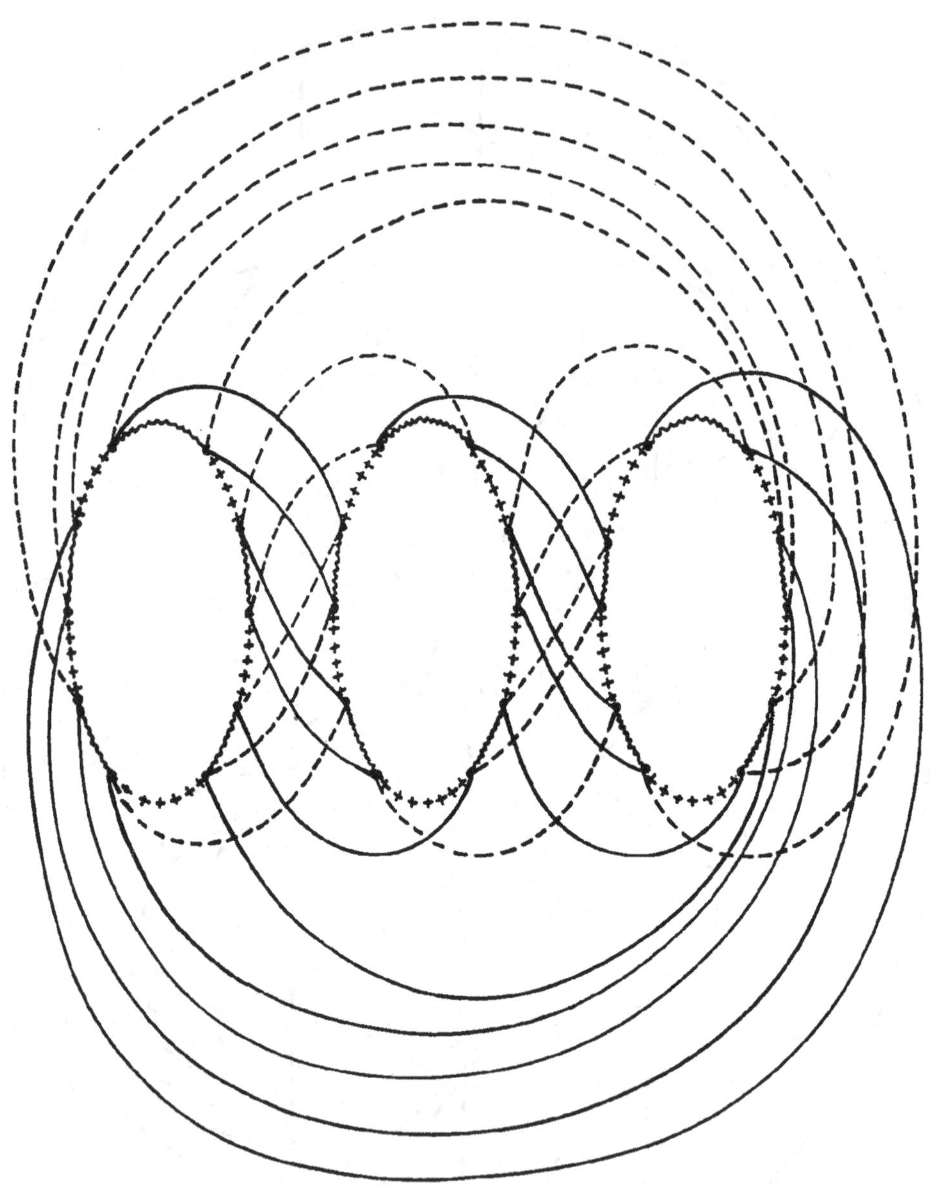

The genus two reduced 4–coloured graph $G^*(2,6)$ representing $M(2,6)$

fig. 13

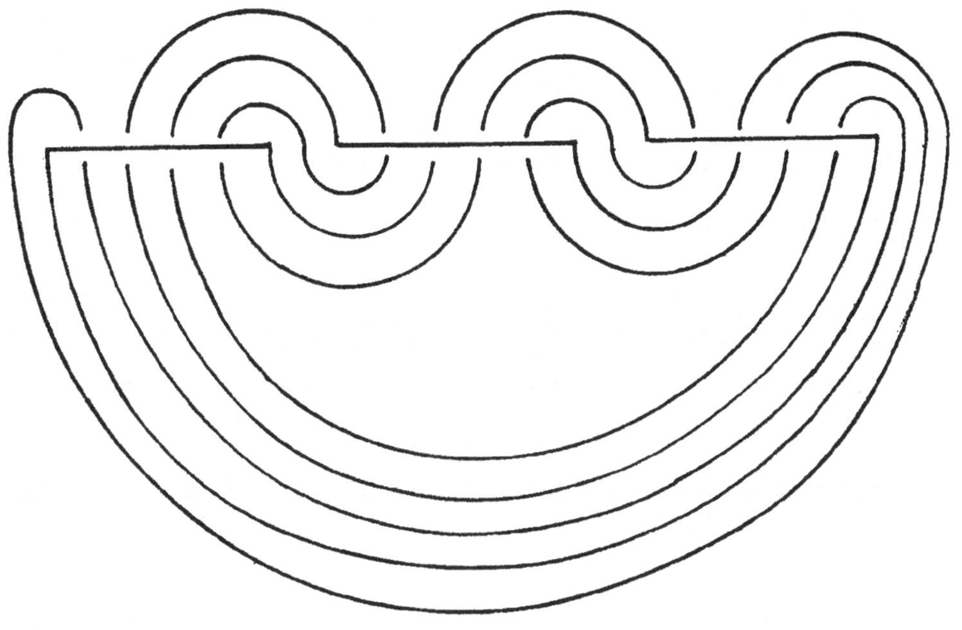

$M(2,6)$ is the double cyclic covering of S^3 branched over this link

fig. 14

5. Prism manifolds.

In this section we consider the Fibonacci groups $F(r,2)$, $r \geq 2$. It is well-known that $F(2s,2) \cong Z_{2s-1}$ and $F(2s+1,2)$, $s \geq 1$, is the metacyclic group of order $4s(s+1)$ presented by

$$< a, b / a^2 = b^{s+1}, a^{-1}ba = b^{2s+1}, b^{2s^2+2s} = 1 >$$

(see [**12**]). Here we prove the following results

Theorem 5.

1) *The unique closed orientable prime 3-manifold having $K(2s+1,2)$, $s \geq 1$, as spine is the prism Seifert manifold of Heegaard genus two*

$$M(2s+1,2) = (0\ o\ 0 / -1\ (2,1)\,(2,1)\,(s+1,s)).$$

2) *The lens space $L(2s-1,2)$ is the unique closed orientable prime 3-manifold which admits $K(2s,2)$, $s \geq 1$, as spine.*

Corollary 6.
The Fibonacci group $F(2s+1,2)$ also admits the following presentations

$$F(2s+1,2) \cong \Pi_1(M(2s+1,2)) \cong < x, y / x^2 = y^2 = (xy)^{s+1} >$$

$$\cong < h, q_1, q_2, q_3 / q_1^2 h = 1,\ q_2^2 h = 1,\ q_3^{s+1} h^s = 1,\ q_1 q_2 q_3 = h^{-1} > .$$

Furthermore, we have

$$AbF(2s+1,2) \cong H_1(M(2s+1,2)) \cong \begin{cases} Z_2 \oplus Z_{2s} & s \quad odd \\ Z_{4s} & s \quad even \end{cases}$$

Proofs:

1) Let us consider the standard presentation

$$F(2s+1,2) = < x_1, x_2 / \underbrace{x_1 x_2}_{1} \cdots \underbrace{x_1 x_2}_{s} x_1 x_2^{-1} = 1$$

$$\underbrace{x_2 x_1}_{1} \cdots \underbrace{x_2 x_1}_{s} x_2 x_1^{-1} = 1 >$$

and denote the 1-cells (resp. 2-cells) of the canonical 2-complex $K(2s+1,2)$ by x_1, x_2 (resp. c_1, c_2) as usual. Then ∂c_i , $i = 1,2$, is attached to $K^{(1)}(2s+1,2) = x_1 \vee x_2$ by a map f_i representing the word

$$\underbrace{x_i x_{i+1}}_{1} \cdots \underbrace{x_i x_{i+1}}_{s} x_i x_{i+1}^{-1} = 1 \text{ (indices mod 2)}.$$

We choose a numbering of the elements of $E = E(2s+1,2)$ such that an appropriate closed simple curve parallel to and near ∂c_1 (resp. ∂c_2) intersects $1, \bar{1}, 2, \bar{2}, \ldots, 2s+1, \overline{2s+1}, \overline{2s+2}, 2s+2$ (resp. $2s+3, \overline{2s+3}, 2s+4, \overline{2s+4}, \ldots, 4s+3, \overline{4s+3}, \overline{4s+4}, 4s+4$) in this order. Thus the set $E(2s+1,2)$ consists of $8s+8$ elements as below:

$$E(2s+1,2) = \{i, \bar{i}/i = 1,2,\ldots,4s+4\}.$$

Then we have

$$\mathbf{A}(2s+1,2) = (1\ 2s+2)(2s+3\ 4s+4)(\overline{2s+1}\ \overline{2s+2})(\overline{4s+3}\ \overline{4s+4}) \prod_{i=1}^{4s+2} \{(\bar{i}\ i+1)/i \neq 2s+1, 2s+2\}$$

$$\mathrm{B}(2s+1,2) = \prod_{i=1}^{4s+4} (i\ \bar{i}).$$

The permutations $\mathbf{C}$ cyclically act on the orbit sets $E_1 = \{1,3,\ldots,2s+1,2s+4,2s+6,\ldots,4s+2,4s+4\}$, $E_2 = \{2,4,\ldots,2s+2,2s+3,2s+5,\ldots,4s+1,4s+3\}$, $\overline{E_1}$ and $\overline{E_2}$.

We consider for example the permutation $\mathbf{C}(2s+1,2)$ defined by

$$\begin{aligned}\mathbf{C}(2s+1,2) = &(1\ 2s+4\ 3\ 2s+6\ \ldots\ 2s-1\ 4s+2\ 2s+1\ 4s+4)\\ &(\overline{4s+4}\,\overline{2s+1}\,\overline{4s+2}\,\overline{2s-1}\ \ldots\ \overline{2s+6}\,\bar{3}\,\overline{2s+4}\,\bar{1})\\ &(2\ 2s+5\ 4\ 2s+7\ \ldots\ 2s\ 4s+3\ 2s+2\ 2s+3)\\ &(\overline{2s+3}\,\overline{2s+2}\,\overline{4s+3}\,\overline{2s}\ \ldots\ \overline{2s+7}\,\bar{4}\,\overline{2s+5}\,\bar{2}).\end{aligned}$$

The product $\mathbf{A}(2s+1,2)\mathbf{C}(2s+1,2)$ has exactly two cycles of lenght 4 and $4s$ cycles of length 2. Indeed, we have

$$\begin{aligned}\mathbf{A}(2s+1,2)\mathbf{C}(2s+1,2) = &(1\ 2s+3)(\overline{4s+3}\ \overline{2s+1})(2\ \overline{4s+4}\ \overline{2s}\ 4s+4)\\ &(2s+2\ 2s+4\ \overline{2s+2}\ \overline{4s+2}) \prod_{i=3}^{2s+1} (i\ \overline{i+2s}) \prod_{i=2s+5}^{4s+3} (i\ \overline{i-2s-4}).\end{aligned}$$

Since $|\mathbf{A}(2s+1,2)| = 4s+4$, $|\mathbf{C}(2s+1,2)| = 4$ and $|\mathbf{A}(2s+1,2)\mathbf{C}(2s+1,2)| = 4s+2$, the relation $|\mathbf{A}| - |\mathbf{C}| + 2 = |\mathbf{AC}|$ holds. Further, it is easily checked that $\mathbf{A}$, $\mathbf{C}$ (resp. $\mathbf{AC}$ and $\mathbf{BC}$) generate a transitive group. Thus theorem 1 implies that there exists a closed orientable prime 3–manifold $M(2s+1,2)$ which admits $K(2s+1,2)$ as spine. Let $H(2s+1,2)$ be the Heegaard diagram of $M(2s+1,2)$ induced by the permutations $\mathbf{A}(2s+1,2)$ and $\mathrm{C}(2s+1,2)$. Then we consider the reduced genus two 4–coloured graph $G^*(2s+1,2)$ which represents $M(2s+1,2)$ (also compare [4]). As it is easily seen, there exists an involutory automorphism of G^* which interchanges 0–coloured (resp. 2–coloured) edges with 1–coloured (resp. 3–coloured) edges. Thus G^* represents the 2–fold cyclic covering space of S^3 branched over a 3–bridge link (see [8]). Moreover a finite sequence of isotopic deformation exists which takes the named link to the Montesinos link $\mathbf{m}(-1;2,\,2,\,\frac{s+1}{s})$ (see [2], def. 12.27, pp. 195 – 196 and [**13**], p. 39 – A, fig. 32). Thus $M(2s+1,2)$ is the prism Seifert manifold $(0\ o\ 0/-1\ (2,1)(2,1)(s+1,s))$. The manifold $M(2s+1,2)$ is unique. Indeed, a manifold M with spine $K(2s+1,2)$ also admits a spine associated to the group presentation

$$F(2s+1,2) =< a, b/a^{s+1}b^{-2} = 1,\ ba^{s} = ab^{-1} >$$

obtained from the standard one by the Nielsen transformation of type 2: $a = x_1x_2$ and $b = x_2$. Thus the induced automorphisms of group presentations correspond to geometric transformations of spines (see [**18**], section 5 and [**6**], [**7**]). Hence M must be a Seifert fibered space over S^3 with three exceptional fibers (see theorem 3.1 [**16**], p. 485). Now the result follows from the classification of Seifert manifolds with finite fundamental group (see [**15**],p. 99).

The first presentation of corollary 6 is directly deduced from the graph $G^*(2s+1,2)$ by standard graph–theoretical tools. The second one follows from the theory of Seifert manifolds (see [**2**], prop. 12.30, p. 197 and [**15**], p. 90). The other statements are easily verifiable. In figg. 15 – 16 – 17 – 18 we illustrate the above constructions for the manifold $M(5,2) = (0\ o\ 0/-1\ (2,1)(2,1)(3,2))$.

2) Let us consider the standard presentation

$$F(2s,2) =< x_1, x_2/ \underbrace{x_1x_2}_{1} \cdots \underbrace{x_1x_2}_{s} x_1^{-1} = 1$$

$$\underbrace{x_2x_1}_{1} \cdots \underbrace{x_2x_1}_{s} x_2^{-1} = 1 >$$

and suppose $M(2s,2)$ be a manifold (if exists) having $K(2s,2)$ as spine. As before, $M(2s,2)$ also admits a spine associated to the group presentation

$$F(2s,2) \cong < a, b/a^{s}b^{-1} = 1,\ a^{s-1}b = 1 >$$

obtained from the standard one by setting $a = x_1x_2$ and $b = x_1$ (see [**18**], section 5 and [**6**],[**7**]). Now we can apply theorems 7 and 10 of [**22**], where $m = s$, $n = -1$, $p = s-1$, $q = 1$ according to the notations used in that paper. We operate on the matrix

$$\begin{pmatrix} m & n \\ p & q \end{pmatrix} = \begin{pmatrix} s & -1 \\ s-1 & 1 \end{pmatrix}$$

using the following rules: (1) interchange of rows or columns; (2) multiplication of a row or column by -1; (3) subtraction of one row from the other under the condition that both entries at least one column have the same sign. Then we have

$$\begin{pmatrix} s & -1 \\ s-1 & 1 \end{pmatrix} \underset{(3)}{\longrightarrow} \begin{pmatrix} s & -1 \\ 1 & -2 \end{pmatrix} \underset{(1)}{\longrightarrow} \begin{pmatrix} 1 & -2 \\ s & -1 \end{pmatrix}$$

$$\underset{(2)}{\longrightarrow} \begin{pmatrix} 1 & 2 \\ s & 1 \end{pmatrix} \underset{(3)}{\longrightarrow} \begin{pmatrix} 1 & 2 \\ 1 & 3-2s \end{pmatrix} = \begin{pmatrix} 1 & n' \\ 1 & q' \end{pmatrix},$$

hence $n' = 2$ and $q' = 3-2s$ (compare theorem 11 of [**22**]). Let $\lambda = |mq-np| = |n'-q'| = 2s-1$ and choose k so that $0 \leq k \leq \lambda$ and $k \equiv n' \pmod{\lambda}$, i.e. $k = 2$ in our case. Thus theorems 7 and 10 of **[22]** imply that the unique closed orientable 3–manifold with spine $K(2s, 2)$ is the lens space $L(\lambda, k) \cong L(2s-1, 2)$. The proofs are completed.

Added in revision. As pointed out by the referee, J.L. Mennicke announced in the note "On Fibonacci groups and some other groups" (Proc. Conference "Groups-Korea 1988", Lect. Note Math. 1398) that $F(2, 2n)$ acts freely as a group of isometries on a certain tesselation of H^3. Thus $F(2, 2n)$ is the fundamental group of the quotient hyperbolic manifold. Moreover, it is remarked there (and it was first discovered by J. Howie and J. Montesinos) that this manifold is the n–fold cyclic cover of S^3 branched over the figure-eight knot.

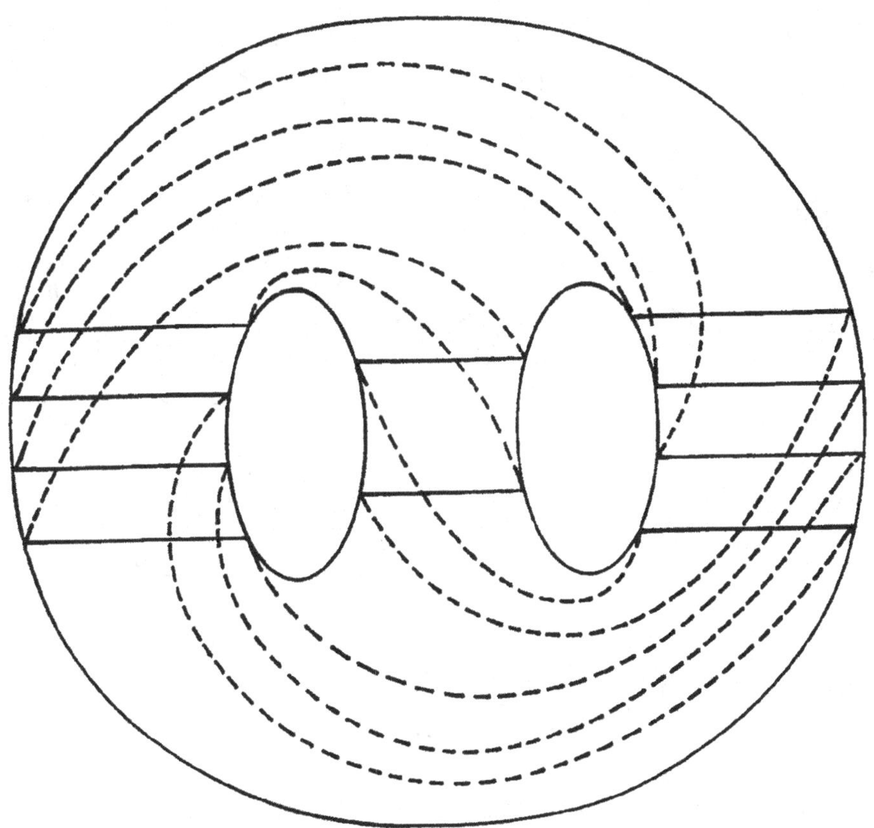

The Heegaard diagram $H(5, 2)$ of $M(5, 2)$

fig. 15

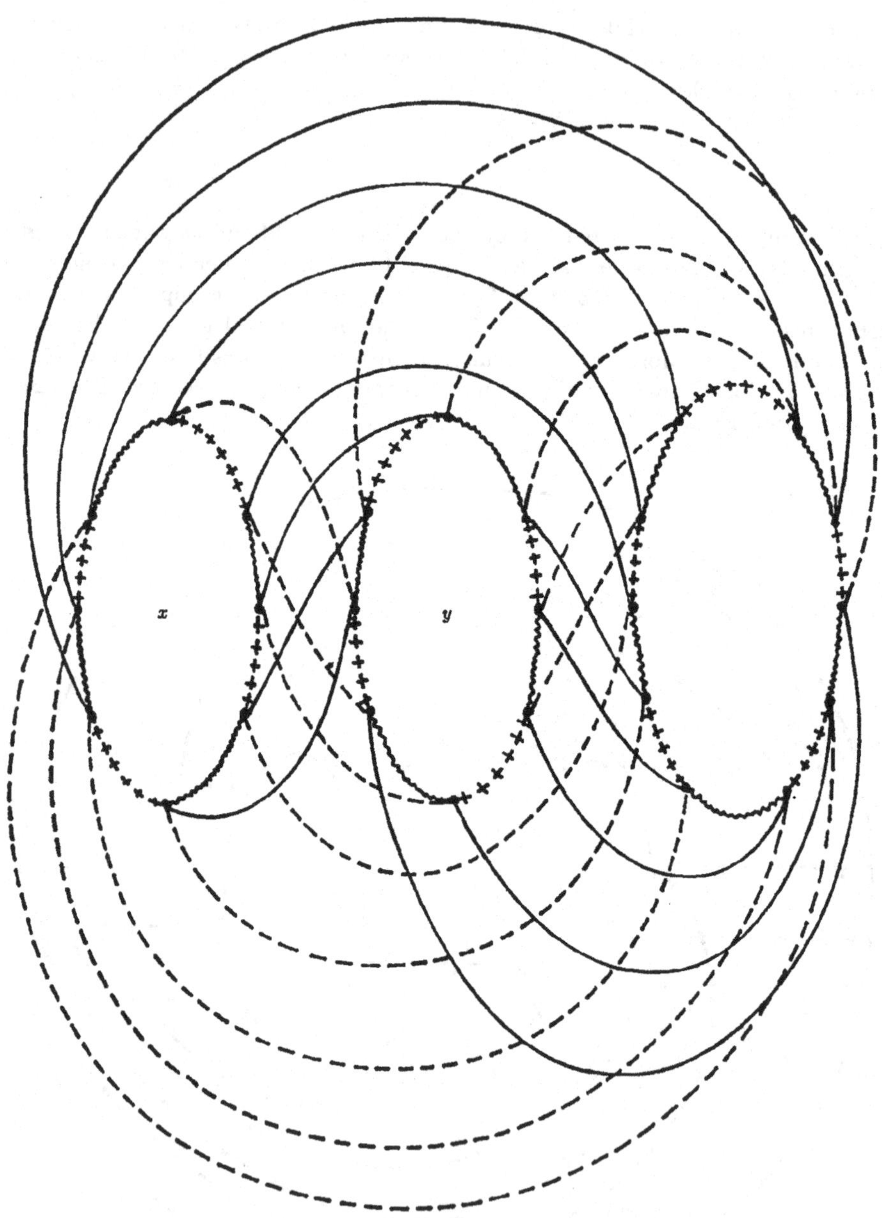

The genus two reduced 4–coloured graph $G^*(5,2)$ representing $M(5,2)$.
Here we have $F(5,2) \cong \Pi_1(M(5,2)) \cong < x, y / x^2 = y^2 = (xy)^3 >$

fig. 16

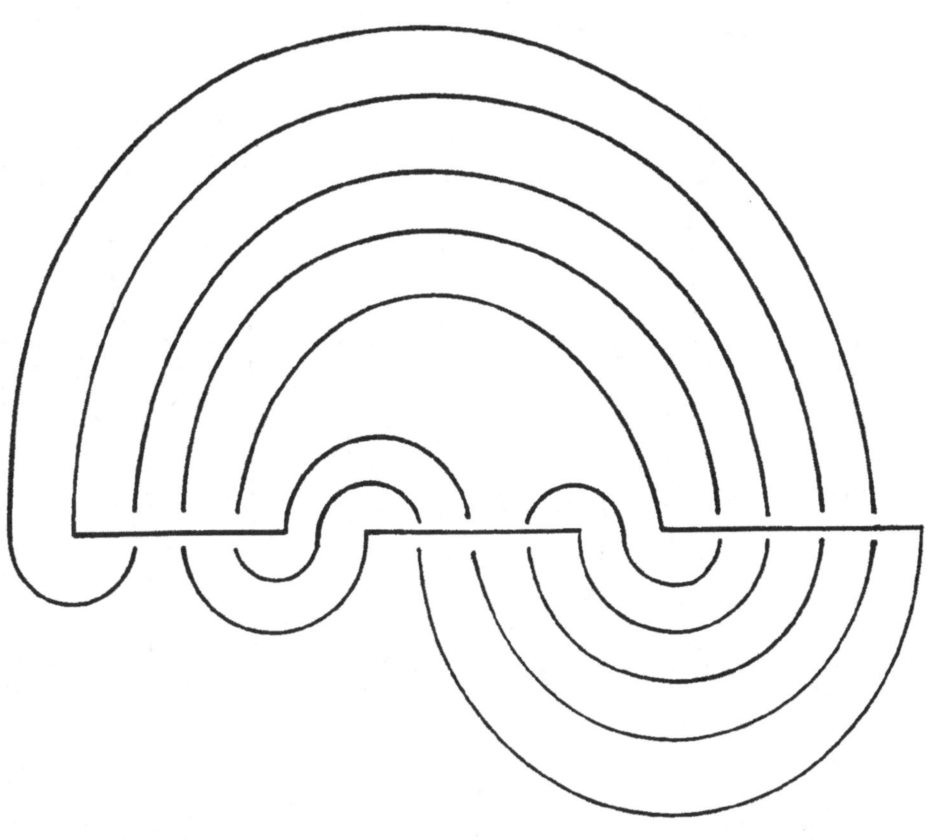

$M(5,2)$ is the double cyclic covering of S^3 branched over this link

fig. 17

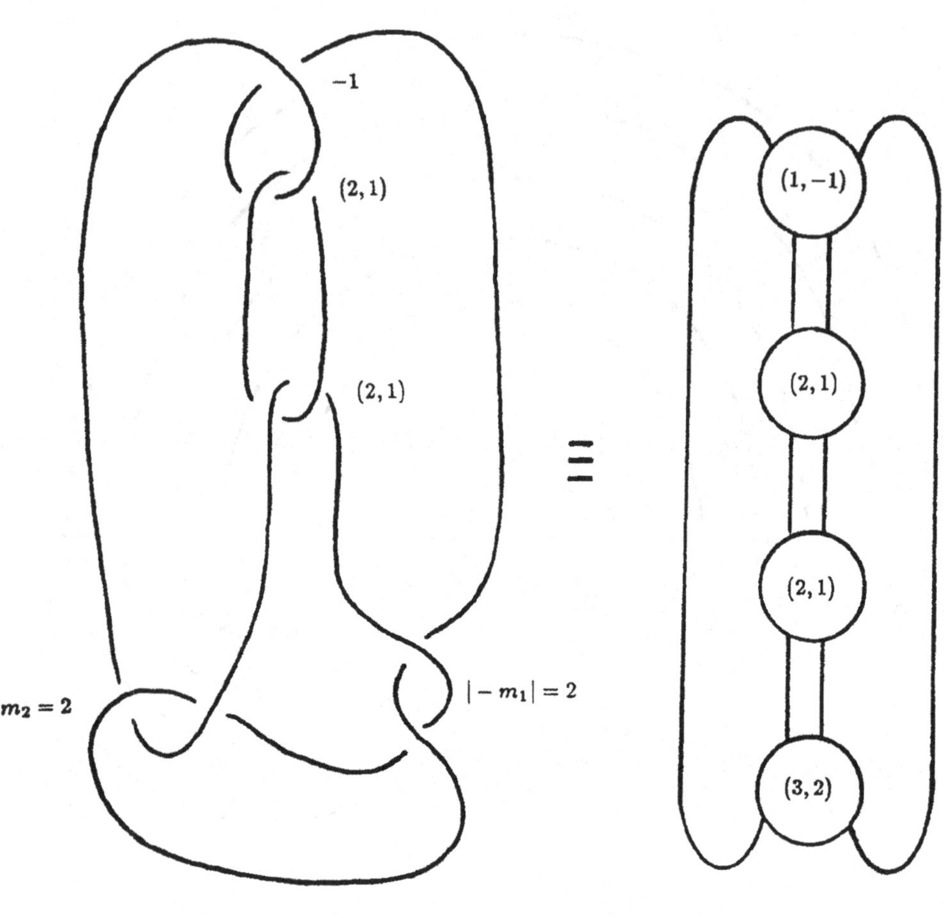

$$\frac{p}{-q} = m_1 + \frac{1}{m_2} \quad = \quad -2 + \tfrac{1}{2} = \tfrac{-3}{2}$$

The link of fig. 17 drawn according to the Montesinos representation

fig. 18

REFERENCES

[1] A. Brunner, *The determination of Fibonacci Groups*, Bull. Austral. Math. Soc. 11(1974), 11 – 14.

[2] G. Burde–H. Zieschang, *Knots*, Walter de Gruyter Ed., Berlin–New York, 1985.

[3] A. Cavicchioli, *Imbeddings of polyhedra in 3-manifolds*, Annali di Mat. Pura ed Appl., to appear.

[4] A. Cavicchioli, *Neuwirth manifolds and colourings of graphs*, to appear.

[5] M. Culler–P. B. Shalen, *Varieties of group representations and splittings of 3-manifolds*, Ann. of Math. 117(1983), 109 – 146.

[6] R. Craggs, *Free Heegaard diagrams and extended Nielsen transformations*, I, Michigan Math. J. 26 (1979), 161–186.

[7] R. Craggs, *Free Heegaard diagrams and extended Nielsen transformations*, II, Illinois J. of Math., 23 (1979), 101–127.

[8] M. Ferri, *Crystallisations of 2-fold branched coverings of S^3*, Proc. Amer. Math. Soc. 73(1979), 271 – 276.

[9] J. Hempel, *3-manifolds*, Ann. of Math. Studies 86, Princeton Univ. Press, Princeton, New Jersey, 1976.

[10] D. L. Johnson, *Topics in the theory of Group Presentations*, London Math. Soc. Lect. Note Series 42, Cambridge Univ. Press, Cambridge–London–New York, 1980.

[11] D. L. Johnson, *A note on the Fibonacci Groups*, Israel J. Math. 17(1974), 277–282.

[12] D. L. Johnson–J. W. Wamsley–D. Wright, *The Fibonacci Groups*, Proc. London Math. Soc. 29(1974), 577 – 592.

[13] J. M. Montesinos, *Sobre la representación de variedades tridimensionales*, Mimeographed Notes, 1977.

[14] L. Neuwirth, *An algorithm for the construction of 3-manifolds from 2-complexes*, Proc. Camb. Phil. Soc. 64(1968), 603 – 613.

[15] P. Orlik, *Seifert manifolds*, Lect. Note in Math. 291, Springer–Verlag Ed., Berlin–Heidelberg–New York, 1972.

[16] R. Osborne, *Simplifying spines of 3-manifolds*, Pacific J. of Math. 74(1978), 473–480.

[17] R. Osborne, *The simplest closed 3-manifolds*, Pacific J. of Math. 74(1978), 481 – 495.

[18] R. Osborne–R.S. Stevens, *Group presentations corresponding to spines of 3-manifolds* II, Trans. Amer. Math. Soc. 234 (1977), 213–243.

[19] D. Rolfsen, *Knots and Links*, Math. Lect. Series 7, Publish or Perish Inc., Berkeley, 1976.

[20] P. Scott, *The classification of compact 3-manifolds*, in "Proc. Conf. on Topology in Low Dimension, Bangor 1979", London Math. Soc. Lect. Note Series 48, Cambridge Univ. Press, Cambridge–London–New York (1982), 3 – 7.

[21] J. Singer, *Three-dimensional manifolds and their Heegaard diagrams*, Trans. Amer. Math. Soc. 35(1933), 88 – 111.

[22] R. S. Stevens, *Classification of 3-manifolds with certain spines*, Trans. Amer. Math. Soc. 205(1975), 151 – 166.

[23] M. Takahashi, *On the presentations of the fundamental groups of 3-manifolds*, Tsukuba J. Math. 13(1989), 175 – 189.

[24] R. M. Thomas, *The Fibonacci Groups* $F(2, 2m)$, Bull. London Math. Soc. 21(1989), 463 – 465.

[25] R. M. Thomas, *Some infinite Fibonacci Groups*, Bull. London Math. Soc. 15(1983), 384 – 386.

[26] F. Waldhausen, *On irreducible 3-manifolds which are sufficiently large*, Ann. of Math. 87(1968), 56 – 88.

Dipartimento di Matematica
Universitá di Modena
Via Campi 213/B
41100 Modena
Italy.

1990 Barcelona Conference
on Algebraic Topology.

ALGORITHM FOR THE COMPUTATION OF THE COHOMOLOGY OF $\mathcal{J}$–GROUPS

B. CENKL AND R. PORTER

Let $G = G_k$ be a finitely generated torsion free nilpotent group with k generators. Such groups are called $\mathcal{J}$–groups by Hall, [4]. The goal is to compute the cohomology groups $H^i(G, \mathbf{Z})$, $i = 0, 1, 2, \cdots, k$.

In [**1**] we proved that the cohomology of the group G can be computed as the cohomology of a differential graded algebra $M(G) = M \otimes \cdots \otimes M$ (k–times), $M = M^0 \oplus M^1$, with a differential D and product $\cdot$.

Here we give algorithms for the computation of the differential D and for the product $\cdot$. The algorithms are given in terms of the group structure of G.

If $\{e_k^{u_k} \otimes \cdots \otimes e_1^{u_1}\}, u_j = 0$ or 1, is an additive basis for $M(G)$ then our algorithms compute the integers

$$a_{v_k \cdots v_1}^{u_k \cdots u_1}, \; p_{l_k \cdots l_1}^{i_k \cdots i_1 \,,\, j_k \cdots j_1}$$

such that

$$D(e_k^{u_k} \otimes \cdots \otimes e_1^{u_1}) = \sum a_{v_k \cdots v_1}^{u_k \cdots u_1} e^{v_k} \otimes \cdots \otimes e^{v_1},$$

$$(e_k^{i_k} \otimes \cdots \otimes e_1^{i_1}) \cdot (e_k^{j_k} \otimes \cdots \otimes e_1^{j_1}) = = \sum p_{l_k \cdots l_1}^{i_k \cdots i_1 \,,\, j_k \cdots j_1} e_k^{l_k} \otimes \cdots \otimes e_1^{l_1}$$

The cohomology of the group G can be computed as the cohomology of a graded algebra of *polynomial cochains* $P(G)$ whose differential d contains all the information about the group structure of G [**1**]. Although $P(G)$ is much smaller than the algebra of all cochains it is still much too large to be suitable for any explicit computation, except in some very special cases. The reduction from $P(G)$ to the *tight complex* $M(G)$ uses the maps in the following diagram

$$(1) \qquad \begin{array}{ccc} P(G) & d & H \\ {\scriptstyle I}\uparrow \; \downarrow{\scriptstyle P} & & \\ M(G) & D & \end{array}$$

The maps satisfy the identities:

(i) $ID = dI$, $Pd = DP$; I, P are degree preserving and d, P are differentials, H lowers the degree

(ii) $PI = 1_{M(G)}$,
(iii) $IP = 1_{P(G)} + dH + Hd$,
(iv) $HI = 0$,
(v) $PH = 0$,
(vi) $HH = 0$.

D is the desired differential on $M(G)$, and the product $\cdot$ on $M(G)$ is induced from the product on $P(G)$ by the formula

$$a \cdot b = P((Ia)(Ib)).$$

The maps in the above diagram are either induced by or composed from the maps in the *Gugenheim tower*. And those maps are in turn completely determined by *elementary maps*.

Polynomial Cochains.

The algebra of polynomial cochains $P(G)$ for the group G with k generators is a graded module

$$P(G) = P^0(G) \oplus P^1(G) \oplus \cdots \oplus P^n(G) \oplus \cdots ,$$

where

$$P^0(G) = \mathbb{Z}$$

$P^n(G)$, $n \geq 1$, is free $\mathbb{Z}$–module with basis consisting of $k \times n$ matrices of nonnegative integers with nonzero columns, together with a *cup product*. The cup product of a $k \times m$ matrix A with a $k \times n$ matrix B is the $k \times (m+n)$ matrix AB obtained by a juxtaposition. The *commutator* $[a, b]$ stands for the element $AB + BA$.

The *differential* d on $P(G)$ is defined as follows. First we introduce a graded $\mathbb{Z}$–module of chains on G;

$$C(G) = C_0 \oplus C_1 \oplus \cdots \oplus C_n \oplus \cdots .$$

$$C_0 = \mathbb{Z} ,$$

C_n, $n \geq 1$, is the free $\mathbb{Z}$–module with basis consisting of all $k \times n$ matrices with integer entries. Let $\mathsf{X} = (x_{ij})$ be such a $k \times n$ matrix.

Let $\mathsf{B} = (p_{ij})$ be a $k \times n$ matrix which belongs to $P^n(G)$. We define a pairing

$$\langle \mathsf{B}, \mathsf{X} \rangle = \prod_{\substack{1 \leq i \leq k \\ 1 \leq j \leq n}} \binom{x_{ij}}{b_{ij}}$$

where $\binom{x}{b}$ is the binomial coefficient. Let X^j be the j–th column of the $k \times n$ matrix X.

$$\partial \mathsf{X} = (\mathsf{X}^2, \ldots, \mathsf{X}^n) + \sum_{j=1}^{n-1} (-1)^j \left(\mathsf{X}^1, \ldots, \rho\left(\mathsf{X}^j, \mathsf{X}^{j+1}\right), \ldots, \mathsf{X}^n\right)$$
$$+ (-1)^n \left(\mathsf{X}^1, \ldots, \mathsf{X}^{n-1}\right) \quad \text{for} \quad n \geq 2$$
$$\partial \mathsf{X} = 0 \qquad \text{for} \quad n = 1.$$

The column $\rho\left(\mathbf{X}^j, \mathbf{X}^{j+1}\right)$ is determined by the group G (see [1]). The differential $d\mathbf{B} \in P^n(G)$ of an element $\mathbf{B} \in P^{n-1}(G)\,, n \geq 2$ is given by the formula

$$\langle d\mathbf{B}, \mathbf{X}\rangle = \langle \mathbf{B}, \partial\mathbf{X}\rangle\,,$$

where $\mathbf{X}$ is any element of C_n.

Because any element of $P^n(G)$ is the cup product of n elements from $P^1(G)$, see [1], and because in applications mostly certain special elements occur we introduce some new symbols and formulas.

$$f_j^{(r)} = \begin{pmatrix} 0 \\ \vdots \\ 0 \\ r \\ 0 \\ \vdots \\ 0 \end{pmatrix} \begin{matrix} (k) \\ \\ \\ (j) \\ \\ \\ (1) \end{matrix}$$

is a $k \times 1$ matrix representing an element of $P^1(G)$. $r \in \mathbf{Z}_+$.

$$f_j = f_j^{(1)}$$

$$f_j^{(r)} f_{j+1}^{(s)} = \begin{pmatrix} 0 & 0 \\ \vdots & \vdots \\ 0 & 0 \\ 0 & s \\ r & 0 \\ 0 & 0 \\ \vdots & \vdots \\ 0 & 0 \end{pmatrix} \begin{matrix} \\ \\ \\ (j+1) \\ (j) \\ \\ \\ \\ \end{matrix}$$

is the cup product of $f_j^{(r)}$ and $f_{j+1}^{(s)}$.

The cup–1 product of $f_i^{(r)}$ and $f_j^{(s)}$, $i < j$ is the element

$$f_i^{(r)} \cup_1 f_j^{(s)} = \begin{pmatrix} 0 \\ \vdots \\ 0 \\ s \\ 0 \\ \vdots \\ 0 \\ r \\ \vdots \\ 0 \end{pmatrix} \begin{matrix} \\ \\ \\ (j) \\ \\ \\ \\ (i) \\ \\ \\ \end{matrix}$$

belonging to $P^1(G)$.

The rule for the differential of the cup–1 product $f \cup_1 g$; $f, g \in P^1(G)$ is

$$d(f \cup_1 g) = df \cup_1 g - f \cup_1 dg - (fg + gf).$$

Let

$$G > G^1 > \cdots > G^{k-1} > G^k = 1$$

be a sequence of normal subgroups of G such that G^s/G^{s+1} is infinite cyclic. Such a system exists by Malcev [5]. Let G_s be the quotient of G by the normal subgroup G^s of G. Then we can consider the algebras of polynomial cochains $P(G_s)$. The structure and the operators from $P(G)$ carry over to $P(G_s)$. Note that the inclusions $G^s > G^{s+1}$ give surjective homomorphism $G_{s+1} \to G_s$ which in turn induce injective maps of algebras

$$P(G_s) \to P(G_{s+1}).$$

Hence an element of $P(G_s)$ can be looked at as an element of $P(G_{s+1})$. Let c_s be such a monomial in $P(G_{s+1})$, i.e., c_s is the image of $c \in P(G_s)$ under the inclusion. Then c_s is represented by a matrix whose top row consists entirely of zeros.

A monomial in $P(G_{s+1})$ is said to be *organized* if it does not contain consecutive pairs of the form $c_s f_{s+1}^{(r)}$. We define an *organizing operator*

$$O_{s+1} : P(G_{s+1}) \to P(G_{s+1})$$

to be the linear operator which acts on a nonorganized monomial w by acting on each consecutive pair $c_s f_{s+1}^{(r)}$ of w according to the rule

$$O_{s+1}\left(c_s f_{s+1}^{(r)}\right) = \left[f_{s+1}^{(r)}, c_s\right] - f_{s+1}^{(r)} c_s\,;$$

the commutator $\left[f_{s+1}^{(r)}, c_s\right]$ stands for $f_{s+1}^{(r)} c_s + c_s f_{s+1}^{(r)}$. If w is an organized monomial then $O_{s+1} w = w$.

Perturbation Technique.

The construction of the maps in (1) is based on the general results about perturbations of differentials and maps. Given an inclusion $i : X \to Y$ of differential graded modules that induces an isomorphism on cohomology and a second (perturbed) differential D_Y, then under appropriate conditions D_Y induces a perturbed differential D_X on X and a perturbation I of i such that $I : X \to Y$ induces an isomorphism on cohomologies computed with respect to the perturbed differential. If Y is a differential algebra then X can be given the structure of a differential algebra such that I is homotopy multiplicative. Our approach [1] is based on the work of Gugenheim [3].

Initial Strong Deformation Retration data.

(ISDR–data) consists of differential graded modules $(X, d_X), (Y, d_Y)$ and maps $i : X \to Y$, $p : Y \to X$ and $h : Y \to Y$ satisfying the properties

(i) $i d_X = d_Y i$, $p d_Y = d_X p$;

(ii) $pi = 1_X$;

(iii) $ip = 1_Y + d_Y h + h d_Y$;

(iv) $hi = 0$;
(v) $ph = 0$;
(vi) $hh = 0$.

ISDR–data is denoted by

$$\text{(I)} \qquad \begin{array}{ccc} Y & d_Y & h \\ {\scriptstyle i}\uparrow\downarrow{\scriptstyle p} & & \\ X & d_X & \end{array}$$

Note that (i) through (vi) imply

(vii) $d_X = pd_Y i$, $hd_Y h = -h$,
$hd_Y i = 0$, $pd_Y h = 0$.

Perturbation data.

(P–data) consists of ISDR–data (I) together with a differential D_Y on Y such that for each fixed dimension $((D_Y - d_Y)h)^j = 0$ for sufficiently large j. A P–data is denoted by

$$\text{(P)} \qquad \begin{array}{cccc} Y & d_Y & h & D_Y \\ {\scriptstyle i}\uparrow\downarrow{\scriptstyle p} & & & \\ X & d_X & & \end{array}$$

Final Strong Deformation Retraction data.

(FSDR–data) consists of differential graded modules $(X, D_X), (Y, D_Y)$ and maps $I : X \to Y$, $P : Y \to X$ and $H : Y \to Y$ satisfying (i)–(vi). The operators are defined by the formulas [**3**]

$$D_X = pD_Y i + pD_Y \sum_{j\geq 0}(hD_Y + 1)^j hD_Y i\,,$$

$$I = i + \sum_{j\geq 0}(hD_Y + 1)^j hD_Y i\,,$$

$$P = p + pD_Y \sum_{j\geq 0}(hD_Y + 1)^j h\,,$$

$$H = \sum_{j\geq 0}(hD_Y + 1)^j h\,,$$

where the sums are finite. A FSDR–data is denoted by

$$\text{(F)} \qquad \begin{array}{ccc} Y & D_Y & H \\ {\scriptstyle I}\uparrow\downarrow{\scriptstyle P} & & \\ X & D_X & \end{array}$$

Elementary Maps.

As mentioned earlier the maps in diagram (1) are constructed from certain elementary maps. Here we define those maps.

First of all we define ISDR–data

$$\text{(A)}\qquad \begin{array}{c} P(G_{s+1}) \\ \alpha_s \big\uparrow \big\downarrow \beta_s \\ P(G_1) \otimes P(G_s) \end{array} \qquad \begin{array}{c} \partial_{s+1} \quad \xi_s \\ \\ d_\otimes \equiv 0, \end{array}$$

$\alpha_s\left(f_1^{(r)} \otimes c\right) = f_{s+1}^{(r)} c_s$,

$\alpha_s(1 \otimes c) = c_s$,

ξ_s is a linear map such that $\xi_x(w) = 0$ if w is an organized monomial which does not contain any commutators or if a cup–1 product precedes a commutator in w;

$\xi_s\left(u\left[f_{s+1}^{(r)}, c_s\right] v\right) = \dim u(-1)u\left(f_{s+1}^{(r)} \cup_1 c_s\right) v$, when u does not contain any commutators,

$\xi_s(v) = \xi_s O_{s+1}(v)$ for any $v \in P(G_{s+1})$,

$\partial_{s+1}\left(f_{s+1}^{(r)} \cup_1 c_s\right) = -\left[f_{s+1}^{(r)}, c_s\right]$ on $P^1(G_{s+1})$

$\partial_{s+1} = 0$ otherwise on $P^1(G_{s+1})$ and on $P^0(G_{s+1})$

∂_{s+1} is a derivation on $P^j(G_{s+1})$, $j \geq 2$.

On organized monomials

$\beta_s = 0$ on all the elements containing commutators of cup–1 products of the form $f_{s+1}^{(r)} \cup_1 c_s$,

$\beta_s(l_1 \cdots l_r a_1 \cdots a_t) = l_1 \cdots l_r \otimes a_1 \cdots a_t$ where l_i's are of the form $f_{s+1}^{(r)}$ and a_j are of the form c_s,

$\beta_s = \beta_s O_{s+1}$ otherwise.

It is proved in [**1**] that all these maps satisfy the required properties.

We get a new set of ISDR–data by tensoring with M. Namely

$$\text{(I)}_1\qquad \begin{array}{c} \overset{k-s-1}{\otimes} M \otimes P(G_{s+1}) \\ \alpha_s \big\uparrow \big\downarrow \beta_s \\ \overset{k-s-1}{\otimes} M \otimes P(G_1) \otimes P(G_s) \end{array} \qquad \begin{array}{c} \partial_{s+1} \quad \xi_s \\ \\ d_\otimes \equiv 0, \end{array}$$

where the same symbols are used for all the maps.

Let D be a "new" differential such that

$$\text{(P)}_1\qquad \begin{array}{c} \overset{k-s}{\otimes} M \otimes P(G_s) \\ \alpha_{s-1} \big\uparrow \big\downarrow \beta_{s-1} \\ \overset{k-s}{\otimes} M \otimes P(G_1) \otimes P(G_{s-1}) \end{array} \qquad \begin{array}{c} \partial_s \quad \beta_{s-1} \quad D \\ \\ d_\otimes \equiv 0. \end{array}$$

is a P–data.

We get the FSDR–data.

Lemma 1.

$$(\mathrm{F})_1 \qquad \begin{array}{ccc} \overset{k-s}{\otimes} M \otimes P(G_s) & D & H \\ I \uparrow \; \downarrow P & & \\ \overset{k-s}{\otimes} M \otimes P(G_1) \otimes P(G_{s-1}) & \widehat{D} & \end{array}$$

where

$$\widehat{D} = \beta_{s-1} D \alpha_{s-1} + \beta_{s-1} D \sum_{j \geq 0} (\xi_{s-1} D + 1)^j \, \xi_{s-1} D \alpha_{s-1} \, ,$$

$$I = \alpha_{s-1} + \sum_{j \geq 0} (\xi_{s-1} D + 1)^j \, \xi_{s-1} D \alpha_{s-1} \, ,$$

$$P = \beta_{s-1} + \beta_{s-1} D \sum_{j \geq 0} (\xi_{s-1} D + 1)^j \, \xi_{s-1} \, ,$$

$$H = \sum_{j \geq 0} (\xi_{s-1} D + 1)^j \, \xi_{s-1} \, .$$

Proof: Follows from the FSDR–data (F). Another proof is in [1].

The second set of ISDR–data is based on the diagram

$$(\mathrm{B}) \qquad \begin{array}{ccc} P(G_1) & \delta_1 & \eta_0 \\ \iota_0 \uparrow \; \downarrow \pi_0 & & \\ M & \delta_M \equiv 0 & \end{array}$$

where the maps are defined by the following formulas:

$\delta_1 f_1^{(1)} = 0 \, ,$

$\delta_1 f_1^{(i)} = - \sum_{j=1}^{i-1} f_1^{(j)} f_1^{(i-j)} \, , \; i \geq 2 \, ,$

$\delta_1 = 0$ on $P^0(G_1)$,

δ_1 extends as a derivation to $P(G_1)$,

$\eta_0 = 0$ on $P^0(G_1)$ and on $P^1(G_1)$,

$\eta_0 \left(f_1^{(1)} f_1^{(i)} \right) = f_1^{(i+1)} \, , \; i \geq 1 \, ,$

$\eta_0 \left(f_1^{(i)} f_1^{(l)} \right) = 0 \, , \; i > 2 \, ,$

$\eta_0 (uv) = \eta_0(u) v \, , u \in P^2(G_1) \, ,$

$\iota_0(1) = 1 \, , \qquad 1 \in M^0 \, ,$

$$\iota_0(e) = f_1^{(1)}, \qquad e \in M^1,$$
$$\pi_0(1) = 1,$$
$$\pi_0\left(f_1^{(1)}\right) = e,$$
$$\pi_0\left(f_1^{(i)}\right) = 0,\ i > 1,$$
$$\pi_0(u) = 0, u \in P^j(G_1),\ j \geq 2.$$

Again we get a new set of ISDR–data by tensoring with M:

$$(\mathrm{I})_2 \qquad \begin{array}{c} \overset{k-s}{\otimes} M \otimes P(G_1) \otimes P(G_{s-1}) \\ {\scriptstyle \iota_{s-1}}\uparrow \ \downarrow{\scriptstyle \pi_{s-1}} \\ \overset{k-s}{\otimes} M \otimes M \otimes P(G_{s-1}) \end{array} \qquad \begin{array}{c} \delta_s \qquad \eta_{s-1} \\ \\ \delta_M \equiv 0, \end{array}$$

where

$$\delta_s = \overset{k-s}{\otimes} 1 \otimes \delta_1 \otimes 1, \quad \pi_{s-1} = \overset{k-s}{\otimes} 1 \otimes \pi_0 \otimes 1,$$
$$\iota_{s-1} = \overset{k-s}{\otimes} 1 \otimes \iota_0 \otimes 1, \quad \eta_{s-1} = \overset{k-s}{\otimes} 1 \otimes \eta_0 \otimes 1.$$

From a P–data

$$(\mathrm{P})_2 \qquad \begin{array}{c} \overset{k-s}{\otimes} M \otimes P(G_1) \otimes P(G_{s-1}) \\ {\scriptstyle \iota_{s-1}}\uparrow \ \downarrow{\scriptstyle \pi_{s-1}} \\ \overset{k-s}{\otimes} M \otimes M \otimes P(G_{s-1}) \end{array} \qquad \begin{array}{c} \delta_s \qquad \eta_{s-1} \qquad D \\ \\ \delta_M \equiv 0, \end{array}$$

We get a second set of FSDR–data.

Lemma 2.

$$(\mathrm{F})_2 \qquad \begin{array}{c} \overset{k-s}{\otimes} M \otimes P(G_1) \otimes P(G_{s-1}) \\ {\scriptstyle I}\uparrow \ \downarrow{\scriptstyle P} \\ \overset{k-s}{\otimes} M \otimes M \otimes P(G_{s-1}) \end{array} \qquad \begin{array}{c} D \qquad H \\ \\ \widehat{D}, \end{array}$$

is an FSDR–data, where

$$\widehat{D} = \pi_{s-1} D \iota_{s-1}$$
$$I = \iota_{s-1}$$
$$P = \pi_{s-1} + \pi_{s-1} D \sum_{j=0}^{i-1} (\eta_{s-1} D + 1)^j \eta_{s-1},$$
$$H = \sum_{j=0}^{i-1} (\eta_{s-1} D + 1)^j \eta_{s-1},$$

where the operators H and P act on $\overset{k-s}{\otimes} M \otimes P^i(G_1) \otimes P(G_{s-1})$.

Proof: Follows from the FSDR–data (F).

Another proof can be found in [**1**], see Lemma 8.12.

The third set of ISDR–data starts with the diagram (B). When tensored with M it gives ISDR–data

$$(\mathrm{I})_3 \qquad \begin{array}{c} \overset{k-1}{\otimes} M \otimes P(G_1) \\ \iota_0 \big\uparrow \ \big\downarrow \pi_0 \\ \overset{k-1}{\otimes} M \otimes M \end{array} \qquad \begin{array}{c} \delta_1 \qquad \eta_0 \\ \\ \delta_M \equiv 0\,, \end{array}$$

and with a "new" differential we get the second P–data

$$(\mathrm{P})_3 \qquad \begin{array}{c} \overset{k-1}{\otimes} M \otimes P(G_1) \\ \iota_0 \big\uparrow \ \big\downarrow \pi_0 \\ \overset{k-1}{\otimes} M \otimes M \end{array} \qquad \begin{array}{c} \delta_1 \qquad \eta_0 \qquad D \\ \\ \delta_M \equiv 0\,, \end{array}$$

And finally

Lemma 3. *The following is FSDR–data*

$$(\mathrm{F})_3 \qquad \begin{array}{c} \overset{k-1}{\otimes} M \otimes P(G_1) \\ I \big\uparrow \ \big\downarrow P \\ \overset{k-1}{\otimes} M \otimes M \end{array} \qquad \begin{array}{c} D \qquad H \\ \\ D\,, \end{array}$$

where

$$\widehat{D} = \pi_0 D \iota_0 + \pi_0 D \sum_{j \geq 0} (\eta_0 D + 1)^j \, \eta_0 D \iota_0\,,$$

$$I = \iota_0 + \sum_{j \geq 0} (\eta_0 D + 1)^j \, \eta_0 D \iota_0\,,$$

$$P = \pi_0 + \pi_0 D \sum_{j \geq 0} (\eta_0 D + 1)^j \, \eta_0\,,$$

$$H = \sum_{j \geq 0} (\eta_0 D + 1)^j \, \eta_0\,.$$

Proof: Follows from FSDR–data (F). A detailed proof can be found in [1].

Gugenheim's Tower.

The maps in the diagram (1) are induced from the maps in the following tower (called Gugenheim's Tower) of chain homotopy equivalences

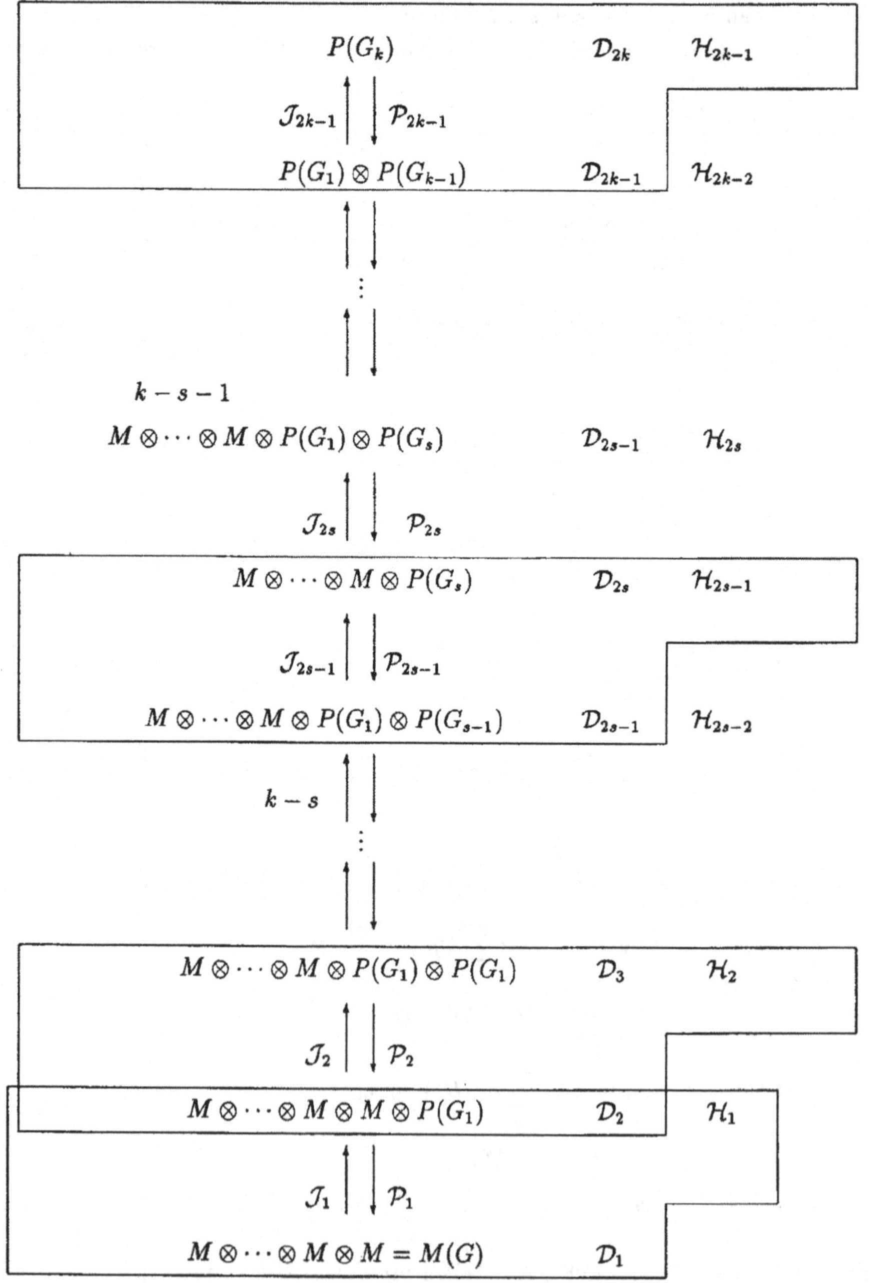

The blocks form FSDR–data. Therefore they satisfy (i)–(vi) of the diagram (I).

$\mathcal{D}_{2k} = d$ is the differential on $P(G_k)$ given by the group G_k.

$D = \mathcal{D}_1$ is the "final" differential on $M(G)$. This differential is given inductively in terms of the elementary operations which are defined in the previous section. Using the notation

$$D_s = \mathcal{D}_{2s}\,, \quad s = 1, 2, \cdots, k,$$

we get the formula for D.

Theorem 1. *The differential D on $M(G)$ in (1) is given by*

$$D = \pi_0 D_1 \iota_0 + \pi_0 D_1 \sum_{j\geq 0} (\eta_0 D_1 + 1)^j \, \eta_0 D_1 \iota_0\,,$$

$$D_s = \pi_s \beta_s D_{s+1} \alpha_s \iota_s + \pi_s \beta_s D_{s+1} \sum_{j\geq 0} (\xi_s D_{s+1} + 1)^j \, \xi_s D_{s+1} \alpha_s\,, s = 1, 2, \cdots, k-1.$$

Remark. Because the sums are finite [1], we don't give an upper bound for the sums explicitly.

Remark. The formula for the differential D can be written in a closed form. It is that form of the differential which is used in actual computations.

Denote

$$K_0 = 1 + \sum_{j\geq 0} (\eta_0 D_1 + 1)^j \, \eta_0\,,$$

$$K_s = 1 + \sum_{j\geq 0} (\xi_s D_{s+1} + 1)^j \, \xi_s\,, \; s = 1, 2, \cdots, k-1\,,$$

$$\iota = \alpha_{k-1} \iota_{k-1} \alpha_{k-2} \iota_{k-1} \cdots \alpha_2 \iota_2 \alpha_1 \iota_1 \iota_0\,,$$

then

$$D = \pi_0 K_0 \pi_1 \beta_1 K_1 \pi_2 \beta_2 K_2 \cdots \pi_{k-1} \beta_{k-1} K_{k-2} d\iota \qquad \text{(D)}$$

The main point of this formula is that the computation of the differential D on a particular element starts with an inclusion

$$\iota : M(G) \to P(G).$$

This inclusion does not commute with the differentials in general and it does not depend on the group structure of G. ι maps the elements of the basis for $M(G)$ as follows:

$$\iota\,(e_k^{u_k} \otimes \cdots \otimes e_1^{u_1}) = (f_k)^{u_k} \cdots (f_1)^{u_1}\,,$$

where

$$(f_j)^1 = f_j \quad \text{and} \quad (f_j)^0 = 1.$$

Proof: The formulas in the theorem are simple compositions of the formulas

$$\mathcal{D}_1 = \pi_0 \mathcal{D}_2 \iota_0 + \pi_0 \mathcal{D}_2 \sum_{j\geq 0} (\eta_0 \mathcal{D}_2 + 1)^j \eta_0 \mathcal{D}_2 \iota_0 ,$$

$$\mathcal{D}_2 = \pi_1 \mathcal{D}_3 \iota_1 ,$$

$$\vdots$$

$$\mathcal{D}_{2s-1} = \beta_{s-1} \mathcal{D}_{2s} \alpha_{2s-1} + \beta_{s-1} \mathcal{D}_{2s} \sum_{j\geq 0} (\xi_{s-1} \mathcal{D}_{2s} + 1)^j \xi_{s-1} \mathcal{D}_{2s} \alpha_{s-1} ,$$

$$\mathcal{D}_{2s} = \pi_s \mathcal{D}_{2s+1} \iota_s ,$$

which in term follow from $(F)_1, (F)_2$ and $(F)_3$.

Theorem 2. *The map I in* (1) *is the composition of the maps in the Gugenheim Tower*

$$I = \mathcal{J}_{2k-1} \mathcal{J}_{2k-2} \cdots \mathcal{J}_2 \mathcal{J}_1 ,$$

where

$$\mathcal{J}_1 = \iota_0 + \sum_{j\geq 0} (\eta_0 D_1 + 1)^j \eta_0 D_1 \iota_0 ,$$

$$\mathcal{J}_{2s} = \iota_s$$

$$\mathcal{J}_{2s+1} = \alpha_s + \sum_{j\geq 0} (\xi_s D_{s+1} + 1) \xi_s D_{s+1} \alpha_s$$

for $s = 1, 2, \cdots, k-1$.

Proof: Follows from $(\mathrm{F})_1$, $(\mathrm{F})_2$ and $(\mathrm{F})_3$.

The formulas in $(\mathrm{F})_1$, $(\mathrm{F})_2$ and $(\mathrm{F})_3$ also give the following two theorems:

Theorem 3. *The map P in* (1) *is the composition of the projection operators in the Gugenheim Tower*

$$P = \mathcal{P}_1 \mathcal{P}_2 \cdots \mathcal{P}_{2k-1} ,$$

where

$$\mathcal{P}_1 = \pi_0 + \pi_0 \mathcal{D}_2 \sum_{j\geq 0} (\eta_0 \mathcal{D}_2 + 1)^j \eta_0 ,$$

$$\mathcal{P}_{2s} = \pi_s + \pi_s \mathcal{D}_{2s+1} \sum_{j=0}^{i-1} (\eta_s \mathcal{D}_{2s+1} + 1)^j \eta_s ,$$

$$\mathcal{P}_{2s+1} = \beta_s + \beta_s \mathcal{D}_{2s+2} \sum_{j\geq 0} (\xi_s \mathcal{D}_{2s+2} + 1)^j \xi_s$$

for $s = 1, 2, \cdots, k-1$.
$\mathcal{P}_{2s}$ *acts on* $M \overset{k-s-1}{\otimes \cdots \otimes} M \otimes P^i(G_1) \otimes P(G_s)$ *and the operators* $\mathcal{D}$ *are given by the formulas in the proof of the Theorem 1.*

Theorem 4. *The homotopy* H *in* (1) *is given in terms of the maps in the tower of Gugenheim by*

$$H = \mathcal{H}_{2k-1} + \mathcal{J}_{2k-1}\mathcal{H}_{2k-2}\mathcal{P}_{2k-1} +$$

$$+\mathcal{J}_{2k-1}\mathcal{J}_{2k-2}\mathcal{H}_{2k-3}\mathcal{P}_{2k-2}\mathcal{P}_{2k-1} + \cdots + \mathcal{J}_{2k-1}\cdots\mathcal{J}_2\mathcal{H}_1\mathcal{P}_2\cdots\mathcal{P}_{2k-1},$$

where

$$\mathcal{H}_1 = \sum_{j\geq 0}(\eta_0\mathcal{D}_2 + 1)^j\,\eta_0,$$

$$\mathcal{H}_{2s} = \sum_{j=0}^{i-1}(\eta_s\mathcal{D}_{2s+1} + 1)^j\,\eta_s,$$

$$\mathcal{H}_{2s+1} = \sum_{j\geq 0}(\xi_s\mathcal{D}_{2s+2} + 1)^j\,\xi_s,$$

for $s = 1, 2, \cdots, k-1$.
$\mathcal{H}_{2s}$ *acts on* $M \overset{k-s-1}{\otimes \cdots \otimes} M \otimes P^i(G_1) \otimes P(G_s)$, *and the operators are given by the formulas in the proof of the Theorem 1.*

Example. Consider the group G of matrices

$$\begin{pmatrix} 1 & a_1 & a_3/k \\ 0 & 1 & a_2 \\ 0 & 0 & 1 \end{pmatrix}$$

where k is a fixed positive integer and $a_j \in \mathbf{Z}$. Let g_i be such a matrix with $a_i = 1,\ a_j = 0,\ i \neq j$. Then any element x of G can be written in the form

$$x = g_1^{x_1} g_2^{x_2} g_3^{x_3},$$

for some integers x_1, x_2, x_3. In fact the map $e : x \to x_2, x_2, x_3$ gives an identification of the sets G and $\mathbf{Z}^3 = \mathbf{Z} \times \mathbf{Z} \times \mathbf{Z}$. The map e induces a group struxture on $\mathbf{Z}^3$ by

$$(x_1, x_2, x_3) \cdot (y_1, y_2, y_3) = (x_1 + y_1, x_2 + y_2, x_3 + y_3 - kx_2y_1).$$

Let G^1 be the subgroup of G generated by g_1, g_2, and G^2 the subgroup generated by g_1. Then $G_3 = G$, $G_2 = G/G^2$ is generated by the projections of g_3, g_2 and $G_1 = G/G^1$ is generated by the projection of g_3.

Our first task is to compute the differential D on the elements $e_3^{u_3} \otimes e_2^{u_2} \otimes e_1^{u_1}$, $u_j = 0, 1$, of the basis for $M(G)$.

The formula for D takes the form

$$D = \pi_0 \left(1 + \sum_{j \geq 0} (\eta_0 D_1 + 1)^j \, \eta_0 \right) \cdot \pi_1 \beta_1 \left(1 + \sum_{j \geq 0} (\xi_1 D_2 + 1)^j \, \xi_1 \right) \cdot$$

$$\cdot \pi_2 \beta_2 \left(1 + \sum_{j \geq 0} (\xi_2 d + 1)^j \, \xi_2 \right) d\iota \, ,$$

where

$$D_1 = \pi_1 \beta_1 \left(1 + \sum_{j \geq 0} (\xi_1 D_2 + 1)^j \, \xi_1 \right) D_2 \alpha_1 \iota_1 \, ,$$

$$D_2 = \pi_2 \beta_2 \left(1 + \sum_{j \geq 0} (\xi_2 d + 1)^j \, \xi_2 \right) d\alpha_2 \iota_2 .$$

When applied to the elements of the basis we get

Theorem 5.

$$\begin{aligned} &De_1 = De_2 = 0 \, , \; De_3 = ke_2 \otimes e_1 \, , \\ &D(e_2 \otimes e_1) = D(e_3 \otimes e_1) = D(e_3 \otimes e_2) = 0 \, , \\ &D(e_3 \otimes e_2 \otimes e_1) = 0. \end{aligned}$$

Proof: Although the result is obtained by a straightforward application of the formula for D it is instructive to go through the computation in one typical case, say De_3

$$\iota e_3 = f_3 .$$

Using the formulas for d in $P(G_3)$ we get

$$df_3 = kf_2 f_1 .$$

Then $\xi_2(f_2 f_1) = 0$ according to the formulas following (A). The same formulas give that $\beta_2(f_2 f_1) = 1 \otimes f_2 f_1$. According to $(\mathrm{I})_2$ $\pi_2(1 \otimes f_2 f_1) = 1 \otimes f_2 f_1$. From (A) again $\xi_1(1 \otimes f_2 f_1) = 0$ and $\beta_1(1 \otimes f_2 f_1) = 1 \otimes f_1 \otimes f_1$. From $(\mathrm{I})_2$ it follows that $\pi_1(f_2 \otimes f_1) = 1 \otimes e_2 \otimes f_1$. (B) gives $\eta_0(1 \otimes e_2 \otimes f_1) = 1 \otimes e_2 \otimes \eta_0(f_1) = 0$ and $\pi_0(1 \otimes e_2 \otimes f_1) = 1 \otimes e_2 \otimes e_1$. This expression is written in a shorter form $e_2 \otimes e_1$ in the statement of the theorem. Similar shorter notation is used in other formulas.

The algebra structure on $M(G)$ is given by the formula

$$a \cdot b = P((Ia)(Ib)).$$

Since the images, under I, of the basis elements have to be computed first we present here the closed expression for I. According to the Theorem 2

$$I = \mathcal{J}_5\mathcal{J}_4\mathcal{J}_3\mathcal{J}_2\mathcal{J}_1 = \iota + \sum_{j\geq 0} (\xi_2 d + 1)^j \xi_2 d\iota +$$

$$+ \left(1 + \sum_{j\geq 0} (\xi_2 d + 1)^j \xi_2 d\right) \alpha_2 \iota_2 \cdot$$

$$\cdot \left[\sum_{j=0} (\xi_1 D_2 + 1)^j \xi_1 + \right.$$

$$+ \alpha_1\iota_1 \left(\sum_{j\geq 0} (\eta_0 D_1 + 1)^j \eta_0\right) \pi_1\beta_1 \left(1 + \sum_{j\geq 0} (\xi_1 D_2 + 1)^j \xi_1\right) +$$

$$+ \left(\sum_{j\geq 0} (\xi_1 D_2 + 1)^j \xi_1\right) \pi_2\beta_2 \left(1 + \sum_{j\geq 0} (\xi_2 d + 1)^j \xi_2\right) d\alpha_2\iota_2\alpha_1\iota_1 \cdot$$

$$\left. \cdot \left(\sum_{j\geq 0} (\eta_0 D_1 + 1)^j \eta_0\right) \pi_1\beta_1 \left(1 + \sum_{j\geq 0} (\xi_1 D_2 + 1)^j \xi_1\right)\right] \cdot$$

$$\cdot \pi_2\beta_2 \left(1 + \sum_{j\geq 0} (\xi_2 d + 1)^j \xi_2\right) d\iota.$$

D_1, D_2 are given by the formulas preceding the Theorem 5. A direct verification shows that I is simply identical with the inclusion ι on the elements in dimensions $0, 1$ and 2. In other words

$$Ie_3 = f_3\,,\ Ie_2 = f_2\,,\ Ie_1 = f_1\,,\ I(e_3 \otimes e_2) = f_3 f_2$$
$$I(e_3 \otimes e_1) = f_3 f_1\,,\ I(e_2 \otimes e_1) = f_2 f_1.$$

From here we can compute the products in $M(G)$ by applying the operator P from the Theorem 3 to the products in $P(G)$.

Theorem 6. *The products in $M(G)$ are given by the formulas*

$$e_3 \cdot e_1 = e_3 \otimes e_1\,,$$
$$e_1 \cdot e_3 = -e_3 \otimes e_1 + ke_2 \otimes e_1\,,$$
$$e_3 \cdot e_2 = e_3 \otimes e_2$$
$$e_2 \cdot e_3 = -e_3 \otimes e_2 + ke_2 \otimes e_1\,,$$
$$e_3 \cdot e_3 = -\binom{k+1}{2} e_2 \otimes e_1\,,$$
$$e_1 \cdot e_1 = 0\,,$$
$$e_2 \cdot e_2 = 0.$$

References

1. B. Cenkl and R. Porter, *"Polynomial cochains on nilmanifolds"* (preprint).
2. B. Cenkl and R. Porter, *"The spectral sequence for polynomial cochains"* (preprint).
3. V.K.A.M. Gugenheim, *"On the chain complex of a fibration"*, Illinois J. Math. **3** (1972), 398–414.
4. P. Hall, *"Nilpotent groups"*, Queen Mary College, Mathcmatics Notes, 1969.
5. A. Malcev, *"On a class of homogeneous spaces"*, Izv. Akad. Nank SSSR Ser. Math. **13** (1949), 9–32. English transl. Math. USSR -Izv. **39** (1949).

Department of Mathcmatics
Norteastern University
521 Lake IIall
Boston, MA 02115
U.S.A.

1990 Barcelona Conference
on Algebraic Topology.

REMARKS ON THE HOMOTOPY THEORY ASSOCIATED TO PERFECT GROUPS

F.R. COHEN*

Let π be a perfect group and $(B\pi)^+$ the Quillen plus construction for the classifying space $B\pi$. In this note we record a question concerning the homotopy theoretic properties of $(B\pi)^+$ which we posed at this conference; some partial answers to our question are also given here. The question addressed here is whether the 1-connected cover of $\Omega(B\pi)^+$ can be fibred by spheres and their loop spaces. We give immediate examples of groups π for which $\Omega(B\pi)^+\langle 1\rangle$ is "resolvable" by spheres and their loop spaces some of which are $A_5, A_6, A_7, SL(3,\mathbb{Z})$, M_{11}, and some free products. The examples which we consider here are studied directly from their cohomology and the remarks here are immediate consequences of these cohomology groups. A fuller account of a wider class of groups and different techniques will appear in the future.

We would like to thank the CRM, especially Manuel Castellet and Jaume Aguadé for providing an excellent mathematical environment and for their hospitality. We would also like to thank Jon Berrick for his interest and Dave Benson for tutorials in finite group theory.

§1. Introduction.

Let $\mathcal{P}$ denote the full subcategory of the category of groups where the objects are perfect groups π. Let $\mathcal{S}$ denote the full subcategory of the pointed homotopy category of topological spaces which (1) contains spheres and (2) is closed under fibrations and weak products of finite type.

We say that a space X is *spherically resolvable of weight* $\leq n$ provided there are spaces and maps

$$\begin{array}{ccccccc} X & \xleftarrow{f_0} & X_1 & \xleftarrow{f_1} & X_2 & \xleftarrow{f_2} & --- \longleftarrow & X_{n-1} \\ \downarrow{\scriptstyle p_0} & & \downarrow{\scriptstyle p_1} & & \downarrow{\scriptstyle p_2} & & & \downarrow{\scriptstyle p_{n-1}} \\ B_0 & & B_1 & & B_2 & & & B_{n-1} \end{array}$$

*Partially supported by the NSF.

such that (1) p_i is a fibration with fibre X_{i+1}, (2) $B_i = \Omega^{n_i} S^{m_i}$ for $0 \leq n_i \leq m_i$, $0 \leq i \leq n-1$ and (3), $p_{n-1} = 1$. A spherically resolvable space of weight $\leq n$ is, of course, in $\mathcal{S}$.

Here is a rash conjecture which we made at this conference: If π is a finite perfect group, then the 1-connected cover of $\Omega(B\pi)^+$ is spherically resolvable of finite weight. We shall give some examples of this conjecture below. Two immediate but useful remarks are given below.

Proposition 1. *Let π be a perfect group such that $H_q(\pi;\mathbb{Z})$ is a finite group for all $q > 0$ where $\mathbb{Z}$ is a trivial π-module. Then there are natural homotopy equivalences*

$$(B\pi)^+ \longrightarrow \prod_{p \text{ prime}} (B\pi)^+_{(p)} \quad \textit{and}$$

$$\bigvee_{p \text{ prime}} (B\pi)^+_{(p)} \longrightarrow \prod_{p \text{ prime}} (B\pi)^+_{(p)} \,.$$

Proposition 2. *Let π be a perfect group which is isomorphic to a discrete subgroup of $SU(n)$. Then $\Omega(B\pi)^+$ is the homotopy theoretic fibre of the natural map $SU(n) \to (SU(n)/\pi)^+$. Thus the $\bmod$-p cohomology ring of $\Omega(B\pi)^+$ is locally nilpotent.*

Let $[k] : S^n \to S^n$ denote a map of degree k. Write $S^n\{k\}$ for the homotopy fibre of $[k]$. The next remark will follow directly from an elementary cohomology argument, part of which appeared in [K].

Proposition 3. *Let π be a perfect finite group.*

(i) If the p-Sylow subgroup of π is isomorphic to $\mathbb{Z}/p^r$, then there is a homotopy equivalence

$$\Omega(B\pi)^+_{(p)} \longrightarrow S^{2n-1}\{p^r\}$$

where $2n$ is the p-period of $H^*(\pi;\mathbb{F}_p)$.

(ii) If the 2-Sylow subgroup of π is isomorphic to the generalized quaternion group of order 2^t, Q_t, then there is a homotopy equivalence

$$\Omega(B\pi)^+_{(2)} \longrightarrow S^3\{2^t\} \,.$$

Furthermore, $S^3\{2^t\}$ is not a homotopy abelian H-space.

We list some examples of this last proposition which follow immediately from Quillen's calculations for $SL(2,\mathbb{F}_q)$ [Q, FP].

Proposition 4. *Let $q = p^d > 3$ for p an odd prime. There is a homotopy equivalence*

$$\Omega BSL(2,\mathbb{F}_q)^+_{(\frac{1}{q})} \longrightarrow S^3\{q^2-1\} \,.$$

If $d = 1$, then there is a homotopy equivalence

$$\Omega BSL(2, \mathbf{F}_q)^+ \longrightarrow S^3\{q^2 - 1\} \times S^{p-2}\{p\}.$$

There are fibrations

$$\Omega BSL(n, \mathbf{F}_q)^+ \longrightarrow \Omega BPSL(n, \mathbf{F}_q)^+ \longrightarrow K(\mathbf{Z}/2, 1)$$

and so the 1-connected cover of $\Omega BPSL(2, \mathbf{F}_q)^+$ is spherically resolvable of weight ≤ 4 if $q = p$, and $p > 3$.

We list some further examples which follow immediately from Proposition 3 and the facts that (1) $SL(2, \mathbf{F}_5)$ is the double cover of A_5, and (2) $SL(2, \mathbf{F}_{3^2})$ is the double cover of A_6 [D].

Proposition 5. *If $\pi = A_5, A_6$, or A_7, the 1-connected cover of $\Omega(B\pi)^+$ is spherically resolvable of finite weight.*

A theorem of Gorenstein and Walter [G] gives that if π is a finite simple group with dihedral Sylow 2-subgroup, then π is isomorphic to $PSL(2, \mathbf{F}_q)$ for $q > 3$ and q odd or to A_7.

Corollary 6. *If π is a finite simple group with dihedral Sylow 2-subgroup, then the 1-connected cover of $\Omega(B\pi)^+_{(\ell)}$ is spherically resolvable of finite weight for all primes ℓ except possibly for $\ell = p$ where $\pi = PSL(2, \mathbf{F}_{p^d}), d > 1$.*

A more complicated example is given by one of the sporadic finite simple groups, the Mathieu group M_{11}. Details will appear elsewhere.

Proposition 7. *The 1-connected cover of $\Omega(BM_{11})^+$ is spherically resolvable.*

Proposition 8. *The space $\Omega(BSL(3, \mathbf{Z})^+_{(3)})$ is in $\mathcal{S}$. Furthermore,*

$$9[\pi_* BSL(3, \mathbf{Z})^+_{(3)}] = 0.$$

We include some remarks concerning the homotopy groups of $BSL(3, \mathbf{Z})^+_{(3)}$ which follow immediately from the proof of Proposition 8. There are infinitely many elements of order 3^2 in the homotopy of $BSL(3, \mathbf{Z})^+$. Furthermore, the image of the map in homotopy obtained from stabilization, $BSL(3, \mathbf{Z})^+_{(3)} \to BSL(6, \mathbf{Z})^+$ is annihilated by 3. It is not the case that $\Omega BSL(3, \mathbf{Z})^+$ is spherically resolvable of finite weight. We do not know whether $\Omega BSL(3, \mathbf{Z})^+$ is in $\mathcal{S}$, but by work of Soulé [S], the homology of $SL(3, \mathbf{Z})$ is all 2 and 3 torsion. Thus there is a homotopy equivalence

$$\Omega BSL(3, \mathbf{Z})^+ \to \Omega BSL(3, \mathbf{Z})^+_{(2)} \times \Omega BSL(3, \mathbf{Z})^+_{(3)}.$$

We will return to the question of whether $\Omega BSL(3, \mathbf{Z})^+_{(2)}$ is in $\mathcal{S}$ elsewhere.

Corollary 9. *Let $\pi_1, \ldots, \pi_k$ be finite perfect groups with periodic* mod *-p cohomology. Then $\Omega(B\pi)^+_{(p)}$ is in $\mathcal{S}$ where π is the free product $\pi_1 * \cdots * \pi_k$.*

Questions.

(1) What is the structure of $\Omega(B\pi)^+$ for finite perfect groups π?
(2) What is the structure of $H_*(\Omega(B\pi)^+; \mathsf{F}_p)$?

§2: Proofs of Propositions 1–4.

To prove Proposition 1, notice that $(B\pi)^+$ is simply-connected and of finite type by assumption. Thus the product of the localization maps

$$(B\pi)^+ \longrightarrow \prod_{p \text{ prime}} (B\pi)^+_{(p)}$$

induces a homology isomorphism as $\overline{H}_q(\pi; \mathbf{Z})$ is all torsion by assumption. Since the connectivity of $(B\pi)^+_{(p)}$ is greater than that of $(B\pi)^+_{(q)}$ for any fixed prime q and all but finitely many primes p by assumption, $\prod_{p \text{ prime}} (B\pi)^+_{(p)}$, the weak infinite product $\prod_{p \text{ prime}} (B\pi)^+_{(p)}$, and the wedge $\bigvee_{p \text{ prime}} (B\pi)^+_{(p)}$ are homotopy equivalent. The proposition follows.

To prove Proposition 2, observe that there is a map of fibrations

$$\begin{array}{ccccc} SU(n)/\pi & \longrightarrow & B\pi & \longrightarrow & BSU(n) \\ \downarrow & & \downarrow & & \downarrow 1 \\ (SU(n)/\pi)^+ & \longrightarrow & (B\pi)^+ & \longrightarrow & BSU(n) \end{array}$$

by [B]. The result follows since $\Omega(B\pi)^+$ is the homotopy fibre of $SU(n) \to (SU(n)/\pi)^+$.

We now give

Proof of Proposition 3: If the p-Sylow subgroup of π is $\mathbf{Z}/p^r$, $p > 2$, then there is an isomorphism of algebras

$$H^*(\pi; \mathsf{F}_p) \cong \Lambda[u_{2n-1}] \otimes \mathsf{F}_p[v_{2n}]$$

where $|u_{2n-1}| = 2n-1$, the rth Bockstein is defined and given by $\beta_r(u) = v$ and $n|(p-1)$. The value of $2n$ is given by a generator least degree in $H^{2n}(\mathbf{Z}/p^r; \mathbf{Z})$, $n > 0$, which is left invariant by the action of the normalizer of $\mathbf{Z}/p^r$ in π [S]. Thus there is an isomorphism of algebras

$$H_*(\Omega(B\pi)^+; \mathsf{F}_p) \cong \mathsf{F}_p[x_{2n-2}] \otimes \Lambda[y_{2n-1}]$$

with $H_{2n-2}(\Omega(B\pi)^+; \mathbf{Z}_{(p)}) \cong \mathbf{Z}/p^r\mathbf{Z}$. Thus there is a map

$$\alpha : P^{2n}(p^r) \to (B\pi)^+$$

inducing an isomorphism on $H_{2n-1}(\ ;\mathbb{Z}_{(p)})$, where $P^{2n}(p^r)$ denotes the cofibre of the map $[p^r]: S^{2n-1} \to S^{2n-1}$. Finally, notice that there is a map

$$\phi : S^{2n-1}\{p^r\} \to \Omega(B\pi)^+$$

given by the composite

$$S^{2n-1}\{p^r\} \xrightarrow{i} \Omega P^{2n}(p^r) \xrightarrow{\Omega\alpha} \Omega(B\pi)^+$$

where i is obtained by passage to fibres in the homotopy commutative diagram

$$\begin{array}{ccc} S^{2n-1} & \longrightarrow & * \\ \downarrow{\scriptstyle [p^r]} & & \downarrow \\ S^{2n-1} & \longrightarrow & P^{2n}(p^r)\,. \end{array}$$

Notice that ϕ_* is a map of $H_*(\Omega S^{2n-1};\mathbb{F}_p)$-modules and ϕ_* induces an isomorphism on $H_j(\ ;\mathbb{F}_p)$ for $j = 2n-1$ and $2n-2$. Thus ϕ_* is an isomorphism and the result 3(i) follows.

The proof of 3(ii) is similar. The mod -2 cohomology of π is periodic of period 4 [S,CE] and there are isomorphisms

$$H^n(BQ_t;\mathbb{F}_2) \cong \begin{cases} \mathbb{F}_2 \oplus \mathbb{F}_2 & \text{if } n \equiv 1,2 \mod 4 \\ \mathbb{F}_2 & \text{if } n \equiv 0,3 \mod 4, \text{ and} \end{cases}$$

$$\overline{H}^n(BQ_t;\mathbb{Z}) \cong \begin{cases} \mathbb{Z}/2 \oplus \mathbb{Z}/2 & \text{if } n \equiv 2 \mod 4 \\ \mathbb{Z}/2^t & \text{if } n \equiv 0 \mod 4, \text{ and} \\ 0 & \text{otherwise.} \end{cases}$$

Since π is perfect, and $H^*(B\pi;\mathbb{F}_2)$ is a summand of $H^*(BQ_t;\mathbb{F}_2)$, $H^1(B\pi;\mathbb{F}_2) = 0$ and hence $H^2(B\pi;\mathbb{F}_2) = 0$ by applying Sq^1. Thus there is an isomorphism of algebras

$$H^*(B\pi;\mathbb{F}_2) \cong \Lambda[u_3] \otimes \mathbb{F}_2[v_4]$$

with $\beta_t u = v$. The result follows as in 3(i).

Proof of Proposition 4: The proof here is essentially that given above as the ℓ-Sylow subgroup of $SL(2,\mathbb{F}_q)$ is cyclic if $\ell > 2$ and generalized quaternion if $\ell = 2$ and $(\ell, q) = 1$: There is an isomorphism

$$\overline{H}^n(SL(2,\mathbb{F}_q);\mathbb{Z}\left[\frac{1}{p}\right]) \cong \begin{cases} \mathbb{Z}/q^2-1 & \text{if } n \equiv 0 \mod 4, n > 0, \\ 0 & \text{otherwise} \end{cases}$$

by [FP,T]. Thus the first part of the proposition follows.

If $q = p$, then the p-Sylow subgroup of $SL(2,\mathsf{F}_p)$ is isomorphic to $\mathbb{Z}/p$. The p-periodicity of $SL(2,\mathsf{F}_p)$ is $p-1$ by inspecting the action of the normalizer of $\mathbb{Z}/p$ in $SL(2,\mathsf{F}_p)$ or by [Q,T]. Thus there is a homotopy equivalence

$$\Omega BSL(2,\mathsf{F}_p)^+_{(p)} \to S^{p-2}\{p\}\ .$$

Thus the second part of the proposition follows. To finish, observe that there is a central extension

$$1 \to \mathbb{Z}/2 \to SL(2,\mathsf{F}_q) \to PSL(2,\mathsf{F}_q) \to 1$$

giving the fibration in the proposition.

§3: Proof of Proposition 5.

There is an extension [D],

$$1 \to \mathbb{Z}/2 \to SL(2,\mathsf{F}_5) \to A_5 \to 1\ .$$

Thus there is a fibration

$$B\mathbb{Z}/2 \to BSL(2,\mathsf{F}_5)^+ \to BA_5^+$$

together with a homotopy equivalence

$$\Omega BSL(2,\mathsf{F}_5)^+ \to S^3\{2^3\cdot 3\cdot 5\}$$

by Proposition 4.

To study A_6, recall that there is an extension [D],

$$1 \to \mathbb{Z}/2 \to SL(2,\mathsf{F}_{3^2}) \to A_6 \to 1\ .$$

But then there is a homotopy equivalence $\Omega BSL(2,\mathsf{F}_{3^2})^+_{(\frac{1}{3})} \to S^3\{2^4\cdot 5\}$. We omit the 3-primary calculations as other methods are more efficient.

To consider A_7, observe that the natural map $i : A_6 \to A_7$ induces an isomorphism in mod -2 and mod -3 homology. Since an inclusion $\mathbb{Z}/5 \to A_7$ may be chosen to factor through Σ_5, we have a homotopy equivalence $\Omega(BA_7)^+_{(5)} \to S^7\{5\}$. Since the Weyl group for the normalizer of $\mathbb{Z}/7$ in A_7 is $\mathbb{Z}/3$ acting by multiplication by 3^2 on $\mathbb{Z}/7$, there is a homotopy equivalence $\Omega(BA_7)^+_{(7)} \to S^5\{7\}$.

§4: Proofs of Statements 8 and 9.

We need to recall the homotopy theory of mod -p^r Moore spaces developed in [CMN] if $p > 2$ and in [Co] if $p = 2$ and $r \geq 2$. Write $[k] : S^n \to S^n$ for a map of degree k and $S^n\{k\}$ and $P^{n+1}(k)$ for the fibre and cofibre respectively of $[k]$. There are spaces $T^{2n+1}\{p^d\}$ and $T^{n+1}\{2^r\}$, $r \geq 2$, constructed satisfying the following properties:

(1). There are homotopy equivalences

$$\Omega P^{2n+2}(p^r) \longrightarrow S^{2n+1}\{p^r\} \times \Omega A_{2n+2}, \quad p > 2,$$
$$\Omega P^{2n+1}(p^r) \longrightarrow T^{2n+1}\{p^r\} \times \Omega B_{2n+1}, \quad p > 2,$$
$$\text{and} \quad \Omega P^{n+1}(2^r) \longrightarrow T^{n+1}\{2^r\} \times \Omega D_{n+1}, \quad r \geq 2\,,$$

where A_{2n+2} and B_{2n+1} are bouquets of mod -p^r Moore spaces, $p > 2$, and D_{n+1} is a bouquet of mod -2^r Moore spaces. Handy references are [CMN, Co].

(2). There are vector space isomorphisms

$$H_*(T^{2n+1}\{p^r\}; \mathsf{F}_p) \longrightarrow H_*(\Omega^2 S^{2n+1}; \mathsf{F}_p) \otimes H_*(\Omega S^{2n+1}; \mathsf{F}_p), \quad p > 2\,,$$
$$\text{and} \quad H_*(T^{n+1}\{2^r\}; \mathsf{F}_2) \longrightarrow H_*(\Omega^2 S^{n+1}; \mathsf{F}_p) \otimes H_*(\Omega S^{n+1}; \mathsf{F}_2)\,.$$

(3). If $n \geq 2$, there are homotopy equivalences

$$\Omega P^{n+1}(p^r) \longrightarrow S \times T\,, \quad p > 2\,,$$

where S is a weak infinite product of spaces $S^{2k+1}\{p^r\}$ and T is a weak infinite product of spaces $T^{2t+1}\{p^r\}$, and

$$\Omega P^{n+1}(2^r) \longrightarrow T$$

where T is a weak infinite product of spaces $T^{n+1}\{2^r\}$.

(4). $T^{2n+1}\{p^t\}$ and $T^{n+1}\{2^r\}, r \geq 2$, are in $\mathcal{S}$.

To prove Corollary 9, we need to recall that if A and B are connected CW-complexes, then the homotopy theoretic fibre of the natural inclusion

$$i : A \vee B \longrightarrow A \times B$$

is $\Sigma(\Omega A) \wedge (\Omega B)$. Thus we record the next lemma which follows from [CMN, N].

Lemma 4.1. *There are homotopy equivalences*

$$\Sigma S^n\{p^r\} \wedge S^m\{p^s\} \longrightarrow \Sigma\left(\bigvee_{j,i \geq 1} P^{n+j(n-1)}(p^r) \wedge P^{m+i(m-1)}(p^s)\right)$$

if $p \geq 2$, *and*

$$P^n(p^s) \wedge P^m(p^r) \longrightarrow P^{n+m}(p^d) \vee P^{n+m-1}(p^d)$$

if $m, n \geq 2$ *and* $r, s \geq 2$ *if* $p = 2$.

We now prove Corollary 9 with $k = 2$ and leave the evident inductive step to the reader. By combining the above remarks, there is a fibration

$$\Sigma\Omega(B\pi_1)^+ \wedge \Omega(B\pi_2)^+ \longrightarrow B(\pi_1 * \pi_2)^+ \longrightarrow (B\pi_1)^+ \times (B\pi_2)^+$$

for perfect groups π_i. If the mod -p cohomology of the finite perfect group π_i is p-periodic, then $\Omega(B\pi_i)^+_{(p)}$ is homotopy equivalent to $S^{2k_i+1}\{p^{r_i}\}$ for some k_i and r_i by Proposition 3. Furthermore, if $p = 2$, then $r_i \geq 3$. Thus there is a homotopy equivalence

$$\Sigma\Omega(B\pi_1)^+_{(p)} \wedge \Omega(B\pi_2)^+_{(p)} \longrightarrow \Sigma S^{2k_1+1}\{p^{r_1}\} \wedge S^{2k_2+1}\{p^{s_2}\}$$

and, by Lemma 4.1, a homotopy equivalence to a bouquet $\vee P^n(p^d)$. Combining remarks (1)–(4) together with the Hilton-Milnor Theorem [W], we see that Corollary 9 follows.

To finish this section, we give the proof of Proposition 8.

By Soulé's calculation [So], there is a map

$$B\Sigma_3 \vee BD_6 \longrightarrow BSL(3, \mathbb{Z})$$

which induces an isomorphism in mod -3 cohomology where D_6 is the dihedral group of order 12; in addition he proves that the reduced homology of $SL(3, \mathbb{Z})$ is all 2 and 3 torsion. By Proposition 1, we restrict attention to primes 2 and 3 to get a homotopy equivalence

$$BSL(3, \mathbb{Z})^+ \longrightarrow BSL(3, \mathbb{Z})^+_{(2)} \times BSL(3, \mathbb{Z})^+_{(3)} .$$

Since $BSL(3, \mathbb{Z})^+_{(3)}$ is 2-connected, the composites

$$P^2(2) \xrightarrow{\ i\ } B\Sigma_3 \longrightarrow BSL(3, \mathbb{Z})^+ \qquad \text{and}$$

$$X = P^2(2) \vee P^2(2) \xrightarrow{\ j\ } BD_6 \longrightarrow BSL(3, \mathbb{Z})^+$$

are null homotopic where i and j induce isomorphisms on $H_1(\ ;\mathbb{Z})$. Thus we obtain

$$(B\Sigma_3/P^2(2)) \vee (BD_6/X) \longrightarrow BSL(3, \mathbb{Z})^+$$

which induces a mod -3 homology isomorphism. Since $A = B\Sigma_3/P^2(2)$ and $B = BD_6/X$ are both simply-connected, we may localize at the prime 3 to get an equivalence

$$A_{(3)} \vee B_{(3)} \longrightarrow BSL(3, \mathbb{Z})^+_{(3)} .$$

Since the mod -3 cohomology groups of $A_{(3)}$ and $B_{(3)}$ are periodic of period 4, we have homotopy equivalences (by the evident modification of Proposition 3)

$$\Omega A_{(3)} \simeq \Omega B_{(3)} \simeq S^3\{3\} .$$

Thus the proof of Proposition 8 is analogous to that of Corollary 9. Namely, we obtain a homotopy equivalence

$$\Omega BSL(3, \mathbb{Z})^+_{(3)} \longrightarrow S \times T$$

where S is a weak product of spaces $S^{2n+1}\{3\}$ and T is a weak product of spaces $T^{2n+1}\{3\}$. Thus by [N], we have

$$9(\pi_* BSL(3, \mathbb{Z})^+_{(3)}) = 0 .$$

References

[B] J. Berrick, *An Approach to Algebraic K-theory*, Pitman Press, 1984.

[BC] D. Benson and J. Carlson, Diagrammatic methods for modular representations and cohomology, *Comm. Algebra* **15** (1987), no. 1-2, 53–121

[CE] H. Cartan and S. Eilenberg, *Homological Algebra*, Princeton Univ. Press. 1956.

[Co] F.R. Cohen, The homotopy theory of mod -2^r Moore spaces, $r > 1$, in preparation.

[CMN] F.R. Cohen, J.C. Moore, and J.A. Neisendorfer, Exponents in homotopy theory, *Ann. of Math. Stud.* **113** (1987), 3–34.

[D] L.E. Dickson, *Linear Groups with an Exposition of the Galois Field Theory*, New York, Dover 1958.

[FP] Z. Fiedorwicz and S. Priddy, *Homology of Classical Groups Over Finite Fields and their Associated Infinite Loop Spaces*, Lecture Notes in Math. 674, Springer-Verlag, 1978.

[G] D. Gorenstein and P. Walter, The characterization of finite groups with dihedral Sylw 2-subgroups (I), *J. Algebra* **2** (1965), 85–151.

[K] S. Kleinerman, The cohomology of Chevalley groups of exceptional Lie type, *Mem. Amer. Math. Soc.*, **39** (1982), no. 268.

[N] J.A. Neisendorfer, The exponent of a Moore space, *Ann. of Math. Stud.* **113** (1987), 35–71.

[Q] D. Quillen, The spectrum of an equivariant cohomology ring, *Ann. of Math.* **94** (1971), 549–572.

[So] C. Soulé, The cohomology of $SL_3(\mathbb{Z})$, *Topology* **17** (1978), 1–22.

[S] R. Swan, The p-period of a finite group, *Illinois J. Math.* **4** (1960), 341–346.

[T] C.B. Thomas, *Characteristic Classes and the Cohomology of Finite Groups*, Cambridge Univ. Press, 1986.

[W] G.W. Whitehead, *Elements of Homotopy Theory*, Springer-Verlag, Graduate Texts in Math. **61**, 1978.

Department of Mathematics
University of Rochester
Rochester, NY 14627
U.S.A.

1990 Barcelona Conference
on Algebraic Topology.

HOMOTOPY LOCALIZATION AND V_1-PERIODIC SPACES

EMMANUEL DROR FARJOUN

0. Introduction:

Several authors have used techniques of [Cartan–Eilenberg], [Bousfield-1], [Quillen] to construct localization functors associated to certain collections of maps in the topological or simplicial category or categories of similar nature. This localization turns a collection of maps with certain closure properties into equivalences in a universal, minimal fashion. Compare [B-2], [H], [Del]. Here we give a straightforward, naive procedure that proves the existence of a localization functor L_φ associated to any map $\varphi : A \to B$ in the category of topological spaces, CW-complexes, or the simplicial category. One can start with any map or any set of maps (which can be added to form one map). This construction subsumes all existing localization construction which are idempotent functors in view of [F] it seems reasonable to conjecture that every idempotent, co-augmented continuous functor is a special case of L_φ for an appropriate map $\varphi : A \to B$. Therefore, it agrees with Bousfield–Kan localization functor R_∞ only when it is idempotent — i.e. for 'good' spaces. Of course the map $A \to B$ is sometimes a map between 'large' spaces such as the wedge of 'all' countable CW-complexes etc.

It is interesting to note that contrary to [Bousfield-2] if one takes φ to be the degree p-map between circles $p : S^1 \to S^1$ one gets a new construction in homotopy theory that turns the fundamental group into its canonical uniquely p-divisible version.

0.1 Periodic spaces: An important special reason for the interest in φ- localization is the need to consider the v_1-periodic version of an arbitrary space. It is shown in [F] that one must ask for a v_1-local version of X namely $L_{v_1}X$ for which the map of function spaces

$$\mathrm{map}(M^\ell(p), L_{v_1}X) \to \mathrm{map}(M^{\ell+2p-2}(p), L_{v_1}X)$$

is a homotopy equivalence where the map is induced by the v_1-maps between the above mod-p-Moore space.

We would like to draw attention to some elementary properties of v_1-localization and give a criterion for a space to be v_1-periodic (or v_1-local). Using the work on nilpotency [DHS], [HS] it is shown that there is a well defined v_1-localization that does not depend on the choice of v_1 map with range $M^3(p)$. A v_1-local space is characterized in terms of the action of $[M^3(p), X]$ on $\{[M^\ell(p), X]\}_\ell$ (for $p > 2$). This action is the precise analogue of the fundamental group action on higher homotopy groups that comes to play in HR-localization or localization away from a prime p.

1. Definition and construction of L_f

1.1. Definition: We say that Y is f-local for a map $f : A \to B$ between spaces if the f induces a homotopy equivalence on function spaces

$$\mathrm{map}(f, Y) : \mathrm{map}(B, Y) \xrightarrow{\simeq} \mathrm{map}(A, Y).$$

1.1.1 Remark. One might ask why we not define a "homotopy–f–local" space to be a space W with $[f, W] : [B, W] \to [A, W]$ an isomorphism of sets. This is a perfectly good definition however it is not 'complete' in the following sense.

1. For any functor F : {Spaces} $\to$ {Spaces} if $[f, FX]$ is an isomorphism of sets then FX is automatically f-local: $\mathrm{map}(f, FX) : \mathrm{map}(B, FX) \xrightarrow{\sim} \mathrm{map}(A, FX)$ is an equivalence. This is a result of [F] which was written precisely in view of the question.

2. In view of (1.) it is impossible to canonically associate a universal "homotopy–f–local" space with every space X.

Here is a simple example due to G. Mislin: Consider $f : S^1 \to *$. Then a homotopy–f–local space is just a simply connected space (while an f-local space is a homotopically discrete space). Now it is not hard to see by cohomological consideration that these is no universal simply connected space U associated via a map $\mathbf{R}P^2 \to U$ with $\mathbf{R}P^2$. Such a space will have to have $H^2(U, \mathbf{Z}) \cong \mathbf{Z}/2\mathbf{Z}$ which is impossible for a simply connected space.

1.2. Pointed, unpointed: One can work either in the pointed or unpointed category, for convenience we work in the pointed category of spaces (spaces are topological spaces having the homotopy type CW-complexes or semi-simplicial sets).

1.3. Theorem: *For any map of spaces $f : A \to B$ there exists a functor (called f-localization) L_f which is co-augmented and homotopically idempotent. The map $X \to L_f X$ is homotopically universal into f-local spaces.*

Proof: The proof runs through 1.10. We use the following basic lemma [Spanier, B-K]. Given any diagram of spaces D_α over some small category C and given any space Y, then there is a homotopy equivalence:

$$\mathrm{holim}_{\leftarrow} \mathrm{map}(D_\alpha; Y) \simeq \mathrm{map}(\mathrm{holim}_{\rightarrow} D_\alpha; Y).$$

This equation depends only on Y being fibrant (e.g. Kan complex) and expresses a basic duality between homotopy direct and inverse limit.

In particular, if

$$\begin{array}{ccc} L & \longrightarrow & K_1 \\ \downarrow & & \downarrow \\ K_2 & \longrightarrow & W \end{array}$$

is a homotopy pushout square then map (W,Y) is the homotopy pull back of $\operatorname{map}(K_1,Y) \to \operatorname{map}(L,Y) \leftarrow \operatorname{map}(K_2,Y)$.

An immediate consequence of the above is the following:

1.4. Observation: Let the following diagram

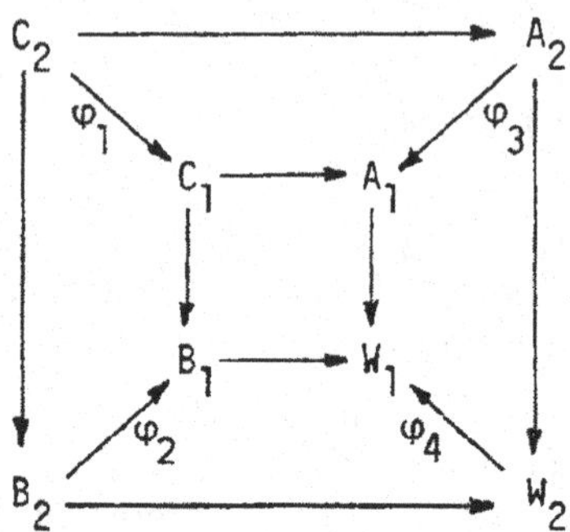

be a map of two pushout squares with W_1 and W_2 being the homotopy pushouts. If a space Y is φ_1-local for $i = 1,2,3$ then Y is φ_4-local.

Proof: In the corresponding pullback squares formed by taking the function complexes of all the spaces in the above diagram into Y we get that three arrows $\hom(\varphi_i, Y)$, $i = 1,2,3$ are homotopy equivalences by assumption on Y so the fourth one $\hom(\varphi_4, Y)$ must also be a homotopy equivalence, due to the homotopy invariance of homotopy pullback, its basic property.

1.5. Derived maps: In order to construct $L_f X$ we glue on to X several families of maps that are defined now:

First let $f \rtimes S^n$ be the map induced by f on the half-smash product: $Y \rtimes S^n .= Y \times S^n \,/\, \{*\} \times S^n$. Thus

$$f \rtimes S^n : A \rtimes S^n \to B \rtimes S^n.$$

Second, let $\operatorname{Dcyl}(f)$ the double mapping cylinder of f be the homotopy pushout of $B \xleftarrow{f} A \xrightarrow{f} B$. Now take $T(f)$ to be the map $\operatorname{Dcyl}(f) \to \operatorname{Dcyl}(id_B)$ induced by the obvious maps. Notice $\operatorname{Dcyl}(id_B) \cong I \times B$.

Define $\bar{f}$ to be the wedge sum of the maps

$$\bar{f} = \left(\bigvee_{n \geq 0} f \rtimes S^n\right) \vee \left(\bigvee_{n \geq 0} T(f \rtimes S^n)\right).$$

Notice that f is a summand of $\bar{f}$ since $f \rtimes S^0 = f \vee f$.

1.6. Now we construct $L_f X$ by repeatedly glueing $\bar{f} : \bar{A} \to \bar{B}$ to X. Let μ be the first infinite limit ordinal with cardinality bigger then the number of cells in $\bar{A} \amalg \bar{B}$.

Define $L^1_f X$ to be the homotopy pushout

$$\begin{array}{ccc} \coprod_{g:\bar{A}\to X} \bar{A} & \longrightarrow & \coprod_g \bar{B} \\ \downarrow & & \downarrow \\ X & \longrightarrow & L^1_f X \end{array}$$

where g runs over the set of maps $\bar{A} \to X$.

Define by transfinite induction

$$L_f^{\alpha+1} X = L_f^1(L_f^\alpha X)$$

and for a limit ordinal β define $L_f^\beta X$ to be the homotopy colimit of the directed diagram $(L_f^\alpha X)_{\alpha<\beta}$.

Finally take $L_f X$ to be $L_f^\mu X$ for μ as above.

1.7. Claim: Let P be a f-local space. Then for each ordinal $\beta \geq 0$ the map

$$\text{map}(j_\beta, P) : \text{map}(L_f^\beta X, P) \to \text{map}(X, P)$$

is a homotopy equivalence for any space X.

Proof: Since map(, P) turn a homotopy direct limit into a homotopy inverse limit; it is sufficient to prove the above for $\beta = 1$.

Notice that $\bar{f} = \vee f_\alpha$ for some collection f_α. It is enough that the claim holds for $X \to Y_\alpha$ where Y_α is the pushout along $f_\alpha : A_\alpha \to B_\alpha$.

Now consider the components f_α. First $f \times S^n : A \ltimes S^n \to B \ltimes S^n$. Clearly hom$(f \times S^n, P)$ is a homotopy equivalence since e.g. for $n = 1$ $A \times S^1$ is the homotopy equalizer of $A \rightrightarrows A$ the two maps being the identity maps, and so map$(A \times S^1, P)$ is the homotopy limit of map$(A, P) \rightrightarrows$ map(A, P). In the diagram,

$$\begin{array}{ccccc} \text{map}(A \times S^1) = \text{holim}_{\leftarrow}(\ ) & \longrightarrow & \text{map}(A, P) & \rightrightarrows & \text{map}(A, P) \\ \uparrow & & \uparrow\simeq & & \uparrow\simeq \\ \text{map}(B \times S^1) = \text{holim}_{\leftarrow}(\ ) & \longrightarrow & \text{map}(B, P) & \rightrightarrows & \text{map}(B, P) \end{array}$$

the map map$(f \times S^1, P)$ appears as a map between two homotopy inverse limits of diagrams which are equivalent for each place by assumption. Therefore this map is also homotopy equivalence. Similarly the map $T(f \rtimes S^n) \to T(1_B \rtimes S^n)$ is a map between homotopy direct limit of diagrams of spaces that turn into homotopy equivalences between them upon taking hom$(-, P)$ since P is f-local.

1.8. If $L_f X$ is local then L_f is idempotent functor. The idempotency is a direct consequence because of the above property of the tower L_f^α. Simply, if X is f-local then by the above map$(X, X) \to$ map$(L_f X, X)$ is a homotopy equivalence and the image of

$id : X \to X$ will give us a map $L_fX \to X$ which is a homotopy inverse to $X \to L_fX$. It follows that if X is f-local $X \to L_fX$ is a homotopy equivalence. Therefore, if L_fX is f-local $L_fL_fX \sim L_fX$.

1.9. L_fX is f-local for all X. We must show that these is a homotopy equivalence

$$\hom(B, L_fX) \to \hom(A, L_fX).$$

These spaces not being connected it must be shown that every component is carried to a corresponding one by a map that induces an isomorphism on the higher homotopy group $\pi_j \quad j > 0$; but first we show that the above map induces an isomorphism on π_0 — the set of path components.

In other words we show that any homotopy class of $A \to L_fX$ has a unique lifting to a class $B \to L_fX$.

The existence is clear from the construction: given any map $g_\mu : A \to L_f^\mu X$, since μ is large enough in comparison to A this map must factor through some map $g_\alpha : A \to L_\mu^\alpha X$ for $\alpha < \mu$. But μ is a limit ordinal so $\alpha + 1 < \mu$ and certainly by construction $g_{\alpha+1}$ can be extended over B where $g_{\alpha+1} : A \to L_f^\alpha X \to L_f^{\alpha+1} X$. So we get an extension of g_μ over B.

Now for uniqueness of the above lift: Given two extensions of $g_\mu : A \to L_fX$ $g_\mu^1, g_\mu^2 : B \to L_f^\mu X = L_fX$ we know that there is a homotopy $g_\mu^1 \circ f \sim g_\mu^2 \circ f$ so there is a map of $\mathrm{Dcyl}(f) \to L_fX$ given by these homotopies. We would like to show that $g_\mu^1 \sim g_\mu^2$. But again for cardinality reasons the map $\mathrm{Dcyl}(f) \to L_fX$ factors through some $L_f^\alpha X$ for some $\alpha < \mu$. We get a map $h_\alpha : \mathrm{Dcyl}(f) \to L_f^\alpha(X)$ whose composition into $L_f^{\alpha+1}X$, the next space in the tower, factor by construction through $B \times I \cong \mathrm{Dcyl}(id_B)$. Therefore the two maps of B into $L_f^\alpha X$ are actually homotopic in $L_f^{\alpha+1}X$ and thus of course in L_fX.

1.10. The higher homotopy groups. Now we choose a component of map $[B, L_fX]$ say of a map, $\psi : B \to L_fX$, and show that f induces an isomorphism of higher homotopy groups of the pointed mapping spaces:

$$\pi_n\big(\mathrm{map}(B, L_fX); \psi\big) \xrightarrow{\cong} \pi_n\big(\mathrm{map}(A, L_fX); \psi \circ f\big).$$

Elements in these homotopy groups are represented by pointed maps of S^n and so by maps of $B \rtimes S^n$ since our glueing process took the map $f \rtimes S^n$ into account these is an isomorphism of sets

$$[B \rtimes S^n, L_fX] \xrightarrow{\circ f} [A \rtimes S^n, L_fX]$$

for each one of these components. So we get the desired result claimed by the theorem.

Examples:

(1) Notice that if f is of the form $f : \Sigma^k A \to A$ then L_fX has a periodic A-homotopy groups. Now if f is nilpotent then $L_f \sim *$ is the cone functor.

(2) If $f_n : \phi \to S^n$ we get $L_{f_n}X = P_nX$, a Postnikov section of X.

(3) $f : A \to *$ or $f : \phi \to A$ one gets an interesting space out of X. In the case $A = S^1 \cup_p e^2$ one gets the Anderson localization of X which is well understood only for a simply connected space.

(4) If the map $f : A \to B$ is $\vee(h_\alpha : A_\alpha \to B_\alpha)$ where h_α runs over all h_*-isomorphism between spaces of cardinality not bigger then $h_*(pt)$ for some homotopy theory $h_*(-)$ one gets the Bousfield h_*- localization. this is one of the origins of the present construction.

(5) It is not true as claimed in [Bou 2] that if $f : S^1 \to S^1$ is the degree p map then L_f is the Anderson localization. L_fX is some sort of uniquely p-divisible space in the sense that the self map $\Omega L_fX \to \Omega L_fX$ of the loop space sending each loop to its p'th power is a homotopy equivalence (not a loop equivalence!). This implies more on π_nX than unique p-divisibility [P].

(6) If $f : \Sigma^{k+\ell}M^3(p) \to \Sigma^\ell M^3(p)$ is an Adams map, a v_1-map, inducing an isomorphism on K-theory, one gets an interesting version of v_1-localization of spaces. $L_{v_1}X = L_fX$ will have a periodic $\mathbb{Z}/p\mathbb{Z}$-homotopy groups, moreover map$(M^3(p), L_{v_1}X)$ is an infinite loop space with the right periodicity property. Again as in case (5) $\pi_*(L_fX, \mathbb{Z}/p\mathbb{Z})$ have definitely stronger periodicity property then the divisibility by the v_1-map (see (2.7) below).

(7) For simply connected spaces L_p for $p : S^2 \to S^2$ is just localizing away from p or with respect to $H_*(\ , \mathbb{Z}[\frac{1}{p}])$. Similarly there is a map $\Pi_\infty S^2 \to \Pi_\infty S^2$ that gives the completion in this case.

(8) In fact we do not know of any localization functor which whenever it is idempotent it does not agree with L_f for some f. It is tempting to ask whether every coaugmented homotopy idempotent functor or spaces $i : X \to FX$ with Fi homotopy equivalence is L_f for some map $f : A \to B$.

Problems: A basic problem is to formulate sufficient conditions on $f : A \to B$ that will guarantee that L_f preserves fibration or cofibration sequences.

In the case $f : \Sigma^{k+\ell}A \to \Sigma^k A$ this implies control over the A-homotopy of L_fX.

It will be interesting to know under what condition $L_{v_1}X \simeq X_K$ the K-theoretical localization of X. Namely, when does inverting v_1 imply inverting all K-theory equivalences.

It can be shown that there exists a map between a large wedge of circles $\omega : \vee_\infty S^1 \to \vee_\infty S^1$ such that ω induces identity on $H_1(\ ; \mathbb{Z})$ and $L_\omega = (\ )_{H\mathbb{Z}} = L_{H\mathbb{Z}}$ the Bousfield localization with respect to integral homology. Similar maps exists $VS^1 \to \vee S^1$ for other connected theories [F].

2. Periodic spaces and periodic actions

When one examines the conditons on the homotopy groups $\pi_*(X, pt)$ that characterize a p-local space (i.e. a space that is local with respect to $p : S^1 \to S^1$) one gets with [P] that, in addition to unique p-divisibility of $\pi_n(X,pt)$ for all $n \geq 1$, an extra condition on the action of $\pi_1(X,pt)$ on $\pi_n(X,pt)$ arises. Namely, the latter must be divisible by all elements of the form:

$$1+\xi+\xi^2+\cdots+\xi^{p-1}=\frac{1-\xi^p}{1-\xi}$$

in the group ring $\mathbf{Z}[\pi, X]$ where ξ is any element in $\pi_1(X,pt)$. This extended divisibility arises algebraically in [Baum] to insure that semi-direct products and wreath-products are uniquely–p–divisible.

In this section an extension of the above divisibility criterion for an arbitrary map $f : \Sigma^k A \to \Sigma A$ between suspensions is given. If f itself is a suspension the condition turns out to be empty.

For the map $v_1 : \Sigma^\ell M^3(p) \to M^3(p)$ $(p > 2)$ this divisibility gives an 'extended' v_1-periodicity that must be satisfied by v_1-periodic spaces (and thus also K-local spaces) on top of the naive divisibility of $\pi_*(X, \mathbf{Z}/p\mathbf{Z})$ by v_1.

We start with an observation about half smash product of suspensions: The quotient space of $X \times Y$ by $X \times pt$ is denoted by $X \ltimes Y$. This construction comes into play since for a given map $\alpha : A \to X$ the nth homotopy group π_n map $(A,X)_\alpha$ is given as $[S^n \ltimes A, X]$ where map (A,X) denotes the space of pointed maps $A \to X$.

2.1. Lemma: *For any pair of pointed spaces A, B there is a natural splitting map which is a homotopy equivalence:*

$$e : B \ltimes \Sigma A \xleftarrow{\simeq} \Sigma(B \wedge A) \vee \Sigma A.$$

Proof: First notice that there is a topological equivalence $B \ltimes A \xrightarrow{\cong} B^+ \wedge A$, where B^+ is the disjoint union of B new base-point. Consider first the special case $A = S^0 = \{0,1\}$. In this case it is easy to give a homotopy equivalence:

$$(S^1 \wedge B) \vee S^1 = \Sigma(B \wedge S^0) \vee \Sigma S^0 \to B \ltimes S^1.$$

This map is defined in the obvious way on each component on the left. For a general space A we simply smash the above equivalence with A:

$$(\Sigma B \wedge A) \vee SigmaA \to B^+ \wedge S^1 \wedge A \simeq B^+ \vee \cong B \ltimes \Sigma A.$$

2.2. The operation $[\Sigma A, X] \times [\Sigma^k A, X] \to [\Sigma^{\ell+k+1} A, X]$ induced by a map $\alpha : \Sigma^{\ell+1} A \to \Sigma A$. Given a map α consider its half-smash with S^k the k-dimensional sphere.

Using the above natural homotopy equivalence

$$\begin{array}{ccc} S^k \ltimes \Sigma^{\ell+1} A & \longrightarrow & S^k \ltimes \Sigma A \\ \downarrow \cong & & \downarrow \cong \\ \Sigma^{k+\ell+1} A \vee \Sigma^{k+1} A & \xrightarrow{S^k \ltimes \alpha} & \Sigma^{k+1} A \vee \Sigma A \end{array}.$$

On the summand $\Sigma^{k+1}A$ the map $S^k \ltimes \alpha$ is the original map, the summand $\Sigma^{k+\ell+1}A$ we get certain Whitehead product of the original map to be discussed below . We denote this map:

$$\mathrm{adj}^k(\alpha) : \Sigma^{\ell+k+1}A \to \Sigma^{k+1}A \vee \Sigma A.$$

2.3. Definition: *Given* $a \in [\Sigma^{k+1}A, X]$ $b \in [\Sigma A, X]$ *one gets a composition* $a^b := (a \vee b) \circ \mathrm{adj}^k(a) \in [\Sigma^{k+\ell+1}A, X]$. This gives the desired operation of b on a.

2.4. Example: Consider $\alpha = p : S^1 \to S^1$ the degree p-map. By smashing with S^k one gets a map

$$S^{k+1} \to S^1 \vee S^{k+1}$$

which sends ι_{k+1} to the group ring element $(1 + \iota_1 + \iota_1^2 + \cdots + \iota_1^{p-1}) \cdot \iota_{n+1}$. Therefore it defined an operation of $\pi_1 X$ on $\pi_n X$ for any space X which sends $u \in \pi_n X$ to $(1 + \xi + \xi^2 + \cdots + \xi^{p-1})u$ for $\xi \in \pi_1 X$.

2.5. Example: If the map $\alpha : \Sigma^{k+1}A \to \Sigma A$ is itself a suspension it is not hard to see that $S^k \ltimes \alpha$ splits as a of α and its suspension, and so the operation in this case is trivial sending each element to the composition.

2.6. Example: For the case $v_1 : \Sigma^{2p-2}M^3(p) \to M^3(p)$ where $p > 2$ a prime, we know that v_1 is not a suspension in this dimension [C-N] and in fact we get a non-trivial action

$$\pi_3(X, \mathbb{Z}/p\mathbb{Z}) \times \pi_n(X, \mathbb{Z}/p\mathbb{Z}) \to \pi_{n+2p+1}(X, \mathbb{Z}/p\mathbb{Z}),$$

where $\pi_k(X, \mathbb{Z}/p\mathbb{Z}) := [M^k(p), X]_{pt}$, for $k \geq 2$.

Using the Hilton–Milnor decomposition one can compute the operation for the universal case $X = M^3(p)$ and express it as a Whitehead product (2.9). As special case of the next proposition one gets that a space is v_1-local in the sense of (1.1) above if and only if every element in $\pi_3(X, \mathbb{Z}/p\mathbb{Z})$ acts by an isomorphism between $\pi_n(X, \mathbb{Z}/p\mathbb{Z})$ and $\pi_{n+2p+1}(X, \mathbb{Z}/p\mathbb{Z})$.

2.7. Proposition: *Let* $\sigma : \Sigma^{\ell+1}A \to \Sigma A$ *be a self map. A space* X *is* σ*-local in the sense of (1.1) if and only if for every element* $u \in [\Sigma A, X]$ *the induced map defined in above* $\mathrm{adj}^k(u) = \theta(u) : [\Sigma^k A, X] \to [\Sigma^{k+\ell+1}A, X]$ *is an isomorphism.*

Proof: For the proof we need only to interprete the weak equivalence map$(\Sigma A, X) \to$ map$(\Sigma^{\ell+1}A, X)$ that defines σ-local spaces in terms of homotopy groups. The above spaces are not connected and therefore we must show that the map between them $\sigma^* =$ map(σ, X) permutes components and send each component of the range map$(\Sigma A, X)$ into the corresponding component of map$(\Sigma^{k+1}A, X)$ by a homotopy equivalence.

But as we have seen in (1.5) while proving (1.3) above the homotopy groups of the mapping spaces above are precisely a subset of components of $[S^n \ltimes \Sigma^k A, X]$; the map $\sigma^* = map(\sigma, X)$ induces a map on the homotopy groups of corresponding components sends a class $S^n \ltimes \Sigma A \to X$ in the mapping space map$(\Sigma A, X)$ into its composition with $S^n \ltimes \sigma$. But by our assumption all this compositions (2.1) are isomorphisms. Thus we get the desired weak equivalence between components.

2.8. Periodic action in terms of Whitehead product. We now express the homotopy operation defined in terms of the decomposition of the half smash, in terms of Whitehead products and Hilton–Milnor invarinats. This will allow us t in the concrete case of the Adams map.

Given A, B, C spaces and given a map $f : \Sigma B \to \Sigma C$ we would like to express $\Sigma A \ltimes f$ in terms of familiar constructions. Let $H_\omega : \Omega\Sigma C \to \Omega\Sigma C^{(|w|)}$ where $\omega^\wedge(\iota^1, \iota^2) : \Omega\Sigma C^{(|w|)} \to \Omega\Sigma(C \vee C)$ is the associated universal Whitehead product on $\Sigma C \vee \Sigma C$. Recall from above that $\Sigma A \ltimes \Sigma B \simeq \Sigma B \vee (\Sigma A \wedge \Sigma B)$. We can write $\Sigma A \ltimes f \equiv \mathrm{Id}(\Sigma A) \ltimes f = f + (\Sigma A \ltimes f)_t$ where the latter map is the twisted part goes from $\Sigma A \wedge \Sigma B \to \Sigma C \vee (\Sigma A \wedge \Sigma C)$.

2.9. Proposition:

$$\begin{aligned}(\Sigma A \ltimes f)_t &= \Sigma A \wedge f + \Sigma_\omega \omega(i, \eta \circ \Sigma A \wedge H_\omega(f) \\ &= \Sigma A \wedge f + \Sigma_k Ad^k(\iota_{\Sigma C})(\iota_{\Sigma A \wedge \Sigma C}) \circ \Sigma A \wedge H_{k+1}(f)\end{aligned}$$

We hope to publish a justification of the formulas above in a future paper with J.Harper.

3. Uniqueness of localization with respect to v_1

In this section the localization L_f when f is an Adams map between Moore spaces mod p is discussed. It is known that for $p > 2$ prime, the lowest dimension for which the Adams map exist is when the range is three-dimensional Moore space $M^3(p) = S^2 \vee C^3$. But there are several map $\Sigma^q M^3 \to M^3$ for $q = 2p - 2$, that induce an isomorphism on K-theory and all are indistinguishably called v_1-maps.

3.1. Proposition: *Let α, β be two v_1-maps $M^{3+q}(p) \to M^3(p)$. Then there is a natural homotopy equivalence of functors $L_\alpha \sim L_\beta$.*

Proof: First it is clear that for any map $f : \Sigma^k A \to A$, $L_{f^2} \sim L_f$. This means that a space Y is f-periodic if and only if it is f^2-periodic. Consider the sequence of maps induced by f

$$[A, Y] \xrightarrow{f_1^*} [\Sigma A, Y] \xrightarrow{f_2^*} [\Sigma^2 A, Y] \to \dots .$$

If we assume that $f_{n+1}^* f_n^*$ is an isomorphism, for all n, then f_n^* must be both injective and surjective and thus f_n^* is an isomorphism.

Now notice that by construction there is a natural transformation, for every $g : \Sigma A \to A$, $L_{g \circ \Sigma g} \to L_g$ between the corresponding localization functor simply because the maps $\Sigma g : \Sigma^2 A \to \Sigma A$ are amongst the map which are used in constructing $L_g X$. Therefore $L_{g \circ \Sigma g}$ is universal and must be equivalent to L_g itself. To proceed we notice that by the nilpotency theorem there exists some $\ell > 0$ for which $\alpha^\ell = \beta^\ell$ as homotopy classes, where α^ℓ, β^ℓ are the obvious composition with higher suspensions. But now $L_\alpha \sim L_{\alpha^\ell} \sim L_{\beta^\ell} \sim L_\beta$ as needed.

3.3. Proposition: *The action of* id : $M^3(p) \to M^3(p)$ *on* $M^k(p) \to M^2(p)$ *is given by Whitehead products where the twisted term is non-trivial here.*

Proof: Use 2.5 above.

References

[Baum] G. Baumslag: Some aspects of groups with unique roots. Acta Math. 104 (1960), pp. 217–303.

[Baum] G. Baumslag: Roots and wreath products. Proc. Comb. Phil. Soc. 56 (1960), pp. 109–117.

[B-1] A.K. Bousfield: Localization of spaces with respect to homology. Topology 14, (1975), pp. 133–150.

[B-2] A.K. Bousfield: Construction of factorization systems in categories. J. of Pure and Appl. Alg., vol. 9 (1976), pp. 207–220.

[B-K] A.K. Bousfield–D.M. Kan: Homotopy limits, completions and localizations. Lecture Notes in Math., No. 304, Springer-Verlag (1972).

[C-N] F. Cohen–J. Neisendorfer: On desuspension of maps. (1989) to appear

[C-E] H. Cartan–S. Eilenberg: Homological algebra. Princeton Univ. Press (1956).

[Del] A. Deleanu: Existence of Adams completions for objects in cocomplete categories. J. of Pure and Appl. Alg., vol. 6 (1975), pp. 31–39.

[DHS] E. Devinatz, M.J. Hopkins and J.H. Smith: Nilpotence and stable homotopy theory I. Ann. of Math. 128 (1988), pp. 207–241.

[F] E. Dror Farjoun: Higher homotopies of natural construction. preprint. (1990) to appear.

[H] A. Heller: Homotopy theories. Memoire AMS 383, 1988.

[HS] M.J. Hopkins and J.H. Smith: Nilpotence and stable homotopy theory II. to appear.

[P] G. Peschke: Localizing groups with action. Pub. Math. de la Univ. Autonoma, Barcelona, vol. 33 (1989), pp. 227–234.

[Q] D. Quillen: Homotopical algebra. Lecture Notes in Math. 43, Springer-Verlag, 1967.

[S] E. Spanier: Algebraic topology. McGraw-Hill (1966).

Hebrew University,
Jerusalem,
Israel.

1990 Barcelona Conference
on Algebraic Topology.

THE MODULO 2 COHOMOLOGY ALGEBRA OF THE WREATH PRODUCT $\Sigma_\infty \int X$

NGUYEN VIET DUNG

Introduction.

Let $X = (X, *)$ be a topological space with base point. According to the results of J.P. May **[3]**, G. Segal [6], M.G. Barratt and P.J. Eccles [1], ... studying the cohomology algebra of the infinite loop space $QX = \Omega^\infty S^\infty X$ reduces to determining the cohomology algebra of the wreath product

$$\Sigma_\infty \int X = \varinjlim E\Sigma_n \underset{\Sigma_n}{x} X^n.$$

Here Σ_n denotes the symmetric group on n letters and limit is given by the canonical inclusions $\Sigma_n \hookrightarrow \Sigma_{n+1}$ (see e.g. [5]). In [5] the homology Hopf algebra $H_*\Sigma_\infty \int X$ was computed clearly in terms of the homology operation derived from the Dickson's coinvariants. In his thesis [7] R.J. Wellington tried to describe H^*QX. But he has not managed to give a description of the algebra H^*QX with concrete generators except in the case when X is a connected space such that H_*X has the structure of a trivial coalgebra.

The purpose of this paper is to determine the cohomology algebra $H^*\Sigma_\infty \int X$. The case when $X = \{pt\}$, the one point space, is computed by N.H.V. Hung in **[2]**. The case of finite loop space will be treated in a sequent paper.

The paper is divided into two sections. Some remarks on ξ–modules of finite type will be given in §1. Combining these remarks and Mui's result on the homology Hopf algebra $H_*\Sigma_\infty \int X$ we will determine the cohomology algebra $H^*\Sigma_\infty \int X$ in §2.

Throughout the paper the coefficients will be taken in the finite field of two elements $Z_2 = Z/2Z$.

§1. ξ–modules.

1.1 Definition

An abelian restricted Lie algebra over the field Z_2 is a graded Z_2 vector space L such that

i) $L_0 = 0$.

ii) there is a vector space homomorphism $\gamma : L_n \longrightarrow L_{2n}$ for all $n \geq 0$.

If H is a commutative, connected Hopf algebra over Z_2, then the augmentation ideal IH is an abelian restricted Lie algebra with $\gamma(x) = x^2$.

If L is an abelian restricted Lie algebra, define the enveloping algebra $U(L)$ of L by the equation

$$U(L) = S(L)/(\gamma(x) + x^2)$$

where $S(\cdot)$ is the symmetric algebra functor. Note that Borel's structure theorem for commutative, connected Hopf algebra of finite type implies that if H is such then there is a sub–abelian restricted Lie algebra $L \subseteq IH$ for which there is an induced isomorphism of algebras

$$U(L) \xrightarrow{\cong} H.$$

1.2 Definition

A ξ–module over the field Z_2 is a graded Z_2 vector space V such that

i) $V_0 = 0$

ii) there is a vector spacc homomorphism $\xi : V_{2n} \longrightarrow V_n$ for all $n \geq 0$.

We routinely extend ξ to all V by demanding that it be 0 on elements of odd degree.

If C is a cocommutative coalgebra, the root map $\xi : C_{2n} \longrightarrow C_n$ gives C the structure of a ξ–module.

1.3 Lemma

Suppose that H is a connected, cocommutative, commutative Hopf algebra of finite type over Z_2 and that $L \subseteq IH$ is a sub–abelian restricted Lie algebra so that

a) $U(L) = H$, and

b) the root map $\xi : IH \longrightarrow IH$ restricts to a map $\xi : L \longrightarrow L$.

Then

i) the inclusion of ξ–modules $L \longrightarrow IH$ is split, and

ii) any spliting $IH \longrightarrow L$ induces an isomorphism of algebras

$$U(L^*) \cong H^*.$$

The $*$ denotes the vector spaces duals; notice that the dual of ξ–module is an abelian restricted Lie algebra.

Proof: The argument is a suitable modification of that of Wellington [**7**]. The splitting is obtained as in [**7**, 1.7] and the isomorphism is obtained as in [**7**, Thm. 3.7].

1.4 Definition

Let n be an integer $n \geq 0$ or let $n = \infty$. Let j be an integer $j \geq 1$. Define a ξ–module $V(j,n)$ by

$$V(j,n)_i = \begin{cases} Z_2\,, & \text{if } i = 2^t j\,,\ 0 \leq t \leq n \\ 0 & \text{otherwise} \end{cases}$$

and $\xi : V(j,n)_{2i} \longrightarrow V(j,n)_i$ an isomorphism for $i = 2^t j$, $t < n$ and $\xi = 0$ otherwise.

$V(j,n)$ could be called cyclic of depth $n+1$ one generator of degree j.

1.5 Lemma

Let V be a ξ–module of finite type. Then there exists a set of pairs of integers (j_i, n_i) so that there is an isomorphism of ξ–modules

$$V = \oplus V(j_i, n_i)\,.$$

That is, every ξ–module is a sum of *cyclic* modules.

Proof: We have that V^* is an abelian restricted Lie algebra of finite type. Then Theorem 3 of [4] implies that there is an isomorphism

$$V^* = \oplus V(j_i, n_i)^*\,.$$

Now dualizing again we get the lemma.

§2. Algebra $H^*\Sigma_\infty \int X$.

Let $X = (X, *)$ be a connected pointed space. According to [5] for each sequence of non–negative integers $K = (k_0, k_1, \dots, k_{n-1}), n \geq 0$, we have a homology operation derived from Dickson's coinvariants

$$D_K : H_*X \longrightarrow H_*\Sigma_\infty \int X$$

with the convention that D_ϕ is the inclusion.

Let $\{x_i;\ i \in I\}$ be a basis for X with $x_{i_0} = 1 \in H_0X$. Then the Hopf algebra $H_*\Sigma_\infty \int X$ is determined in [5] as follows.

2.1 Theorem (H. Mui)

i) We have the isomorphism of Hopf algebras

$$H_*\Sigma_\infty \int X = Z_2\,[D_K x_i; K = (k_0, \dots, k_{n-1}),\ n \geq 0,\ k_0 > 0,\ i \in I,\ (K; i) \neq (\emptyset; i_0)]$$

where the comultiplication is given by

$$\Delta D_K(x) = \sum_{K=K'+K''} \sum_{(x)} D_{K'}(x') \otimes D_{K''}(x'')$$

with $\Delta x = \sum_{(x)} x' \otimes x''$.

ii) the above formula can be reduced by the relation

$$D_{(0,k_1,\dots,k_{n-1})}(x) = (D_{(k_1,\dots,k_{n-1})}(x))^2$$

Now assume that X is connected and of finite type. We have

$$L = \text{Span}\,\{D_K(x); x \in H_*X,\ (K; x) \neq (\emptyset, 1)\} \subseteq IH_*\Sigma_\infty \int X.$$

This is an abelian restricted Lie algebra with

$$\gamma(D_K(x)) = D_{(0,K)}(x)$$

and L is a sub–abelian restricted Lie algebra of $IH_*\Sigma_\infty \int X$.

In fact, Mui's result may be interpreted to say that

$$H_*\Sigma_\infty \int X \cong U(L).$$

By Theorem 2.1 (i),

$$\xi D_K(x) = \begin{cases} D_{(i_1,\ldots,i_n)}(\xi x), & \text{if } K = (2i_1, \ldots, 2i_n) \\ 0 & \text{otherwise.} \end{cases}$$

So, the hypothese of Lemma 1.3 are satisfied and we have

$$H^*\Sigma_\infty \int X = U(L^*). \tag{2.2}$$

Now, by Lemma 1.5,

$$H_*X = \oplus V(j_i, n_i).$$

Let $\{x_k\}$ be the basis for X obtained from the evident basis for the summands $V(j_i, n_i)$. Call x_k odd if

$$x_k \in V(j_i, n_i)_{j_i}$$

for some i and define the depth $h(k)$ of an element x_k of this basis by the formula

$$h(k) - 1 = \max\{n\,;x_k \in \operatorname{Im} \xi^n\}.$$

The depth can be infinite. Now Mui's result and the formula for ξD_K given above immediately imply the following result

2.3 Lemma

There is an isomorphism of ξ–modules

$$L = \oplus_{(x_k,K)} V(j(x_k, K), n(x_k, K))$$

where the sum is over all pairs (x_k, K) such that $(x_k, K) \neq (1, \emptyset)$ and either

i) x_k is odd, or

ii) $K = (k_0, \ldots, k_{n-1})$ with some k_t odd.

In addition, $D_K(x_k)$ is the non–zero element of $V(j(x_k, K), n(x_k, K))_{j(x_k,K)}$ and

$$j(x_k, K) = \deg(D_K(x_k))$$

and

$$n(x_k, K) = h(k) - 1\,.$$

Now if $W_K(x_k)$ is dual to $D_K(x_k)$, then (2.2) and Lemma 2.3 imply the main result of the paper

2.4 Theorem

Assume that X is connected and of finite type and that $\{x_k\}$ is the basis for X given above. Then we have the isomorphism of algebras

$$H^*\Sigma_\infty \int X = Z_2 \left[W_k(x_k);\ K \in Z_+^n,\ n \geq 0, \quad x_k \text{ is odd or } K = (k_0, \ldots, k_{n-1}) \text{ with some } k_t \text{ odd}\right] / ((W_K(x_k))^{2^{h(k)}}).$$

2.5 Remark

If X is not of finite type, write $X = \operatorname{colim} X_\alpha$ where $X_\alpha \subseteq X$ is of finite type. Then

$$H^*\Sigma_\infty \int X = \lim H^*\Sigma_\infty \int X_\alpha$$

will be, in general, a completed polynomial algebra.

Acknowledgements. The author would like to thank Prof. H. Mui for proposing the problem and for useful discussions. He is also grateful to the referee who has critically read the first manuscript of this paper and helped him to bring it to the present form.

References.

[1] M.G. Barrat and P.J. Eccles, "T^+*-structures I*", Topology **13** (1974), 23–45.

[2] N.H.V. Hung, *"The modulo 2 cohomology algebras of symmetric group"*, Japan. J. Math. **13**, 1(1989).

[3] F.R. Cohen, T.J. Lada and J.P. May, *"The homology of iterated loop spaces"*, Springer Lect. notes in Math. **Vol. 533**.

[4] J.P. May, *"Some remarks on the structure of Hopf algebras"*, Proc. A.M.S. **23** (1969), 708–713.

[5] H. Mui, *"Homology operations derived from modular coinvariants"*, Göttingen preprint.

[6] G. Segal, *"Configuration spaces and iterated loop spaces"*, Invent. Math. **21** (1973), 213–222.

[7] R.J. Wellington, *"The unstable Adams spectral sequence for iterated loop spaces"*, Memoirs A.M.S. **258**, March (1982).

Section of Topology – Geometry
Institute of Mathematics
P.O. Box 631 Bo Ho
10000 Hanoi
Vietnam

1990 Barcelona Conference
on Algebraic Topology.

LANNES' DIVISION FUNCTORS ON SUMMANDS OF $H^*(B(\mathbf{Z}/p)^r)$

JOHN C. HARRIS AND R. JAMES SHANK *

Introduction

Lannes has constructed a family of functors which he refers to as division functors. These functors are defined on either the category of unstable modules over the Steenrod algebra or the category of unstable algebras over the Steenrod algebra. We denote the unstable module category by $\mathcal{U}$ and the unstable algebra category by $\mathcal{K}$. We are primarily interested in division in $\mathcal{U}$. If B is a $\mathcal{U}$-object of finite type then division by B in $\mathcal{U}$, $A \mapsto (A : B)$, is the left adjoint to $A \mapsto A \otimes B$. There is particular interest in division by $H^*(B\mathbf{Z}/p)$. Following Lannes we denote $(A : H^*(B\mathbf{Z}/p))$ by $T(A)$. The T functor has a number of useful properties: it is additive, exact, and preserves tensor products. Lannes originally used the T functor in his reformulations of the proofs of the Sullivan conjecture and the Segal conjecture for elementary abelian p-groups (see [L]).

In [HS] , the authors determined the value of the T functor on the indecomposable $\mathcal{U}$-summands of $H^*(BV)$ where V is an elementary abelian p-group. In the current paper we extend the results of [HS] to describe the modules $(A : B)$ where A and B are both summands of $H^*(BV)$. Since the division functor is biadditive, it suffices to consider only the indecomposable summands. These summands were classified in [HK] using the modular representation theory of the semigroup ring $\mathbf{F}_p[\mathrm{end}(V)]$. Let L_λ and L_μ denote indecomposable summands $H^*(BV)$. Our main result, Theorem 3.6, describes $(L_\lambda : L_\mu)$ in terms of representation theory.

It is known that the L_λ are injective in the category $\mathcal{U}$ ([C] , [M] , [LZ]). Another collection of injectives are the dual Brown–Gitler modules $\{J(n) \mid n \geq 0\}$ ([M]). Lannes and Schwartz ([LS]) have shown that all injectives in $\mathcal{U}$ are direct sums of injectives of the form $L_\lambda \otimes J(n)$. As an application of our division functor calculations we give formulas for the dimensions of the spaces $\hom_{\mathcal{U}}(L_\lambda \otimes J(l), L_\mu \otimes J(m))$. We hope these dimensions may be of use in studying injective resolutions in $\mathcal{U}$.

Throughout the paper V and W will be elementary abelian p-groups, H^*V will denote $H^*(BV)$, and R will be the endomorphism ring $\mathrm{end}_{\mathcal{U}}(H^*V)$.

The paper is organized as follows. In Section 1, we recall some of the properties of the division functors and we describe the module $(H^*V : fH^*W)$ where f is an idempotent in $\mathrm{end}_{\mathcal{U}}(H^*W)$. In Section 2, we describe $(e_\lambda H^*V : fH^*W)$, when e_λ is

---

*The second author is a N.S.E.R.C. Postdoctoral fellow
1980 Mathematics Subject Classification (1985 Revision): 55S10

a primitive idempotent in R. In Section 3, we use some algebraic lemmas to give a completely representation theoretic description of $(e_\lambda H^*V : e_\mu H^*V)$ when e_λ and e_μ are primitive idempotents in R. In Section 4 we discuss the computations needed to determine an explicit description of $(e_\lambda H^*V : e_\mu H^*V)$. We also recover the principal algebaic result of [HS] . Section 5 contains the calculation of the dimensions of the spaces $\hom_{\mathcal{U}}(L_\lambda \otimes J(l), L_\mu \otimes J(m))$. Finally, in the appendices, we give some examples.

Some of the results in this paper first appeared in the second author's thesis ([S]) written at the University of Toronto under the direction of Paul Selick.

Section 1

Here we list those properties of the division functors which we will use in the latter sections. We refer the reader to [L] for details.

Let A and B be objects of $\mathcal{U}$ with B of finite type (that is, B is finite dimensional in each degree). Then $A \mapsto (A : B)$ is a covariant functor from $\mathcal{U}$ to $\mathcal{U}$ and $B \mapsto (A : B)$ is a contravariant functor from $\mathcal{U}^{f.t.}$ to $\mathcal{U}$. It follows that $(A : B)$ is both a left $\mathrm{end}_{\mathcal{U}}(A)$-module and a right $\mathrm{end}_{\mathcal{U}}(B)$-module.

Lemma 1.1. *Let $f \in \mathrm{end}_{\mathcal{U}}(B)$ be an idempotent. Then $(A : fB) \cong (A : B)f$.*

Proof. Here $fB = \mathrm{im} f$ and $(A : B)f = \mathrm{im}(A : f)$. Note that $(A : f)$ is an idempotent in $\mathrm{end}_{\mathcal{U}}\big((A : B)\big)$. Let $\pi: B \to fB$ and $i: fB \to B$ be the projection and inclusion; so $i \circ \pi = f$ and $\pi \circ i = id_{fB}$. Since $(A : f) = (A : \pi)\circ(A : i)$ and $id_{(A:fB)} = (A : i)\circ(A : \pi)$, we have $(A : fB) = \mathrm{im}(A : i) \cong \mathrm{im}(A : f) = (A : B)f$. ∎

Remark 1.2. In general, for $g \in \mathrm{end}_{\mathcal{U}}(B)$, $(A : gB)$ and $(A : B)g$ are not isomorphic. See Appendix 3.

Lannes has shown that for any object in $\mathcal{U}$, say M, there is a natural isomorphism $(M \otimes M : H^*W) \xrightarrow{\lambda_{M,W}} (M : H^*W) \otimes (M : H^*W)$ (see [L]). If A is an object in $\mathcal{K}$ with the product given by μ, then $(\mu : H^*W) \circ \lambda^{-1}_{M,W}$ is a product on $(A : H^*W)$. In fact, with this product, $(A : H^*W)$ is an object in $\mathcal{K}$.

Let $F_{\mathcal{K}}$ denote the free functor which is left adjoint to the forgetful functor from $\mathcal{K}$ to the category of non-negatively graded vector spaces. If we identify V and W with H^1V and H^1W, respectively (hence regarding these as graded vector spaces concentrated in degree one), then $H^*V \cong F_{\mathcal{K}}(V)$ and $H^*W \cong F_{\mathcal{K}}(W)$.

The following proposition is a consequence of Lannes construction of the division functors and Lemma 1.1.

Proposition 1.3. ([L])
(i) $(H^*V : H^*W) \cong F_{\mathcal{K}}(V \otimes W^*) \otimes F_{\mathcal{K}}(V)$ as objects in $\mathcal{K}$,
(ii) If $f \in \mathrm{end}_{\mathcal{U}}(H^*W)$ is an idempotent then $(H^*V : fH^*W) \cong F_{\mathcal{K}}(V \otimes W^*)f \otimes F_{\mathcal{K}}(V)$ as objects in $\mathcal{U}$.

Note that the right action of f on W^* comes from the usual left action of f on $W = H^1W$.

Since $V \otimes W^*$ has grading zero, the identity in the Steenrod algebra acts as the p-th power map on $F_{\mathcal{K}}(V \otimes W^*)$. Therefore, $F_{\mathcal{K}}(V \otimes W^*)$ is a free p-Boolean algebra (having $x^p = x$ for all x). If we choose bases $\{v_i\}$ for V and $\{w_j\}$ for W then the elements $a_{ij} = v_i \otimes w_j^*$ span $V \otimes W^*$ and, as algebras, $F_{\mathcal{K}}(V \otimes W^*) \cong \mathbf{F}_p[a_{ij}]/(a_{ij}^p - a_{ij})$.

Section 2

The summands of H^*V are in one to one correspondence with the idempotents in $\mathrm{end}_{\mathcal{U}}(H^*V)$. To get a handle on these idempotents one uses the representation theory of finite dimensional algebras. This is possible because of the following theorem of Adams, Gunawardena, and Miller.

Theorem 2.1. ([AGM] , p. 438) *As vector spaces,* $\hom_{\mathcal{U}}(H^*V, H^*W) \cong \mathbf{F}_p[\hom(V, W)]$ *and, as algebras,* $\mathrm{end}_{\mathcal{U}}(H^*V) \cong \mathbf{F}_p[\mathrm{end}(V)]$.

We continue to write R for the ring $\mathrm{end}_{\mathcal{U}}(H^*V)$. The irreducible left R-modules can be described using Young diagrams ([HK]). We denote the irreducibles by $\{E_\lambda \mid \lambda \in \Lambda\}$, where $\Lambda = \{(\lambda_1, \ldots, \lambda_r) \mid 0 \leq \lambda_i \leq p-1\}$ and the Young diagram associated to λ has $\lambda_1 + \ldots + \lambda_i$ nodes in the i-th row. To each E_λ we can associate a primitive idempotent e_λ so that the indecomposable module Re_λ is the projective cover of E_λ. By general representation theory it can be shown that $R \cong \bigoplus_{\lambda\in\Lambda} \dim_{\mathbf{F}_p}(E_\lambda)\, Re_\lambda$.

Theorem 2.2. ([HK] , A) $H^*V \cong \bigoplus_{\lambda\in\Lambda} \dim_{\mathbf{F}_p}(E_\lambda)\, e_\lambda H^*V$ *and the* $e_\lambda H^*V$ *are indecomposable* $\mathcal{U}$*-modules.*

For an idempotent $f \in \mathrm{end}_{\mathcal{U}}(H^*W)$ it follows from Proposition 1.3 that $(H^*V : fH^*W) \cong F_{\mathcal{K}}(V \otimes W^*)f \otimes H^*V$. If d is the dimension of $F_{\mathcal{K}}(V \otimes W^*)f$, then it follows that $(H^*V : fH^*W)$ is isomorphic to a sum of d copies of H^*V. Hence, for an idempotent e in R, $(eH^*V : fH^*W)$, which is a summand of $(H^*V : fH^*W)$, can be written as a sum of copies of the indecomposables $e_\nu H^*V$, for $\nu \in \Lambda$.

The following result is a generalization of ([HS] , 3.5).

Theorem 2.3. $(e_\lambda H^*V : fH^*W) \cong \bigoplus_{\nu\in\Lambda} a_{\lambda\nu}\, e_\nu H^*V$ *where* $a_{\lambda\nu}$ *is the multiplicity of* E_λ *in* $F_{\mathcal{K}}(V \otimes W^*)f \otimes E_\nu$.

Proof. Let $S = \mathrm{end}_{\mathcal{U}}((H^*V : fH^*W))$. Since the division functors are additive the action of R on H^*V induces a ring homomorphism from R to S. If we denote this homomorphism by γ then $\gamma(e_\lambda)$ is an idempotent in S associated to the summand $(e_\lambda H^*V : fH^*W)$. We want to write the projective module $S\gamma(e_\lambda)$ as a sum of indecomposible S-modules. Let $\{G_\nu\}_{\nu\in N}$ be the collection of irreducible S-modules and, for each ν, let g_ν be a primitive idempotent associated with G_ν. The projective cover of G_ν is Sg_ν. If $\gamma_*(G_\nu)$ is the R-module induced by γ and $b_{\lambda\nu}$ is the multiplicity of E_λ in $\gamma_*(G_\nu)$ then, using standard representation theory (see [HS] , 1.8), $S\gamma(e_\lambda) \cong \bigoplus_{\nu\in N} b_{\lambda\nu}\, Sg_\nu$.

With d equal to the dimension of $F_{\mathcal{K}}(V\otimes W^*)f$ as above, we have $S \cong \mathrm{M}_{d,d}(\mathbf{F}_p)\otimes R$ where $\mathrm{M}_{d,d}(\mathbf{F}_p)$ is the ring of $d \times d$ matrices. By Morita equivalence, we can identify N with Λ and G_ν with $(\mathbf{F}_p)^d \otimes E_\nu$. Here $(\mathbf{F}_p)^d$ is the standard module for $\mathrm{M}_{d,d}(\mathbf{F}_p)$.

Now let $\Delta: R \to R \otimes R$ be the diagonal map, and let $\beta: R \to \mathrm{M}_{d,d}(\mathbf{F}_p)$ be the map induced by the action of R on V. Then $\gamma = \Delta \circ (\beta \otimes 1)$ so

$$\gamma_*((\mathbf{F}_p)^d \otimes E_\nu) \;\cong\; \beta_*((\mathbf{F}_p)^d) \otimes E_\nu \;\cong\; F_{\mathcal{K}}(V \otimes W^*)f \otimes E_\nu$$

as R-modules. Hence $b_{\lambda\nu} = a_{\lambda\nu}$. ∎

Section 3

In this section we consider the composition factors of the left $\mathbf{F}_p[\mathrm{end}(V)]$-module $F_{\mathcal{K}}(V \otimes W^*)f$. Theorem 2.3 is then restated as Theorem 3.5 and specialized to the case

$W = V$ in Theorem 3.6. We continue to use R to denote $\mathbf{F}_p[\text{end}(V)]$. We now denote $\mathbf{F}_p[\text{end}(W)]$ by S and $\mathbf{F}_p[\text{end}(V \otimes W^*)]$ by Q.

The action of $\text{end}(V \otimes W^*)$ on $V \otimes W^*$ defines a left action of Q on $F_{\mathcal{K}}(V \otimes W^*)$. The left R-module structure on $F_{\mathcal{K}}(V \otimes W^*)$ is induced by the homomorphism from R to Q, and the right S-module structure is induced by the anti-homomorphism from S to Q The images of R and S in Q commute making $F_{\mathcal{K}}(V \otimes W^*)$ an R-S bimodule. We start by looking at the Q-module structure on $F_{\mathcal{K}}(V \otimes W^*)$.

Let $U = V \otimes W^*$, so $Q = \mathbf{F}_p[\text{end}(U)]$. (We remark that all of the following results which involve U are valid for any vector space concentrated in degree zero.) Let $\mathcal{C}$ denote the category of commutative algebras over $\mathbf{F}_p$, and let $F_{\mathcal{C}}$ denote the free functor which is left adjoint to the forgetful functor from $\mathcal{C}$ to the category of vector spaces. Then $F_{\mathcal{C}}(U)$ is the polynomial algebra generated by U. Let $\mathcal{N}$ denote the subcategory of $\mathcal{C}$ consisting of 'p-Nilpotent' algebras, i.e., algebras satisfying $x^p = 0$ for all x, and let $F_{\mathcal{N}}$ denote the free functor which is left adjoint to the forgetful functor from $\mathcal{N}$ to the category of vector spaces. Then $F_{\mathcal{N}}(U)$ is the quotient of $F_{\mathcal{C}}(U)$ by the ideal generated by p^{th} powers. Since U is a graded vector space concentrated in degree zero, $F_{\mathcal{K}}(U)$ is the quotient of $F_{\mathcal{C}}(U)$ by the ideal generated by elements of the form $x^p - x$. If we filter these three algebras by degree then the projection maps, $F_{\mathcal{C}}(U) \to F_{\mathcal{N}}(U)$ and $F_{\mathcal{C}}(U) \to F_{\mathcal{K}}(U)$, are filtration preserving Q-module morphisms.

Lemma 3.1. ([HS] , p. 10) *If $F_{\mathcal{K}}(U)$ and $F_{\mathcal{N}}(U)$ are filtered by degree then the associated graded algebras are isomorphic as left Q-modules.*

Proof. If M is a filtered object then we denote the associated graded object by $G_*(M)$. Both $G_*(F_{\mathcal{K}}(U))$ and $G_*(F_{\mathcal{N}}(U))$ are isomorphic to the truncation of $G_*(F_{\mathcal{C}}(U))$. ∎

Let $\mathbf{F}_p[U]$ denote the group ring of the elementary abelian p-group U.

Lemma 3.2. ([K] , p. 199) *If $\mathbf{F}_p[U]$ is filtered by powers of the augmentation ideal and $F_{\mathcal{N}}(U)$ is filtered by degree, then the associated graded algebras are isomorphic as left Q-modules.*

Proof. Consider the algebra map from $F_{\mathcal{N}}(U)$ to $\mathbf{F}_p[U]$ which takes $u \in U$ to $[u] - [0]$. This is a filtration preserving map of left Q-modules, and the induced map from $G_*(F_{\mathcal{N}}(U))$ to $G_*(\mathbf{F}_p[U])$ is an isomorphism. ∎

Lemma 3.3. $\mathbf{F}_p[V \otimes W^*]$ *and* $\mathbf{F}_p[\text{hom}(W, V)]$ *are isomorphic as R-S bimodules.*

Proof. Choosing bases for V and W identifies $V \otimes W^*$ and $\text{hom}(W, V)$ with the set $M_{r,s}(\mathbf{Z}/p)$. This identification preserves the left R action and the right S action. ∎

Proposition 3.4. *Let α be any element of S. Then as left R-modules $F_{\mathcal{K}}(V \otimes W^*)\alpha$ and $\mathbf{F}_p[\text{hom}(W, V)]\alpha$ have the same composition factors with the same multiplicities.*

Proof. Let $U = V \otimes W^*$. The filtrations on $F_{\mathcal{K}}(U)$, $F_{\mathcal{N}}(U)$, and $\mathbf{F}_p[U]$ defined above induce left R-module filtrations on $F_{\mathcal{K}}(U)\alpha$, $F_{\mathcal{N}}(U)\alpha$, and $\mathbf{F}_p[U]\alpha$. Lemmas 3.1 and 3.2 imply that $G_*(F_{\mathcal{K}}(V \otimes W^*)\alpha)$ and $G_*(\mathbf{F}_p[V \otimes W^*]\alpha)$ are isomorphic left R-modules. Thus $F_{\mathcal{K}}(V \otimes W^*)\alpha$ and $\mathbf{F}_p[V \otimes W^*]\alpha$ have the same compostion factors. It follows from Lemma 3.3, that $\mathbf{F}_p[V \otimes W^*]\alpha$ and $\mathbf{F}_p[\text{hom}(W, V)]\alpha$ are isomorphic left R-modules. Therefore $F_{\mathcal{K}}(V \otimes W^*)\alpha$ and $\mathbf{F}_p[\text{hom}(W, V)]\alpha$ have the same composition factors. ∎

Now we can restate Theorem 2.3 using the above results.

Theorem 3.5. *Let $e_\lambda \in R$ be a primitive idempotent and $f \in S$ be an idempotent, then $(e_\lambda H^*V : fH^*W) \cong \bigoplus_{\nu\in\Lambda} a_{\lambda\nu}\, e_\nu H^*V$, where $a_{\lambda\nu}$ is the multiplicity of E_λ in $\mathbf{F}_p[\hom(W,V)]f \otimes E_\nu$.*

Since $H^*B(\mathbf{Z}/p)^m$ is a summand of $H^*B(\mathbf{Z}/p)^{m+n}$ it suffices to take $W = V$ in this theorem.

Theorem 3.6. *Let e_λ and e_μ be primitive idempotents in R, then*

$$(e_\lambda H^*V : e_\mu H^*V) \cong \bigoplus_{\nu\in\Lambda} a_{\lambda\mu\nu}\, e_\nu H^*V,$$

where $a_{\lambda\mu\nu}$ is the multiplicity of E_λ in $Re_\mu \otimes E_\nu$.

Section 4

In this section, we discuss the computations needed to determine the $a_{\lambda\mu\nu}$ in Theorem 3.6. Here we will think of R as the semigroup ring $\mathbf{F}_p[\mathrm{M}_{r,r}(\mathbf{Z}/p)]$.

The calculation of the multiplicity of E_λ in $Re_\mu \otimes E_\nu$ can be broken into two steps: first find the composition factors of Re_μ and then find the composition factors in $E_\sigma \otimes E_\nu$ for the E_σ's occurring in Re_μ. When $r = 1$, these computations are easy: $Re_{(i)} \cong E_{(i)} \cong (Det)^i$, where Det is the determinant representation. When $r = 2$, composition series for the Re_μ and the $E_\sigma \otimes E_\nu$ are given in a paper of D. J. Glover ([G]). In general, however, neither step can be carried out.

Finding the composition factors of Re_μ is a standard problem in representation theory. Let $c_{\sigma\mu}$ be the multiplicity of E_σ in Re_μ. The matrix $(c_{\sigma\mu})$ is called the *Cartan matrix* for R. ([CR] , 83.8). It follows from the discussion above that the Cartan matrices are known for all primes when $r = 1$ or 2. In addition the matrix for $\mathbf{F}_2[\mathrm{M}_{3,3}(\mathbf{Z}/2)]$ has been determined ([HHS] , 4.2).

There are examples of particular indecomposible projectives whose composition factors are known. For example, for $1 \le i \le p-1$, $Re_{(p-1,\dots,p-1,i)} \cong E_{(p-1,\dots,p-1,i)}$. These are called the (twisted) Steinberg representations. Here is another example.

Proposition 4.1. *$Re_{(0,\dots,0)} \cong E_{(0,\dots,0)}$ and, for $1 \le i \le p-1$, $Re_{(i,0,\dots,0)}$ has r composition factors: $\{E_{(i,0,\dots,0)}, E_{(p-1-i,i,0,\dots,0)}, \dots, E_{(0,\dots,0,p-1-i,i)}\}$. Each factor occurs with multiplicity one.*

Before proving the proposition we set up some notation and we show how to recover the main algebraic result of [HS] . For $0 \le d \le r(p-1)$ write $d = a(p-1) + b$ with $0 \le b < p-1$. Let $\lambda[d]$ be the sequence $(\lambda_1, \dots, \lambda_r)$ in Λ having $\lambda_a = p-1-b$, $\lambda_{a+1} = b$ and $\lambda_i = 0$ for $i \ne a, a+1$. The irreducibles occurring in the proposition are the $E_{\lambda[d]}$ for $0 \le d \le r(p-1)$. Note that in the Young diagram associated to $\lambda[d]$, the first a rows each have $p-1$ nodes and the $(a+1)$-st row has b nodes.

The principal algebraic result of [HS] now follows from Theorem 3.6 and Proposition 4.1.

Corollary 4.2. ([HS] , 3.8) *$(e_\lambda H^*V : H^*(B\mathbf{Z}/p)) \cong \bigoplus_{\nu\in\Lambda} a_{\lambda\nu}\, e_\nu H^*V$ where $a_{\lambda\nu}$ is the multiplicity of E_λ in $\bigoplus_{d=0}^{r(p-1)} E_{\lambda[d]} \otimes E_\nu$.*

Proof. $H^*(B\mathbf{Z}/p) \cong \bigoplus_{i=0}^{p-1} e_{\lambda[i]} H^*V$ and the R-modules $\bigoplus_{i=0}^{p-1} Re_{\lambda[i]}$ and $\bigoplus_{d=0}^{r(p-1)} E_{\lambda[d]}$ have the same composition factors. ∎

We now turn to the proof of Proposition 4.1. The first statement is easy: $e_{(0,\dots,0)}$ is the zero matrix, $\mathbf{0}$, and $Re_{(0,\dots,0)} \cong \mathbf{F}_p\{0\}$ is one dimensional as is $E_{(0,\dots,0)}$. The rest of the proposition follows from the following two lemmas, the first of which follows from results of Kuhn and Carlisle.

Lemma 4.3. *The composition factors of $\bigoplus_{i=0}^{p-1} Re_{\lambda[i]}$ are $\{E_{\lambda[d]} \mid 0 \le d \le r(p-1)\}$. Each factor occurs with multiplicity one.*

Proof. Let $f = \sum_{i=0}^{p-1} e_{\lambda[i]}$. The multiplicity of the irreducible E_σ in the R-module Rf equals the dimension of $e_\sigma Rf$ ([CR] , Section 54). So we need to show that this vector space has dimension one when $\sigma = \lambda[d]$ and dimension zero otherwise.

For any idempotent $e \in \mathrm{end}_{\mathcal{U}}(H^*W)$, $\hom_{\mathcal{U}}(H^*V, eH^*W)$ and $\hom_{\mathcal{U}}(H^*V, H^*W)e$ are isomorphic R-modules. Therefore, $e_\sigma Rf$ is isomorphic to $e_\sigma \hom_{\mathcal{U}}(H^*V, fH^*V)$. The $\mathcal{U}$-modules fH^*V and $H^*(B\mathbf{Z}/p)$ are isomorphic ([HK] , Section 6). So $e_\sigma Rf$ is isomorphic to $e_\sigma \hom_{\mathcal{U}}(H^*V, H^*(B\mathbf{Z}/p))$ which is isomorphic to $e_\sigma \mathbf{F}_p[\mathrm{M}_{r,1}(\mathbf{Z}/p)]$ (by [AGM]). This latter space is isomorphic to $e_\sigma(\mathbf{F}_p[x_1,\dots,x_r]/(x_1^p,\dots,x_r^p))$ by ([K] , 1.8, 1.9). Here $\mathbf{F}_p[x_1,\dots,x_r]$ is regarded as a left $R \cong \mathbf{F}_p[\mathrm{M}_{r,r}(\mathbf{Z}/p)]$-module in the usual way. Finally, the space $e_\sigma(\mathbf{F}_p[x_1,\dots,x_r]/(x_1^p,\dots,x_r^p))$ is one dimensional when $\sigma = \lambda[d]$ and zero otherwise ([CK] , 6.1). ∎

Let $n(\lambda) = \lambda_1 + 2\lambda_2 + \cdots + r\lambda_r$, the number of nodes in the Young diagram associated to λ. Note that $n(\lambda[d]) = d$.

Lemma 4.4. *If E_λ is a composition factor of Re_μ, then $n(\lambda) \equiv n(\mu) \pmod{p-1}$.*

Proof. To prove this lemma we regard E_λ and Re_μ as left $\mathrm{GL}_r(\mathbf{F}_p)$ modules and we consider the action of an element $\tau \in \mathrm{GL}_r(\mathbf{F}_p)$ defined as follows. Let ω be a primitive generator for $\mathbf{F}_{p^r}$ over $\mathbf{F}_p$, and let τ be the matrix representing left multiplication by ω in the basis $\{1,\omega,\dots,\omega^{r-1}\}$. The lemma follows from: (i) the eigenvalues for the action of τ on E_λ are all of the form ω^j with $j \equiv n(\lambda) \pmod{p-1}$ and (ii) the eigenvalues for the action of τ on Re_μ are all of the form ω^k with $k \equiv n(\mu) \pmod{p-1}$.

(i) Recall that E_λ is a composition factor of the Weyl module W_λ associated to the Young diagram of λ. The matrix τ is diagonalizable over $\mathbf{F}_{p^r}$ since its eigenvalues $\{\omega,\omega^p,\dots,\omega^{p^{r-1}}\}$ are distinct. Therefore we can apply a result ([JK] , 8.1.18) which computes the eigenvalues of a diagonal matrix acting on a Weyl module. This result shows that the eigenvalues of τ acting on W_λ are of the form $\omega^{n_0}\omega^{pn_1}\cdots\omega^{p^{r-1}n_{r-1}}$, where $n_0 + \cdots + n_{r-1} = n(\lambda)$. Therefore the eigenvalues of τ acting on W_λ are of the form ω^j with $j = n_0 + \cdots + p^{r-1}n_{r-1} \equiv n_0 + \cdots + n_{r-1} = n(\lambda) \pmod{p-1}$. Since E_λ is a composition factor of W_λ, the eigenvalues of τ acting on E_λ also have this property.

(ii) The element τ generates a cyclic group G of order $p^r - 1$. Since G is abelian and its order is relatively prime to p, the group ring $\mathbf{F}_{p^r}[G]$ has (unique) primitive orthogonal idempotents $e_j = -\sum_{k=1}^{p^r-1} \omega^{-kj}\tau^k$, for $1 \le j \le (p^r - 1)$ (see [CR] , 33.8) which sum to the identity. Note that $\mathbf{F}_{p^r}[G]e_j$ is a one dimensional left G module and $\tau e_j = \omega^j e_j$.

For $1 \le i \le p-1$, let $f_i = \sum_{j \equiv i \pmod{p-1}} e_j$. A computation shows that $f_i = -\sum_{l=1}^{p-1} \omega^{\theta l i}\tau^{-\theta l}$, where $\theta = (p^r - 1)/(p-1)$. Since $\omega^\theta \in \mathbf{F}_p$, the f_i lie in $\mathbf{F}_p[G] \subseteq R$. And since τ^θ is a scalar matrix, the f_i lie in the center of R.

Now think of Rf_i as a left G-module. The action of τ on Rf_i has ω^j as an eigenvalue if and only if $e_j(R \otimes \mathbf{F}_{p^r})f_i \neq 0$ ([CR] , 54.12). Since R has an identity and f_i is central, this happens if and only if $e_j f_i \neq 0$. And since the e_j are orthogonal, this happens if and only if $i \equiv j \pmod{p-1}$. Therefore the eigenvalues of τ acting on Rf_i are of the form ω^j with $j \equiv i \pmod{p-1}$.

Since Re_μ is indecomposable, it must be isomorphic to a summand of one of the Rf_i, so the eigenvalues of τ acting on Re_μ are of the form ω^k with $k \equiv i \pmod{p-1}$. But we know that E_μ is a composition factor of Re_μ, so by (i) above, $k \equiv n(\mu) \pmod{p-1}$. ∎

Finding the composition factors in $E_\sigma \otimes E_\nu$ is also a standard problem in representation theory. As mentioned above, these have been determined for $r \leq 2$. In addition when $r = 3$ or when $p = 2$ and $r = 4$ one can determine these factors using the methods in Carlisle's thesis ([Ca]). The proof of Lemma 4.4 gives the following condition on factors of $E_\sigma \otimes E_\nu$.

Corollary 4.5. *If E_λ is a composition factor in $E_\sigma \otimes E_\nu$, then*

$$n(\lambda) \equiv n(\sigma)n(\nu) \pmod{p-1}.$$

Using some of the above computations we list, in Appendix 1, $(e_\lambda H^*V : e_\mu H^*V)$ when $V = (\mathbf{Z}/2)^3$, and in Appendix 2, $(e_\lambda H^*V : e_{(100)}H^*V)$ and $(e_\lambda H^*V : e_{(200)}H^*V)$ when $V = (\mathbf{Z}/3)^3$.

Section 5

In this section we give a description of the spaces $\hom_{\mathcal{U}}(e_\lambda H^*V \otimes J(l), e_\mu H^*V \otimes J(m))$, where $J(n)$ is the dual Brown–Gitler module with top class in degree n. Recall that $J(n)$ is characterized by the property that, for any A in $\mathcal{U}$,

$$\hom_{\mathcal{U}}(A, J(n)) \cong A^n$$

where A^n denotes the degree n elements in A. It can be shown that $J(n)$ is finite dimensional, is zero in degrees greater than n, and has dimension one in degree n. Indeed, the complete structure of $J(n)$ is given in ([M] , 6.17).

It follows from the characterization of $J(m)$ and the adjointness of division that $\hom_{\mathcal{U}}(e_\lambda H^*V \otimes J(l), e_\mu H^*V \otimes J(m))$ is isomorphic to the degree m elements in $(e_\lambda H^*V \otimes J(l) : e_\mu H^*V)$. Recall that a $\mathcal{U}$-module J is called *locally finite*, if every finitely generated submodule is finite.

Lemma 5.1. *If A, B, and J are in $\mathcal{U}$, B is a summand of H^*V, and J is locally finite, then $(A \otimes J : B) \cong (A : B) \otimes J$.*

Proof. Let ϕ_B be the $\mathcal{U}$-morphism, natural in B, from $(A \otimes J : B)$ to $(A : B) \otimes J$ whose adjoint is $\mathrm{ad}(1_{(A:B)}) \otimes 1_J$.

Since division by H^*V commutes with tensor products and $(J : H^*V) \cong J$, we have $(A \otimes J : H^*V) \cong (A : H^*V) \otimes J$. Furthermore this isomorphism is ϕ_{H^*V}.

By the naturality of ϕ_*, the diagram

$$\begin{array}{ccc} (A \otimes J : H^*V) & \xrightarrow{\phi_{H^*V}} & (A : H^*V) \otimes J \\ \downarrow & & \downarrow \\ (A \otimes J : B) & \xrightarrow{\phi_B} & (A : B) \otimes J \end{array}$$

induced by the inclusion of B commutes and ϕ_B is surjective. Similarly, the diagram

$$\begin{array}{ccc} (A\otimes J:B) & \xrightarrow{\phi_B} & (A:B)\otimes J \\ \downarrow & & \downarrow \\ (A\otimes J:H^*V) & \xrightarrow{\phi_{H^*V}} & (A:H^*V)\otimes J \end{array}$$

induced by the projection onto B commutes and ϕ_B is injective. ∎

For a non-negatively graded vector space M, the Poincaré series of M is given by $P.S.(M;t)=\sum_{i\geq 0}\dim(M^i)\,t^i$.

Using the characterization of $J(m)$, Lemma 5.1, and Theorem 3.6, the dimension of $\hom_{\mathcal{U}}(e_\lambda H^*V\otimes J(l), e_\mu H^*V\otimes J(m))$ equals the coefficient of t^m in the Poincaré series of $\sum_{\nu\in\Lambda} a_{\lambda\mu\nu}\, e_\nu H^*V \otimes J(l)$. The Poincaré series of $J(l)$ can be determined from ([M] , 6.17). For the series of $e_\nu H^*V$, we use the following generalization of ([Mit], 1.6).

Proposition 5.2. *$P.S.(e_\nu H^*V;t)=\sum_{i\geq 0}$(multiplicity of E_ν in $H^iV)\,t^i$.*

These series have been computed for $r\leq 3$ and for $p=2$, $r=4$ ([Ca]). They are also known for certain types of ν ([MP] , [CW]).

Appendix 1

Let $V=(\mathbf{Z}/2)^3$. Here we tabulate $(e_\lambda H^*V : e_\mu H^*V)$. As in Theorem 3.6 we let $a_{\lambda\mu\nu}$ be the number of copies of $e_\nu H^*V$ in $(e_\lambda H^*V : e_\mu H^*V)$. We give one table for each value of μ (except $\mu=(000)$, since division by $e_{(000)}H^*V$ is the identity functor). In the tables the λ's label the rows and the ν's label the columns. For example, the third row of the first table indicates that $(e_{(110)}H^*V : e_{(100)}H^*V)\cong e_{(100)}H^*V\oplus e_{(010)}H^*V\oplus 2e_{(110)}H^*V$. We omit the rows with $\lambda=(000)$ since $(e_{(000)}H^*V : e_\mu H^*V)=0$ when $\mu\neq(000)$.

	0	1	0	1	0	1	0	1
$a_{\lambda(100)\nu}$	0	0	1	1	0	0	1	1
	0	0	0	0	1	1	1	1
100	1	1						
010	1	2	1	2				
110		1	1	2				
001	1	1	1	2	1	1	1	2
101		1	2	5	1	2	2	5
011			1	3	1	2	2	5
111				1		1	1	3

	0	1	0	1	0	1	0	1
$a_{\lambda(010)\nu}$	0	0	1	1	0	0	1	1
	0	0	0	0	1	1	1	1
100	1	1						
010	3	2	3	2				
110		3	1	4				
001	3	3	3	6	3	3	3	6
101	2	5	6	15	3	6	6	15
011		4	3	13	3	6	6	15
111			2	5		3	3	9

$a_{\lambda(110)\nu}$	0	1	0	1	0	1	0	1
	0	0	1	1	0	0	1	1
	0	0	0	0	1	1	1	1
100								
010		2		2				
110	1	1	1	1				
001		2	2	6		2	2	6
101	1	4	4	11	1	4	4	11
011	1	3	4	9	1	5	4	11
111		1	1	4	1	2	2	5

$a_{\lambda(001)\nu}$	0	1	0	1	0	1	0	1
	0	0	1	1	0	0	1	1
	0	0	0	0	1	1	1	1
100	1	1						
010	3	2	3	2				
110		3	1	4				
001	5	4	4	8	5	4	4	8
101	3	8	8	20	4	9	8	20
011	1	7	6	18	4	9	9	20
111			3	9		3	4	13

$a_{\lambda(101)\nu}$	0	1	0	1	0	1	0	1
	0	0	1	1	0	0	1	1
	0	0	0	0	1	1	1	1
100								
010	2	2	2	2				
110	1	3	1	3				
001	3	6	7	15	3	6	7	15
101	6	12	12	34	6	12	12	34
011	3	15	9	31	5	17	11	33
111			6	14	1	3	7	17

$a_{\lambda(011)\nu}$	0	1	0	1	0	1	0	1
	0	0	1	1	0	0	1	1
	0	0	0	0	1	1	1	1
100								
010		2		2				
110	1	1	1	1				
001	1	5	4	11	1	5	4	11
101	3	7	10	23	3	7	10	23
011	4	10	8	22	4	12	8	24
111			3	10	1	1	4	11

$a_{\lambda(111)\nu}$	0	1	0	1	0	1	0	1
	0	0	1	1	0	0	1	1
	0	0	0	0	1	1	1	1
100								
010								
110								
001		1	1	4		1	1	4
101		3	2	6		3	2	6
011		2	3	6		2	3	6
111	1	1	1	3	1	1	1	3

Appendix 2

Let $V = (\mathbf{Z}/3)^3$. On the next two pages we tabulate $(e_\lambda H^*V : e_{(100)}H^*V)$ and $(e_\lambda H^*V : e_{(200)}H^*V)$. Since $H^*(B(\mathbf{Z}/3)) \cong \bigoplus_{i=0}^{2} e_{(i00)}H^*V$ and division by $e_{(000)}H^*V$ is the identity functor, these tables can be used to recover the modules $(e_\lambda H^*V : H^*B\mathbf{Z}/3)$ which were tabulated in [HS] .

Appendix 3

Here we give an example of a $g \in \mathbf{F}_p[\mathrm{end}(W)]$ such that $(H^*V : gH^*W) \ncong (H^*V : H^*W)g$. Let $V = \mathbf{Z}/2$, $W = (\mathbf{Z}/2)^2$, and $g = \left(\begin{smallmatrix}0&0\\0&0\end{smallmatrix}\right) + \left(\begin{smallmatrix}1&1\\0&0\end{smallmatrix}\right) + \left(\begin{smallmatrix}1&0\\0&0\end{smallmatrix}\right) + \left(\begin{smallmatrix}0&1\\0&0\end{smallmatrix}\right)$. It can be shown that gH^*W is isomorphic to $(H^*V)^{\geq 2}$, the submodule of elements of H^*V of degree greater than or equal to 2. Here g was chosen as the composition of the projection from H^*W to $e_{(01)}H^*W \cong \widetilde{H}^*(BA_4)$ with the unique non-zero map from $\widetilde{H}^*(BA_4)$ to H^*V.

Let t denote the non-zero element in H^1V and let ι be the non-zero element in $J(1)$. The map $\phi: H^*V \to (H^*V)^{\geq 2} \otimes J(1)$ which sends t^0 to 0, t^1 to 0, and for $i \geq 1$, t^{2i} to 0 and t^{2i+1} to $t^{2i} \otimes \iota$ is a morphism in $\mathcal{U}$. Therefore $\hom_{\mathcal{U}}(H^*V, (H^*V)^{\geq 2} \otimes J(1))$ is non-zero, and it follows by adjunction that $(H^*V : (H^*V)^{\geq 2}) \cong (H^*V : gH^*W)$ is non-zero.

Now consider $(H^*V : H^*W)g \cong F_{\mathcal{K}}(V \otimes W^*)g \otimes H^*V$. The element g acts on the basis elements of $F_{\mathcal{K}}(V \otimes W^*)$ as follows (where $a_{ij} = v_i \otimes w_j^*$):

$$\begin{aligned} 1 &\mapsto (1) + (1) + (1) + (1) = 0 \\ a_{11} &\mapsto (0) + (a_{11} + a_{12}) + (a_{11}) + (a_{12}) = 0 \\ a_{12} &\mapsto (0) + (0) + (0) + (0) = 0 \\ a_{11}a_{12} &\mapsto (0) + (0) + (0) + (0) = 0. \end{aligned}$$

Therefore $(H^*V : H^*W)g = 0$.

$(e_\lambda H^* V : e_{(100)} H^* V)$

$a_{\lambda(100)\nu}$	0 0 0	1 0 0	2 0 0	0 1 0	1 1 0	2 1 0	0 2 0	1 2 0	2 2 0	0 0 1	1 0 1	2 0 1	0 1 1	1 1 1	2 1 1	0 2 1	1 2 1	2 2 1	0 0 2	1 0 2	2 0 2	0 1 2	1 1 2	2 1 2	0 2 2	1 2 2	2 2 2
0 0 0																											
1 0 0	1		1																								
2 0 0		1																									
0 1 0		1			1			1																			
1 1 0	1		3	1		3	1		3																		
2 1 0		1			1			1																			
0 2 0		1			1			1																			
1 2 0			2	1		3	1		3																		
2 2 0					1			1																			
0 0 1			3	2		7	3		13		2			2			7				3	2		7	3		13
1 0 1					1			2		1		2	1		2	1		5					1			2	
2 0 1				1		2			2		1			1			2					1		2			2
0 1 1	1		1			2	2		5		1			1			2		1		1			2	2		5
1 1 1		1			2			5		1		3	1		5	3		11		1			2			5	
2 1 1			1			3	2		6		1			1			3				1			3	2		6
0 2 1				1		2			2		1			1			2					1		2			2
1 2 1					1			3				2	1		3	1		6					1			3	
2 2 1						1			1					1			1							1			1
0 0 2		2			2			7				3	2		7	3		13		2			2			7	
1 0 2			1	1		2	1		5					1			2		1		2	1		2	1		5
2 0 2					1			2					1		2			2		1			1			2	
0 1 2								1		1		1			2	2		5		1			1			2	
1 1 2						2	2		8		1			2			5		1		3	1		5	3		11
2 1 2								2				1			3	2		6		1			1			3	
0 2 2								1					1		2			2		1			1			2	
1 2 2									3					1			3				2	1		3	1		6
2 2 2															1			1					1			1	

$(e_\lambda H^*V : e_{(200)}H^*V)$

$a_{\lambda(200)\nu}$	0	1	2	0	1	2	0	1	2	0	1	2	0	1	2	0	1	2	0	1	2	0	1	2	0	1	2
	0	0	0	1	1	1	2	2	2	0	0	0	1	1	1	2	2	2	0	0	0	1	1	1	2	2	2
	0	0	0	0	0	0	0	0	0	1	1	1	1	1	1	1	1	1	2	2	2	2	2	2	2	2	2
0 0 0																											
1 0 0		1																									
2 0 0	1		1																								
0 1 0			1	1		2			1																		
1 1 0		2			2			2																			
2 1 0			1	1		2			1																		
0 2 0	1		2			1	1		2																		
1 2 0		1			2			2																			
2 2 0			1			1	1		2																		
0 0 1		1			4			6		1		2	1		6	2		10		1			4			6	
1 0 1				1		3	1		4		2			1			3					1		3	1		4
2 0 1								1		1		2			1	2		5								1	
0 1 1		1			1			3				1	2		3			4		1			1			3	
1 1 1			1	2		6	1		8		2			3			6				1	2		6	1		8
2 1 1					2			3				1	1		3			4					2			3	
0 2 1								1		1		2			1	2		5								1	
1 2 1						3	1		4		1			2			3							3	1		4
2 2 1								1				1			1	1		3								1	
0 0 2	1		2	1		6	2		1		1			4			6		1		2	1		6	2		10
1 0 2		1			1			3					1		3	1		4		2			1			3	
2 0 2			1			1	2		5								1		1		2			1	2		5
0 1 2				1		1			3		1			1			3				1	2		3			4
1 1 2					1			4				1	2		6	1		8		2			3			6	
2 1 2						1			3					2			3				1	1		3			4
0 2 2							1		3								1		1		2			1	2		5
1 2 2								1							3	1		4		1			2			3	
2 2 2									1								1				1			1	1		3

References

[AGM] J. F. Adams, J. H. Gunawardena, and H. R. Miller. The Segal conjecture for elementary abelian p-groups. *Topology* **24** (1985), 435-460.

[CS] H. E. A. Campbell and P. S. Selick. Polynomial algebras over the Steenrod algebra. *Comment. Math. Helv.* **65** (1990), 171-180.

[C] G. Carlsson. Equivariant stable homotopy and Segal's Burnside ring conjecture. *Ann. of Math.* (2) **120** (1984), 189-224.

[Ca] D. Carlisle. The modular representation theory of $\mathrm{GL}(n, p)$, and applications to topology. Ph.D. thesis, University of Manchester (1985).

[CK] D. Carlisle and N. J. Kuhn. Subalgebras of the Steenrod algebra and the action of matrices on truncated polynomial algebras. *J. Algebra*, **121** (1989), 370-387.

[CW] D. Carlisle and G. Walker. Poincaré series for the occurrence of certain modular representations of $\mathrm{GL}(n, p)$ in the symmetric algebra. Preprint (1989).

[CR] C. W. Curtis and I. Reiner. *Representation Theory of Finite Groups and Associative Algebras*, (Wiley, 1962).

[G] D. J. Glover. A study of certain modular representations. *J. Algebra* **51** (1978), 425-475.

[HHS] J. C. Harris, T. J. Hunter, and R. J. Shank. Steenrod algebra module maps from $H^*(B(\mathbf{Z}/p)^n)$ to $H^*(B(\mathbf{Z}/p)^s)$. *Proc. Amer. Math. Soc.* To appear in April, 1991.

[HK] J. C. Harris and N. J. Kuhn. Stable decompositions of classifying spaces of finite abelian p-groups. *Math. Proc. Camb. Phil. Soc.* **103** (1988), 427-449.

[HS] J. C. Harris and R. J. Shank. Lannes' T functor on summands of $H^*(B(\mathbf{Z}/p)^s)$. Preprint, 1990.

[JK] G. James and A. Kerber. *The Representation Theory of the Symmetric Group.* Encyclopedia of Math. and its Applications, vol. 16 (Addison-Wesley, 1981).

[K] N. J. Kuhn. The Morava K-theory of some classifying spaces. *Trans. Amer. Math. Soc.* **304** (1987), 193-205.

[L] J. Lannes. Sur la cohomologie modulo p des p-groups abelians elementaires. In *Proc. Durham Symposium on Homotopy Theory 1985*, L. M. S. Lecture Notes vol. 117 (Camb. Univ. Press 1987), 97-116.

[LS] J. Lannes and L. Schwartz. Sur la structure des $\mathcal{A}$-modules instable injectifs. *Topology* **28** (1989), 153-169.

[LZ] J. Lannes and S. Zarati. Sur les $\mathcal{U}$-injectifs. *Ann. Scient. Ec. Norm. Sup.* **19** (1986), 303-333.

[M] H. R. Miller. The Sullivan conjecture on maps from classifying spaces. *Ann. of Math.* **120** (1984), 39-87.

[Mit] S. A. Mitchell. Splitting $B(\mathbf{Z}/p)^n$ and BT^n via modular representation theory. *Math. Z.* **189** (1985), 1-9.

[MP] S. A. Mitchell and S. B. Priddy. Stable splittings derived from the Steinberg module. *Topology* **22** (1983), 285-298.

[S] R. J. Shank. Polynomial algebras over the Steenrod algebra, summands of $H^*(B(\mathbf{Z}/2)^s)$ and Lannes' division functors. Ph.D. thesis, University of Toronto (1989).

J.C. Harris: University of Toronto,
Toronto, Ontario M5S 1A1,
Canada (harris@math.toronto.edu)

R.J. Shank: University of Rochester,
Rochester, New York 14627,
U.S.A. (rojs@cc.rochester.edu)

1990 Barcelona Conference
on Algebraic Topology.

CLASSES HOMOTOPIQUES ASSOCIEES A UNE G–OPERATION

CLAUDE HAYAT–LEGRAND

Let G be a discrete group, and let Π and Π' be two G-modules. Whatever $n > 0$, an interpretation of the elements of the group $\mathrm{Ext}^n_G(\Pi, \Pi')$ *as homotopy classes of sections of a fibre bundle associated with the given data is offered. Thereafter based G-actions on a based space X having Π and Π' as the only non-trivial homotopy groups are studied in terms of suitably defined difference characteristic classes assuming their values in* $\mathrm{Ext}^n_G(\Pi, \Pi')$.

Dans cette étude, on construit des classes caractéristiques associées à une G–opération, G étant un groupe discret. Pour cela, on donne une représentation du n–ième groupe de G–extension $\mathrm{Ext}^n_G(\Pi, \Pi')$, pour deux G–modules Π et Π', comme espace des classes d'homotopie des sections d'un fibré construit à partir des G–modules Π et Π'. Ce fibré joue le rôle de fibré structural, [10], pour le fibré $BE(X)$ quand X est un espace qui a Π et Π' comme seuls groupes d'homotopie non triviaux. On détermine alors des classes différences appartenant à $\mathrm{Ext}^2_G(\Pi, \Pi')$, associéées à deux G–opérations.

Ces classes caractéristiques ont déjà été utilisées, dans des cas particuliers, pour calculer des différentielles de suites spectrales, celle du revêtement par Legrand A. [11], celle d'extension de groupes par Huebschmann J. [9], et celle de Kasparov en KK–théorie par Fieux E. [8]. Dans ce dernier cas, on utilise la construction q de Cuntz [5], permettant de représenter le groupe $KK_i(A, B)$, où A et B sont des C^*-algèbres, comme i-ème groupe d'homotopie de l'espace $\mathrm{Hom}(qA, B \otimes K)$ des morphismes de C^*-algèbres de qA dans $B \otimes K$. La topologie sur $\mathrm{Hom}(qA, B \otimes K)$ est déduite de la topologie de la convergence simple et K est l'algèbre des opérateurs compacts d'un Hilbert séparable.

Dans les cas cités plus haut, et dans le dernier cas quand $A = \mathbf{C}$, les classes sont construites avec un espace X ayant le type d'homotopie d'un groupe abélien, donc d'un produit d'espaces d'Eilenberg-Mac Lane. Ici, nous allons supposer, plus généralement que X est du type d'homotopie d'un produit tordu de deux espaces d'Eilenberg-Mac Lane.

La représentation de $\mathrm{Ext}^n_G(\Pi, \Pi')$ pourrait s'établir à partir d'une suite spectrale instable d' Adams. Ici, on utilise les résultats de Shih W. [13], Didierjean G. [6] sur le calcul des groupes d'homotopie de l'espace des équivalences d'homotopie de X, et les constructions de Cooke G. [4], pour étudier une réciproque, c'est-à-dire associer une G–opération à une classe dans $\mathrm{Ext}^2_G(\Pi, \Pi')$.

1) **Représentation du n–ième groupe de G–extensions.** Par définition, si Π' est un G–module [12]: $\mathrm{Ext}^n_G(\mathbf{Z}, \Pi') = \mathrm{H}^n(\mathrm{G}, \Pi')$, et on sait que, [10] et [2], $H^n(G, \Pi') =$

$\Pi_0\Gamma[EG \times_G K(\Pi', n)]$. Le fibré $G \to EG \to BG$ est le fibré universel associé à G, le foncteur Γ est le foncteur sections pointées appliqué ici au fibré $K(\Pi', n) \to EG \times_G K(\Pi', n) \to BG$, de fibre l'espace d'Eilenberg-Mac Lane $K(\Pi', n)$.

Fixons un entier $p > 1$. Il sera, dans le paragraphe suivant, l'indice du premier groupe d'homotopie non nulle de l'espace sur lequel agira G. Comme le groupe $\mathrm{Hom}(K(\Pi, p), K(\Pi', p+1))$, des homomorphismes de $K(\Pi, p)$ dans $K(\Pi', p+1)$, est un groupe abélien topologique, on peut justifier l'itération $B^{n-1}\mathrm{Hom}(K(\Pi, p), K(\Pi', p+1))$. Le goupe G opère sur $\mathrm{Hom}(K(\Pi, p), K(\Pi', p+1))$ de façon classique, avec un point fixé qui est l'application constante de $K(\Pi, p)$ sur le point fixé par G de $K(\Pi', p+1)$.

Les résultats rappelés plus haut se prolongent par:

Théorème 1. *Soient G un groupe discret,* Π *et* Π' *deux $G-$modules, on a, pour tout $n > 0$, un isomorphisme canonique:*

$$\mathrm{Ext}^n_G(\Pi, \Pi') = \Pi_0\Gamma[EG \times_G B^{n-1}\mathrm{Hom}(K(\Pi, p), K(\Pi', p+1))].$$

On déduit immédiatement:

Corollaire 2. *Si* Π *et* Π' *sont des $G-$modules triviaux, on a:*

$$\mathrm{Ext}^2_G(\Pi, \Pi') = [BG, \mathrm{Hom}(K(\Pi, p), K(\Pi', p+2))].$$

Preuve du Corollaire 2: Si Π et Π' sont des G-modules triviaux, alors

$$EG \times_G B\mathrm{Hom}(K(\Pi, p), K(\Pi', p+1)) = BG \times B\mathrm{Hom}(K(\Pi, p), K(\Pi', p+1)).$$

Or, on a;

$$B\mathrm{Hom}(K(\Pi, p), K(\Pi', p+1)) = \mathrm{Hom}(K(\Pi, p), K(\Pi', p+2))_0,$$

où $\mathrm{Hom}(K(\Pi, p), K(\Pi', p+2))_0$ est la composante connexe de l'application constante sur le pointage de $K(\Pi', p+2)$. Comme BG est connexe, on trouve le résultat du corollaire 2.

Preuve du théorème 1: On utilise la caractérisation du foncteur $n-$ième $G-$extension donnée dans [3, p.145] . On va donc vérifier les trois propriétés caractéristiques à isomorphisme près du foncteur $n-$ième $G-$extension.

On pose pour la démonstration:

$$\begin{aligned} \Gamma(n, \Pi, C) &= \Gamma[EG \times_G B^{n-1}\mathrm{Hom}(K(\Pi, p), K(C, p+1))]; \\ U_n(C) &= \Pi_0\Gamma(n, \Pi, C), \text{ si n} > 0; \\ U_0(C) &= \Pi_0\Gamma[EG \times_G \mathrm{Hom}(K(\Pi, p), K(C, p))]. \end{aligned}$$

Comme il s'agit d'applications pointées, $U_0(C)$ est égal à $\Pi_0\Gamma[EG \times_G K(\mathrm{Hom}(\Pi, C), 0)]$, c'est-à -dire à $\mathrm{Hom}_G(\Pi, C)$.

D'autre part, une suite exacte $0 \to C \to A \to N \to 0$ de G−modules, donne une suite exacte longue d'homotopie du fibré: $\Gamma(n,\Pi,C) \hookrightarrow \Gamma(n,\Pi,A) \to \Gamma(n,\Pi,N)$. Comme $\Pi_i\Gamma(\ell,\Pi,C) = \Pi_0\Gamma(\ell-i,\Pi,C)$, on obtient une suite exacte longue pour $U_n(C)$.

La dernière propriété caractéristique à vérifier, est que $U_n(C) = 0$, $(n > 0)$ quand C est un module G−injectif. En effet C est alors un module injectif. Pour le montrer, on utilise la formule d'associativité de [2, p.165], qui permet d'écrire:

$$\mathrm{Hom}(-,C) = \mathrm{Hom}_{\mathbb{Z}\otimes\mathbb{Z}[G]}(\mathbb{Z}[G]\otimes -, C) = \mathrm{Hom}(\mathbb{Z}[G], \mathrm{Hom}_{\mathbb{Z}[G]}(-,C)).$$

Puisque C est G−injectif, $\mathrm{Hom}_{\mathbb{Z}[G]}(-,C)$ est exact, $\mathrm{Hom}(\mathbb{Z}[G],-)$ est exact car $\mathbb{Z}[G]$ est libre, donc $\mathrm{Hom}(-,C)$ est exact et C est injectif.

Alors $\mathrm{Ext}(\Pi,\mathrm{C})$ est nul, ainsi que $\mathrm{Ext}^n_{\mathrm{G}}(\Pi,\mathrm{C})$. La suite exacte longue [11]:

$$\to H^{n-2}(G,\mathrm{Ext}(\Pi,\mathrm{C})) \to \mathrm{H}^n(\mathrm{G},\mathrm{Hom}(\Pi,\mathrm{C})) \to \mathrm{Ext}^n_{\mathrm{G}}(\Pi,\mathrm{C}) \to \mathrm{H}^{n-1}(\mathrm{G},\mathrm{Ext}(\Pi,\mathrm{C})),$$

donne que $H^n(G,\mathrm{Hom}(\Pi,C))$ est nul si C est un module G−injectif.

D'autre part, comme $\mathrm{Ext}(\Pi,\mathrm{C})$ est nul, $\mathrm{Hom}(K(\Pi,p),K(C,p+1))$ est un espace de type $K(\mathrm{Hom}(\Pi,C),1)$, et comme on l'a rappellé au début de ce paragraphe, $\Pi_0\Gamma(n,\Pi,C) = H^n(G,\mathrm{Hom}(\Pi,C))$. □

2) **Classes de Borel.** Dans ce qui suit X a deux groupes d'homotopie non nuls Π et Π', et c'est l'espace total d'un fibré dont la base est de type $K(\Pi,p)$ et la fibre de type $K(\Pi',p+1)$, classifié par un invariant d'Eilenberg $\eta \in H^{p+2}(K(\Pi,p),\Pi')$. Alors on connait [13] le fibré :

$$\mathrm{Hom}(K(\Pi,p),K(\Pi',p+1)) \to E(X) \to (\mathrm{Aut}\Pi \times \mathrm{Aut}\Pi')_\eta,$$

$(\mathrm{Aut}\Pi \times \mathrm{Aut}\Pi')_\eta$ étant le sous-groupe d'isotropie de η, pour l'opération évidente de $\mathrm{Aut}\Pi \times \mathrm{Aut}\Pi'$ sur $H^{p+2}(K(\Pi,p),\Pi')$ et $E(X)$ étant l'espace de Hopf des *équivalences d'homotopie pointées* de l'espace pointé X. .

Une *G−opération pointée* est un morphisme $\phi : G \to E(X)$ qui est un H-morphisme du H-espace G dans le H-espace $E(X)$ (on précise que ϕ n'est pas un morphisme à homotopie près). Par projection une G-opération pointée donne un morphisme $u : G \to (\mathrm{Aut}\Pi \times \mathrm{Aut}\Pi')_\eta$.

Avec ces hypothèses, on obtient un *théorème de structure* pour $BE(X)$. En effet, considérons le fibré $Bu^*(BE(X))$ induit par Bu du fibré $BE(X)$ de fibre $B\mathrm{Hom}(K(\Pi,p),K(\Pi',p+1))$.

$$\begin{array}{ccc}
B\mathrm{Hom}(K(\Pi,p),K(\Pi',p+1)) & = & B\mathrm{Hom}(K(\Pi,p),K(\Pi',p+1)) \\
\downarrow & & \downarrow \\
Bu^*(BE(X)) & \longrightarrow & BE(X) \\
\downarrow & & \downarrow \\
BG & \xrightarrow{Bu} & B(\mathrm{Aut}\Pi \times \mathrm{Aut}\Pi')_\eta
\end{array}$$

L'opération $B\phi_0$ détermine une section encore notée $B\phi_0$ de ce fibré induit. En se plaçant dans la catégorie des B-fibrés (voir par exemple [11] ou [14]) et en simplicial, on démontre le théorème suivant:

Théorème 3. *Soit ϕ_0 une G-opération pointée sur X espace pointé ayant deux groupes d'homotopie Π et Π' et d'invariant d' Eilenberg η. Si $u : G \to (\mathrm{Aut}\Pi \times \mathrm{Aut}\Pi')_\eta$ est le morphisme déduit de ϕ_0, alors le fibré induit $Bu^*(BE(X))$ est équivalent au fibré $EG \times_G B\mathrm{Hom}(K(\Pi,p), K(\Pi',p+1))$, par une équivalence qui fait de la classe de $B\phi_0$ la classe de la section nulle de $\Gamma[EG \times_G B\mathrm{Hom}(K(\Pi,p), K(\Pi',p+1))]$, l'espace des sections de ce fibré.*

Notons $[BG, BE(X)]_u$ les classes d'homotopie des morphismes $\alpha : BG \to BE(X)$, tels que le composé $\Pi_1 BG \to \Pi_1 BE(X) \to \Pi_1 B(\mathrm{Aut}\Pi \times \mathrm{Aut}\Pi')_\eta$ soit u. Les théorèmes 1 et 3 nous permettent de construire une opération *libre* du groupe $\mathrm{Ext}^2_{\mathrm{G}}(\Pi, \Pi')$ sur l'ensemble $[BG, BE(X)]_u$:

$$\mathcal{L} : \mathrm{Ext}^2_{\mathrm{G}}(\Pi, \Pi') \times [\mathrm{BG}, \mathrm{BE(X)}]_{\mathrm{u}} \to [\mathrm{BG}, \mathrm{BE(X)}]_{\mathrm{u}}.$$

On a donc le résultat suivant:

Théorème 4. *A deux G-opérations pointées $\phi_0 : G \to E(X)$ et $\phi_1 : G \to E(X)$, induisant le même $u : G \to (\mathrm{Aut}\Pi \times \mathrm{Aut}\Pi')_\eta$, on sait associer un élément $\theta(\phi_0, \phi_1)$ appelé classe de Borel de (ϕ_0, ϕ_1) appartenant à $\mathrm{Ext}^2_{\mathrm{G}}(\Pi, \Pi')$. Cette classe est ca ractéristique dans le sens suivant: $\theta(\phi_0, \phi_1) = 0$ si et seulement si $[B\phi_0] = [B\phi_1]$.* Réciproquement, supposons maintenant que le morphisme $u : G \to (\mathrm{Aut}\Pi \times \mathrm{Aut}\Pi')_\eta$ se relève en une G-opération pointée $\phi_0 : G \to E(X)$. Cette hypothèse est vérifiée quand l'obstruction de Cooke [4], qui est ici dans $H^3(G, \mathrm{Hom}(\Pi, \Pi'))$, est nulle. Soit une classe $\theta \in \mathrm{Ext}^2_{\mathrm{G}}(\Pi, \Pi')$. Cette classe θ détermine, à homotopie près, un morphisme $\alpha : BG \to BE(X)$, tel que le composé $\Pi_1 BG \to \Pi_1 BE(X) \to \Pi_1 B(\mathrm{Aut}\Pi \times \mathrm{Aut}\Pi')_\eta$ soit u. Le problème de relèvement de Cooke [3] a une solution ce qui veut dire qu'il existe un espace Y, homotopiquement équivalent à X et une opération de G sur Y. Si les espaces sont pointés, on peut, à homotopie prés, faire de l'opération de G sur Y une opération ayant un point fixe c'est-à-dire une G-opération $\beta : G \to E(Y)$. L'équivalence $f : X \to Y$ obtenue donne une équivalence d'homotopie $Bf : BE(X) \to BE(Y)$, car $BE(X)$ est classifiant [7], [1] et [15] pour les couples (π, f), où π est une fibration et f une équivalence d'homotopie pointée de X sur la fibre $\pi^{-1}(*)$. De plus le diagramme suivant est homotopiquement commutatif:

$$\begin{array}{ccc} BG & \longrightarrow & BE(X) \\ \Big\downarrow{\scriptstyle B\beta} & & \Big\downarrow{\scriptstyle Bf} \\ BE(Y) & = & BE(Y). \end{array}$$

Corollaire 5. *A une classe $\theta \in \mathrm{Ext}^2_{\mathrm{G}}(\Pi, \Pi')$, on sait associer une G-opération $\beta : G \to E(Y)$, telle que $\mathcal{L}(\theta, [B\phi_0]) = [B\beta]$.*

Les démonstrations des théorèmes 3 et 4 font parties d'un article à paraître.

Remarque Dans les cas particuliers connus [8], [9], [10], l'obstruction de Cooke, qui est dans $H^3(G, \mathrm{Hom}(\Pi, \Pi'))$, est nulle. Il s'agit de deux types de situation: ou

$u : G \to (\mathrm{Aut}\Pi \times \mathrm{Aut}\Pi')_\eta$ a pour image l'élément neutre, c'est-à-dire que Π et Π' sont de G−modules triviaux, ou l'invariant d'Eilenberg η est nul, alors $E(X)$ est homotope à $\mathrm{Hom}((K(\Pi,p), K(\Pi',p+1)) \times (\mathrm{Aut}\Pi \times \mathrm{Aut}\Pi')$. Dans ces deux cas, il existe une section "canonique" de BG dans $Bu^*(BE(X))$, dont la classe est la classe de l'élément nul de $\Gamma[EG \times_G B\mathrm{Hom}(K(\Pi,p), K(\Pi',p+1))]$, et les classes de Borel associées à une G−opération sont des classes caractéristiques au sens classique, c'est-à-dire qu'elles sont nulles si et seulement si la G-opération est homotope à l'opération "canonique" sur X.

Références

[1] ALLAUD G. : On the classification of fibre space. Math. Z. 92, 110-125, (1966).

[2] BAUES H. J. : Algebraic Homotopy. Cambridge Studies in Advanced Math. 15, Cambridge University Press, (1990).

[3] CARTAN H. - EILENBERG S. : Homological Algebra. Princeton University Press, (1973).

[4] COOKE G. : Replacing homotopy actions by topological actions. Trans. of A.M.S., 237, (1978).

[5] CUNTZ J. : A New Look at KK-theory, K-Theory, 31-51, (1987).

[6] DIDIERJEAN G. : Homotopie de l'espace des équivalences fibrées. Ann. Inst. Fourier, 35(3), (1985), 33-47.

[7] DOLD A. - LASHOF R. : Principal quasifibrations and fibre homotopy equivalence of bundles. Ill. J. Math. 3, 285-305, (1959).

[8] FIEUX E. : Classes caractéristiques en KK−théorie des C^*−algèbres avec opérateurs. Thèse, Département de Math., URA 1408 Top. et Géom., U.P.S. Toulouse, (1990).

[9] HUEBSCHMANN J. : Change of rings and characteristic classes. Math. Proc. Camb. Phil. Soc.,106, 29, (1989).

[10] LEGRAND A. : Homotopie des espaces de sections. LN 941, (1982).

[11] LEGRAND A. : Caractérisation des opérations d'algèbres sur les modules différentiels. Composito Math., 66, 23-36, (1988).

[12] MAC LANE S. : Homology. Springer Verlag, (1979).

[13] SHIH W. : On the group $\mathcal{E}(X)$ of equivalences maps. Bull. Amer. Math. Soc., 492, 361-365, (1964).

[14] SMITH L. : Lectures on the Eilenberg-Moore Spectral Sequence. LN 134, (1970).

[15] STASHEFF J. D. : A classification theorem for fibre spaces. Topology 2, 239-246, (1963).

U.R.A. 1408
Laboratoire de Topologie et Géomètrie.
Toulouse III,
31062 Toulouse Cedex,
France.

1990 Barcelona Conference
on Algebraic Topology.

A NOTE ON THE BRAUER LIFT MAP[1]

FRIEDRICH HEGENBARTH

1. Introduction.

Let F_q be a finite field of characteristic p and let $Gl(n) = Gl(n, F_q)$ be the general linear group of $n \times n$-matrices over F_q. Furthermore, let $\overline{F}_q$ be the algebraic closure of F_q and let $\theta : \overline{F}_q - 0 \to \mathrm{C} - 0$ be a character. With these data there can be constructed a continuous map $Bl : BGl = \lim_{n\to\infty} BGl(n) \to BU$ depending on θ [7]. In this note we are going to prove the following

1.1 Theorem

If θ is an injective character, then the induced homomorphism

$$(Bl)^* : K(BGl) \to K(BU)$$

is surjective. The precise definition of Bl is given below.

D. Quillen has proved that the classifying space BGl is related to the homotopy fibre $F\Psi^q$ of $\Psi^q -$ Id : $BU \to BU$. Indeed, $F\Psi^q$ is homotopy equivalent to BGl^+, the "+"-construction of BGl (see [7]). A version of Theorem 1 is proved in [3]*. Here, however, we give a completly different proof using representation theory of $Gl(n)$. We construct explicitly a set of generators of $K(BGl)$. Recall that there are natural ring homomorphisms

$$A_n : R(Gl(n)) \to R(\widehat{Gl(n)}) = K(BGl(n))$$

which link representation theory to K-theory [1]. The A_n fit together to give a homomorphism

$$A_\infty : R(Gl(\infty)) = \lim_{n\to\infty} R(Gl(n)) \to K(BGl(\infty))$$

In order to prove Theorem 1 we use the basic results of J.A. Green [2] and define a homomorphism

$$Bl_\infty : R(U(\infty)) \to R(Gl(\infty)).$$

Theorem 1 then will follow from

[1] This work was performed under the auspicies of the G.N.S.A.G.A. of the C.N.R. and financially supported by the M.P.I. of Italy.

1.2 Theorem
If θ is an injective character, then

$$Bl_\infty : R(U(\infty)) \to R(Gl(\infty))$$

is surjective.

In order to prove Theorem 2 we have to compare the induction homomorphisms for the finite groups $Gl(n)$ with those of the compact Lie groups $U(n)$. The latter ones are defined via K-theory [**5**]. Furthermore, we use a double coset formula. We refer to Mitchell and Priddy [**6**] for a beautiful discussion of double coset formulae. Furthermore, we will study the dependence of Bl and Bl_∞ on the character $\theta : \overline{F}_q^* \to (\mathbf{C})^*$.

In particular, the set of all these characters is isomorphic to $\prod Z_l$ where the product goes over all primes $l \neq p$. This will give an explicit splitting of $K(BGl)$ in its l-adic parts.

2. The Induction Homomorphism for the Unitary Groups.

I. Madsen and J. Milgram have defined induction homomorphisms

$$\mathrm{Ind}_H^{U(N)} : R(H) \to R(U(N))$$

for any subgroup H of maximal rank. This homomorphism satisfies the following formula: If $x \in R(H)$, then

$$\mathrm{Ind}_H^{U(N)}(x) = \sum \sigma \cdot x\,,$$

where the sum is over $\sigma \in W(U(N))/W(H)$. Here $W(G)$ denotes the Weyl group of G [**5**]. In particular, $W(U(N)) = \Sigma_N$, the symmetric group on N letters. Let $T(N)$ be a maximal torus of $U(N)$ which is also contained in H and let $j : T(N) \to U(N)$ (resp. H) are its natural inclusions. Then the induced homomorphism

$$j^* : R(U(N)) \to R(T(N)) \cong Z[t_1, t_1^{-1}, \ldots, t_N, t_N^{-1}]$$

is an injection and

$$j^*\mathrm{Ind}_H^{U(N)}(x) = \sum \sigma \cdot j^*(x).$$

Here σ runs through $\Sigma_N/W(H)$. Considering $j^*(x)$ as a character, the action of σ on $j^*(x)$ is given by permuting the eigenvalues. We will now consider the case in which $H = P_{n,m} \subset U(n+m)$ is the subgroup

$$\begin{pmatrix} U(n) & * \\ 0 & U(m) \end{pmatrix}.$$

Recall that the Weyl group of $P_{n,m}$ is $\Sigma_n \times \Sigma_m$. Let now $A \in U(n+m)$ and let $u_1(A), \ldots, u_{n+m}(A)$ be its eigenvalues, then the above formula yields

$$\mathrm{Ind}_{P_{n,m}}^{U(n+m)}(x \otimes y)(A) =$$

$$= \sum x(u_{w(1)}(A), \ldots, u_{w(n)}(A))y(u_{w(n+1)}(A), \ldots, u_{w(n+m)}(A)),$$

where w runs through $\Sigma_{n+m}/\Sigma_n \times \Sigma_m$. In the next section we will compare this formula with the induction formula of finite groups $Gl(n, F_q)$ via the Brauer lift map.

3. Representation Theory of $Gl(n, F_q)$.

The following Theorem of J.A. Green is used to construct complex representations of finite groups [2]: Let $\rho : G \to Gl(N, F_q)$ be a modular representation of the finite group G and let $\theta : \overline{F}_q^* \to \mathbf{C}^*$ be a fixed homomorphism. Then for any symmetric function $f \in Z[t_1, \ldots, t_N]^{\Sigma_N}$ the map $\mathcal{X}_f : G \to \mathbf{C}$ defined by

$$g \to f(\theta u_1(g), \ldots, \theta u_N(g))$$

is the character of a (virtual) complex representation. Here $u_1(g), \ldots, u_N(g)$ denote the eigenvalues of $\rho(g)$. A proof of this Theorem can also be found in [4]. In the following we fix θ and take $\rho = \mathrm{Id} : Gl(N) \to Gl(N)$. Let

$$G_N : Z[t_1, \ldots, t_N]^{\Sigma_N} \to R(Gl(N))$$

be the above map $f \to \mathcal{X}_f$. Recall that

$$R(U(N)) \cong Z[s_1, \ldots, s_N, s_N^{-1}] \subset Z[t_1, t_1^{-1}, \ldots, t_N, t_N^{-1}] \cong R(T(N)),$$

where $s_1, \ldots, s_N$ are the elementary symmetric functions in $t_1, \ldots, t_N$. This yields a map

$$Bl_N : RU(N)) \to R(Gl(N)).$$

It is easy to show that this map is a ring homomorphism. We set $N = n + m$ and consider the parabolic subgroup $P_{n,m} = P_{n,m}(F_q)$, i.e. the subgroup

$$\begin{pmatrix} Gl(n) & * \\ 0 & Gl(m) \end{pmatrix}$$

This group can also be written in the form $Gl(n) \times Gl(m) U_{n,m}$. Let

$$L : R(Gl(n)) \otimes R(Gl(m)) \to R(P_{n,m}),$$

defined by $L(x \otimes y)(g_1 \times g_2 u) = x(g_1) y(g_2)$ for any element $g_1 \times g_2 u \in P_{n,m}$. One considers the following "induction" homomorphism:

$$\mathrm{Ind}_{Gl(n)\times Gl(m)}^{Gl(n+m)} \circ L : R(G(n)) \otimes R(Gl(m)) \to R(G(n+m)).$$

We will make use of the following notation:
If $x \in R(Gl(n))$ and $y \in R(Gl(m))$, then

$$x \circ y = \mathrm{Ind}_{Gl(n)+Gl(m)}^{Gl(m+n)}(x \otimes y).$$

The main step to prove Theorem 1 is

3.1 Proposition

The following diagram is commutative:

$$\begin{array}{ccc} R(U(n))\otimes R(U(m)) & \xrightarrow{\text{Ind}} & R(U(n+m)) \\ {\scriptstyle Bl_n\otimes Bl_m}\downarrow & & \downarrow{\scriptstyle Bl_{n+m}} \\ R(Gl(n))\otimes R(Gl(m)) & \xrightarrow{\text{Ind}} & R(Gl(n+m)) \end{array}$$

The horizontal maps denote the induction homorphisms and the left and right vertical homomorphisms are $Bl_n \otimes Bl_m$ and Bl_{n+m} respectively.

Proof:

Let $f_1 \in R(U(n))$ and $f_2 \in R(U(m))$ be symmetric functions. It follows from section 2 that for any $g \in Gl(n+m)$,

$$Bl_{n+m}\mathrm{Ind}_{U(n)\times U(m)}^{U(n+m)}(f_1\otimes f_2)(g) =$$

$$= \sum f_1(\theta u_{w(1)}(g),\dots)f_2(\theta u_{w(n+1)}(g),\dots \theta u_{w(n+m)}(g))$$

where the sum is over $w \in \Sigma_{n+m}/(\Sigma_n \times \Sigma_m)$. Now we use the fact that two characters of $Gl(n+m)$ coincide if and only if their restrictions to the parabolic subgroups $P_a(F_q) = P_a$ coincide. Here $a = (a_1,\dots,a_r)$ is a partition of $n+m$ and P_a is the subgroup of type [4]

$$\begin{pmatrix} Gl(a_1) & * & \cdots & * \\ 0 & Gl(a_2) & \cdots & * \\ \vdots & \vdots & & \vdots \\ 0 & 0 & \cdots & Gl(a_r) \end{pmatrix}.$$

The Mackey formula yields

$$\mathrm{Res}_{P_i}^{Gl(n+m)}\mathrm{Ind}_{P_{n,m}}^{Gl(n+m)}(Bl_n(f_1)\otimes Bl_m(f_2))(g) =$$

$$= \sum \mathrm{Ind}_{wP_{n,m}w^{-1}\cap P_a}^{P_a}(f_1\otimes f_2))(w^{-1}gw),$$

where the sum runs over $\overline{w} \in P_a \setminus Gl(n+m)/P_{n,m}$ and $w \in \overline{w}$. Fixing w and $g \in P_a$, one obtains

$$\mathrm{Ind}_{wP_{n,m}w^{-1}\cap P_a}^{P_a}(Bl_n(f_1)\otimes Bl_m(f_2))(g) =$$

$$= R^{-1}\sum f_1(\theta u_{w(1)}(hgh^{-1}),\dots \theta u_{w(n)}(hgh^{-1}))\cdot$$

$$\cdot f_2(\theta u_{w(n+1)}(hgh^{-1}),\dots \theta u_{w(n+m)}(hgh^{-1})).$$

Moreover, R denotes the order of the group $wP_{n,m} \cap P_a$, where the sum is over $h \in P_a$ such that $hgh^{-1} \in wP_aw^{-1}$. The conjugation of g by $h \in P_a$ defines a permutation $v_h \in \Sigma_a = \Sigma_{a_1} \times \Sigma_{a_2} \times \cdots \times \Sigma_{a_r}$ of the eigenvalues of g (recall that P_a can be uniquely decomposed into a product of unipotent, diagonal and permutation matrices). The above expression can therefore be written as

$$R^{-1}\sum f_1(\theta u_{vw(1)}(g),\ldots,\theta u_{vw(n)}(g))\cdot f_2(\theta u_{vw(n+1)}(g),\cdots,\theta u_{vw(n+m)}(g)),$$

where $v = v_h$ and h runs through P_a.

If h and h' are in P_a such that $h'h^{-1} \in wP_{n,m}w^{-1} \cap P_a$, the corresponding summands $f_1(\ldots)f_2(\ldots)$ are equal. The above formula therefore becomes

$$\mathrm{Ind}^{P_a}_{wP_{n,m}w^{-1}\cap P_a}(Bl_n(f_1)\otimes Bl_m(f_2))(g) =$$

$$= \sum f_1(\theta u_{vw(1)}(g),\ldots \theta u_{vw(n)}(g))\cdot f_2(\theta u_{vw(n+1)}(g),\ldots \theta u_{vw(n+m)}(g)),$$

where v runs through Σ_a. The natural inclusions $\Sigma_{n+m} \subset Gl(n+m)$ and $\Sigma_a \subset P_a$ induce an identification

$$\Sigma_a \setminus \Sigma_{n+m}/\Sigma_n \times \Sigma_m = P_a \setminus Gl(n+m)/P_{n,m}$$

(*Bruhat decomposition*). We have therefore the following formula

$$\mathrm{Res}^{Gl(n+m)}_{P_a}\mathrm{Ind}^{Gl(n+m)}_{P_{n,m}}(Bl_n(f_1)\otimes Bl_m(f_2))(g) =$$

$$= \sum\sum f_1(\theta u_{w(1)}(g),\ldots \theta u_{w(n)}(g))\cdot f_2(\theta u_{w(n+1)}(g),\ldots \theta u_{w(n+m)}(g)),$$

where the first sum is over $\overline{w} \in \Sigma_a \setminus \Sigma_{n+m}/\Sigma_n \times \Sigma_m$ and the second is over $w \in \overline{w} \in \Sigma(n+m)/\Sigma_n \times \Sigma_m$. But this is equal to $Bl_{n+m}\mathrm{Ind}^{U(n+m)}_{U(n)\times U(m)}(f_1 \otimes f_2)(g)$, proving the proposition.

4. Proof of the Theorems.

In section 3 we have considered the homomorphisms

$$Bl_N : R(U(N)) \to R(Gl(N)).$$

We can take the inverse limit for $N \to \infty$ and obtain a homomorphism

$$Bl_\infty : R(U(\infty)) \to R(Gl(\infty, F_q)).$$

The map $Bl : BGl(\infty, F_q) \to BU$ is obtained as follows:

Let $s_1 \in R(U(\infty))$ be the first elementary symmetric function. Then we have $A_\infty \circ Bl_\infty(s_1) \in K(BGl(\infty, F_q))$ and its component in the reduced K-group defines the map Bl. In order to prove surjectivity of the induced homomorphism

$$Bl^* : K(BU) \to K(BGl(\infty, F_q))$$

it suffices to prove sujectivity of Bl_∞. Since the $\lim^1$-term vanishes, it is enough to show that the compsition

$$\mathrm{Res} \circ Bl_\infty : RU(\infty) \to R(Gl(\infty)) \to R(Gl(N))$$

is surjective. Let $x_j = Bl_N(s_j)$. It is well-known that $R(Gl(N))$ is generated (additively) by the elements $x_{j_1} \circ x_{j_2} \circ \cdots x_{j_r}$, $j_1 + j_2 + \cdots + j_r = N$. For this one has to assume that $\theta : \overline{F}^*_q \to \mathbf{C}^*$ is injective (see [4, p.147ff]). Theorem 1 and Theorem 2 now follow immediately from Proposition 3.1.

5. Final Remarks.

The construction of the map Bl depends on the choice of the character $\theta : \overline{F}_q^* \to \mathbf{C}^*$. Since the set of all such characters is isomorphic to $\prod \widehat{\mathbf{Z}}_l$, where the product is over all primes $l \neq p$, we can write $\theta = (\theta_l)_{l \neq p}$. The relation between $Bl = Bl(\theta)$ and the $Bl(\theta_l)$, l a prime different from p, can be easily deduced from the following well-known fact:

The kernel of $A_N : R(Gl(N)) \to K(BGl(N))$ consist of those characters which vanish on elements $g \in Gl(N)$ of prime power order [**1**]. This implies

$$Bl(\theta) = \prod_{l \neq p} Bl(\theta_l).$$

Moreover, if l and k are different primes, then $Bl(\theta_l)Bl(\theta_k) = 0$. Hence the induced homomorphism

$$Bl(\theta_l)^* : K(BU) \to K(BGl(\infty, F_q))$$

maps onto the l-adic part.

References.

1. M.F. Atiyah: *"Characters and Cohomology of Finite Groups"*, Publ. Math. IHES **9** (1961), 23–64.
2. J.A. Green: *"The Characters of the Finite General Linear Groups"*, Trans. Amer. Math. Soc. **80** (1953), 402–447.
3. F. Hegenbarth: *"On the K-Theory of the Classifting Spaces of the General Linear Groups over Finite Fields"*. Proc. Differential Topology Symposium, Siegen 1987, Springer Lect. Notes **1350**, 259–265.
4. I.G. MacDonald: *"Symmetric Functions and Hall Polynomials"*, Clarendon Press. Oxford, 1979.
5. I. Madsen, J. Milgram: *"The Classifying Spaces for Surgery and Cobordism of Manifolds"* Ann. of Math. Studies **92**, Princeton University Press, 1979.
6. S.A. Mitchell, S.B. Priddy: *"A Double Coset Formula for Levi Subgroups and Splitting BGl_n"*, Proc. Algebraic Topology, Arcata 1986, Springer Lect. Notes **1370**, 325–334.
7. D. Quillen: *"On the Cohomology and K-theory of the General Linear Groups over a Finite Field"*, Ann. of Math. **96** (1972), 552–586.

* Addendum.

The proof of Theorem 1.1 given in reference [**3**] is not correct, because Theorem 2 of [**3**] is only true if l is an odd prime and q generates the units of the l-adic integers. To prove Theorem 2 of [**3**] one has to compare the two spaces $BGlF_q$ and J. These two spaces coincide only under the above assumption. Under this assumption Theorem 2 of **[3]** was already proved by V. Snaith (see: V. Snaith: Dyer–Lashof Operations in K–Theory, Springer Lecture Notes 496 (1975), 103–294).

The author would like to thank the referee for pointing out this error.

Dipartimento di Matematica II
Università di Roma
Via Fontanile di Carcaricola
00133 Roma
Italia

1990 Barcelona Conference
on Algebraic Topology.

CATEGORICAL MODELS OF N–TYPES FOR PRO–CROSSED COMPLEXES AND $\mathcal{J}_n$–PROSPACES*

L.J. HERNÁNDEZ AND T. PORTER

Abstract: As part of a program to study Pro n-types and Proper n-types of locally finite simplicial complexes, in this paper we give notions of n-fibrations, n-cofibrations and weak n-equivalences in the category of crossed complexes Crs that satisfy the axioms for a closed model structure in the sense of Quillen. The category obtained by formal inverting the weak n-equivalences $\mathrm{Ho}_n(\mathrm{Crs})$ is said to be the category of n-types of crossed complexes. We also extend the notions above to pro-crossed complexes, pro-simplicial sets and prospaces to obtain the categories $\mathrm{Ho}_n(\mathrm{proCrs})$, $\mathrm{Ho}_n(\mathrm{proSS})$ and $\mathrm{Ho}_n(\mathrm{proTop})$.

We consider the notions of $\mathcal{J}_n$-space (a slight modification of the notion of J_n-space given by J.H.C. Whitehead), $\mathcal{J}_n$-crossed complex and their generalizations to the categories of prospaces and pro-crossed complexes and we prove that the category of n-types of $\mathcal{J}_n$-prospaces is equivalent to the category of n-types of $\mathcal{J}_n$-pro-crossed complexes. We also use the properties of skeleton, coskeleton and truncation functors to compare categories of n-types to categories of homotopy types. This analysis allows one to give new category models for the categories of n-types of prospaces and pro-crossed complexes.

A.M.S. classification: 55P15, 55P99.
Key words: n-type, proper n-type, pro-n-type, closed model category, prospace, pro-simplicial set, crossed complex, pro-crossed complex, $\mathcal{J}_n$-prospace and $\mathcal{J}_n$-pro-crossed complex.

---

* The authors acknowledge the finantial help given by the British-Spanish joint research program 'British Council-M.E.C., 1988-89, 51/18' and the research project P587-0062 of the DGYCYT.

0. Introduction

In 1949, J.H.C. Whitehead [W.1, W.2] introduced the following notion of (n+1)-type: Two CW-complexes X and Y were said to have the same (n+1)-type if there are continuous maps $f: X^{n+1} \longrightarrow Y^{n+1}$ and $g: Y^{n+1} \longrightarrow X^{n+1}$ (X^m denotes the m-skeleton of X) such that gf/X^n is homotopic to the inclusion of X^n into X^{n+1} and similarly on the other side. In later papers that notion was changed by one dimension; that is, two complexes satisfying the property above were said to have the same n-type.

Whitehead was looking for a purely algebraic equivalent of n-type satisfying realisability properties in two senses. He hoped not only that for each algebraic model G there would be a CW-complex X such that the algebraic model associated with X is isomorphic to G but also for the realisability of the morphisms between algebraic models (c.f. [Ba.1] where this is interpreted in categorical terms).

Associated with a reduced CW-complex X, Whitehead considered the sequence of groups,

$$\cdots \longrightarrow \pi_{i+1}(X^{i+1}, X^i) \xrightarrow{\delta_{i+1}} \pi_i(X^i, X^{i-1}) \longrightarrow \cdots \longrightarrow \pi_2(X^2, X^1) \longrightarrow \pi_1(X^1, *),$$

where δ_{i+1} is defined by the composition

$$\pi_{i+1}(X^{i+1}, X^i) \longrightarrow \pi_i(X^i) \longrightarrow \pi_i(X^i, X^{i-1}).$$

Any sequence of groups satisfying certain properties of the sequence above was called a homotopy system.

Using maps and homotopies of homotopy systems Whitehead gave an algebraic characterization of the homotopy types of 3-dimensional CW-complexes and of finite simply-connected 4-dimensional CW-complexes. A description of 2-types in terms of 3-dimensional cohomology classes was given by Mac Lane and Whitehead [M-W] in 1950. They defined an algebraic 2-type by a triplet (π_1,π_2,k) where π_1 is a group, π_2 an abelian group which admits π_1 as a group of operators and k is an element of $H^3(\pi_1;\pi_2)$. Mac Lane and Whitehead proved that two CW-complexes have the same 2-type if and only if the corresponding algebraic 2-types are isomorphic and that algebraic 2-types and morphisms between them can be realised by spaces and continuous maps. When Whitehead considered the realisability of maps between algebraic models of certain n-types for n>2, he found that the machinery he had developed could only handle the problem for a subclass of n-types, those corresponding to the J_n-spaces that will be considering later on.

Homotopy systems are good algebraic models for characterising some types and n-types of 0-connected CW-complexes, but they do not have enough structure to characterize all types and n-types of CW-complexes. One of Whitehead's problems was that their requirements for realisability were too strong (see also [Ba.1, Ba.2]).

Recently, Brown and Higgins [B-H.1, B-H.2, B-H.3] have developed the notion of crossed complex which is a slight generalization of homotopy systems. The crossed complex associated with a filtered space $X=\{X^k\}$ takes in dimension 0, the set X^0, in dimension 1 the fundamental groupoid of X^1 and in higher dimensions the family of relative homotopy groups $\{\pi_k(X^k, X^{k-1}, p) | p \in X^0\}$, $k \geq 2$.

One of the advantages of this approach is that any crossed complex, C, can be realised by a CW-complex X=(BC see below) together with a filtered space structure $\{X^k\}$ defined by subcomplexes. This filtration, though, is not usually the skeletal filtration which can only give crossed complexes of "free type".

The algebraic and homotopy properties of the category of crossed complexes have been studied by N. Ashley, H.J. Baues, R. Brown, P.J. Higgins, M. Golasiński, and N.D. Gilbert, amongst others. Recently Brown and Higgins have defined a classifying functor which associates a CW-complex BG with a crossed complex G. They define BG as the realisation of the cubical nerve of G, see [B-H.2]. Letting πX denote the crossed complex associated with the CW-complex X as above, they have proved that π:Ho(CCW) $\longrightarrow$ Ho(Crs) is left adjoint to B: Ho(Crs) $\longrightarrow$ Ho(CCW) where Ho(CCW) denotes the homotopy category of CW-complexes and cellular maps and homotopies and Ho(Crs) is the homotopy category of Crs obtained by inverting the weak equivalences of the closed model structure of Crs given by Brown and Golasiński, [B-G].

One of the aims of the present paper is to develop some algebraic models for other homotopy categories, related to problems in proper homotopy and shape theory. It is well known that the proper homotopy category of σ-compact spaces and the strong shape category can be considered as full subcategories of well structured homotopy category of prospaces or pro-simplicial sets, see [E-H]. For the proper homotopy category we can use the embedding theorems of Edwards and Hastings, see [E-H], and an embedding for the strong shape is given by the Vietoris functor, see [P]. A method to give algebraic models for proper or strong shape types thus consists in giving firstly algebraic models for the homotopy types or n-types of prospaces or pro-simplicial sets and afterwards, by using suitable embeddings for proper or strong shape n-types to obtain the desired algebraic models. As an example, the authors have proved that the functor $\operatorname{cosk}_{n+1}\varepsilon:(P_\sigma)_\infty \longrightarrow$ proTop induces an embedding of the category of proper n-types of σ-compact simplicial complexes into the homotopy category of prospaces, see [H-P.3].

This paper develops some tools which are necessary for the extension to the homotopy category of prospaces of the Whitehead, Mac Lane and Brown-Higgins theorems on algebraic characterizations of types and n-types that are obtained by using homotopy systems and crossed complexes. The methods we use depend strongly on the functoriality of certain constructions and the naturality of various isomorphisms. Thus to a large extent the main part of the paper is concerned with a detailed examination of the "building blocks" of a proof of results of Whitehead and Brown-Higgins making explicit much of the implicit functoriality and naturality involved.

We begin the paper by giving the definition and some properties of the category of crossed complexes. For a more complete description of this category we refer the reader to the papers [B-H.1, B-H.2, B-H.3]. We have also included the closed model structure given to Crs by Brown-Golasiński, see [B-G].

The method that we have used to define the category of n-types of spaces, simplicial sets or crossed complexes consists in the "formal inversion" of weak n-equivalences; that is, given the category $\mathcal{C}$ and Σ the class of weak n-equivalences we consider the category $\mathrm{Ho}_n(\mathcal{C}) = \mathcal{C}\,\Sigma^{-1}$, see [Q]. The existence of this category is no problem for the cases that we are going to consider. If $\mathcal{B}$ is a full subcategory of $\mathcal{C}$ we shall denote by $\mathrm{Ho}_n(\mathcal{C})|\mathcal{B}$ the full subcategory of $\mathrm{Ho}_n(\mathcal{C})$ determined by the objects of $\mathcal{B}$ and by $\mathrm{Ho}_n(\mathcal{B})$ the category $\mathcal{B}(\Sigma \cap \mathrm{Mor}\mathcal{B})^{-1}$; that is, the category obtained by inverting the weak n-equivalences of $\mathcal{B}$.

A morphism f of crossed complexes is said to be a weak n-equivalence if f induces isomorphisms on the k-th homotopy groups, $0 \leq k \leq n$, for any choice of base point. The homotopy groups that we are using in Crs are those defined in [B-G].

In section 3 we give appropiate notions of n-fibrations and n-cofibrations for crossed complexes in such a way that for each non negative integer n, Crs admits a closed model structure in the sense of Quillen. Therefore for the category of n-types of crossed complexes we can use all the results obtained by Quillen [Q]. Afterwards we use the method of Edwards Hastings to extend this closed model structure to the category of pro-crossed complexes. In this way we have obtained a good algebraic and homotopy structure to deal with certain proper n-types. Similar arguments can be given for prospaces and pro-simplicial sets.

In [W.1], as part of his study of n-types, Whitehead introduced the notion of J_n-complex. The property of being a J_n-complex is an invariant of (n-1)-type and for J_n-complexes, the

Hurewicz homomorphism of the universal covering is an isomorphism in dimensions≤n and an epimorphism in dimension n+1 (In fact a J_n-complex is precisely one on which Whitehead's Γ_k-functor vanishes for k=1,...,n.). Thus the simpler structure of J_n-complexes allows a much easier classification for such spaces. We have considered the notions of J-space and $\mathfrak{J}_n$-space and their generalization to the categories of prospaces, pro-simplicial sets and pro-croosed complexes. The notion of $\mathfrak{J}_n$-space is a slight modification of Whitehead's notion of J_n-space (For the Hurewicz homomorphism of the universal covering we remove the condition of being an epimorphism in dimension n+1) and the notion of J-space agrees with Whitehead's notion when $n=+\infty$.

Using properties of closed model categories we finally prove that the category of n-types of $\mathfrak{J}_n$-prospaces is equivalent to the category of n-types of $\mathfrak{J}_n$-pro-crossed complexes (see section 7) and using the properties of the skeleton, coskeleton and truncation functors we give other models for these n-types. We thus not only recover Whitehead's realisability result for morphisms between J_n-spaces, but obtain "pro"versions of this, thus making it calculable for strong shape and proper homotopy. In the case of 2-types if we use the representation of elements of $H^3(G;H)$ given by Huebschmann, Holt, Lue and Mac Lane (see [M]) as reduced crossed complexes C of dimension 2 such that $\Pi_1(C)\cong G$, $\Pi_2(C)\cong H$, we have that Corollary 7.10(iii) is the extension of the Mac Lane-Whitehead Theorem to the category of tower of spaces.

The next step to obtain a more complete description of the n-types of prospaces will be to use richer algebraic models (cat^n groups, n-crossed cubes, hypercrossed complexes, etc) to obtain the corresponding new algebraic models. The applications of this paper to give algebraic models for proper and strong shape n-types or to compute proper homotopy and strong shape classes will be developed in later papers.

1. The category of crossed complexes.

Recently, Brown and Higgins [B-H.1, B-H.2, B-H.3] have made an algebraic and homotopy study of the category of crossed complexes. We will start by giving a brief survey of the relevant parts of their results.

Recall the definition of crossed complex given by Brown and Higgins in [B-H.3].

Let C_1 be a groupoid with object set C_0. If p, q$\in C_0$, $C_1(p,q)$ will denote the set of arrows from p to q. A "non-Abelian" C_1-module consists of a family of groups

$A = \{A(p) \mid p \in C_0\}$ on which C_1 operates, obeying the laws for a C_1-module except possibly commutativity. Additive and multiplicative notation will be used for A and C_1, respectively. Such a module is a crossed module over C_1 if it is equipped with a morphism $\delta: A \longrightarrow C_1$ sending $A(p)$ to $C_1(p, p)$ (for $p \in C_0$) which satisfies the laws

$$\delta(a+a') = \delta a \cdot \delta a' \qquad a, a' \in A(p),$$
$$\delta(a^c) = c^{-1}\,\delta a\, c \qquad a \in A(p), \quad c \in C_1(p, q),$$
$$a^{\delta a'} = -a' + a + a' \qquad a, a' \in A(p).$$

A crossed complex is a sequence

$$\cdots \longrightarrow C_n \xrightarrow{\delta} C_{n-1} \longrightarrow \cdots \longrightarrow C_2 \longrightarrow C_1 \underset{\delta^1}{\overset{\delta^0}{\rightrightarrows}} C_0$$

such that:

(i) $C_1 \rightrightarrows C_0$ is a groupoid;
(ii) $C_2 \longrightarrow C_1$ is a crossed module over C_1;
(iii) C_n is a C_1-module (Abelian) for $n \geq 3$;
(iv) $\delta: C_n \longrightarrow C_{n-1}$ is an operator morphism for $n \geq 3$;
(v) $\delta\delta: C_n \longrightarrow C_{n-2}$ is trivial for $n \geq 3$;
(vi) $\delta\, C_2$ acts trivially on C_n for $n \geq 3$.

The indexing map $\beta: C \longrightarrow C_0$ is defined by $\beta(x) = q$ if $x \in C_0$ and $x = q$, $x \in C_1(p,q)$ or $x \in C_k(q)$ $(k \geq 2)$.

A morphism $f: C \longrightarrow D$ of crossed complexes consists of a family of morphisms $f_n: C_n \longrightarrow D_n$ which preserve all the structure; that is, (f_1, f_0) is a groupoid morphism and each f_k $(k \geq 2)$ is a (f_1, f_0)-operator morphism, ($f_k(x^g) = (f_k x)^{f_1 g}$, $\beta(f_k x) = f_0\, \beta x$) that commutes with δ $(\delta f_k = f_{k-1}\delta,\ k \geq 2)$.

By a pointed crossed complex we mean a crossed complex C and a base point $* \in C_0$ and by a pointed morphism a morphism that preserves base points.

We will denote by Crs and Crs* the categories of crossed complexes and pointed crossed complexes respectively.

A 1-fold (left) homotopy $(h, f): f' \simeq f: B \longrightarrow C$ is a family of maps $h_k: B_k \longrightarrow C_{k+1}$ $(k \geq 0)$ such that $\beta(h_k(b)) = \beta(f_k(b))$, $b \in B_k$; h_1 is a derivation over f_1; h_k $(k \geq 2)$ is an operator morphism over f_1; and for $b \in B$,

$$f'_k(b) = \begin{cases} [f_k b + h_{k-1}\delta b + \delta h_k b]^{(h_0 \beta b)^{-1}} & \text{if } k \geq 2, \\ (h_0 \delta^0 b)(f_1 b)(\delta h_1 b)(h_0 \delta^1 b)^{-1} & \text{if } k = 1, \\ \delta^0 h_0 b & \text{if } k = 0. \end{cases}$$

An n-fold (left) homotopy $(h, f): B \longrightarrow C$ is a family of maps $h_k: B_k \longrightarrow C_{k+n}$ $(k \geq 0)$ such that $\beta(h_k(b)) = \beta(f(b)$ for all $b \in B$; h_1 is a derivation over f_1; and $h_k (k \geq 2)$ is an operator morphism over f_1.

An n-fold homotopy (h, f) is pointed if f is pointed and $h_0(*) = 1_*$ if $n = 1$, or $h_0(*) = 0_*$ if $n \geq 2$.

For a given 1-fold homotopy $(h, f): f' \simeq f$, we define $\delta^0(h, f) = f'$ and $\delta^1(h, f) = f$ and for an n-fold homotopy, $(n \geq 2)$, (h, f), define $\delta(h, f) = (\delta h, f)$ where

$$\delta h(b) = \begin{cases} \delta(h(b)) + (-1)^{n+1} h(\delta(b)) & \text{if } b \in B_k \ (k \geq 2), \\ (-1)^{n+1} (h(\delta^0 b))^{fb} + (-1)^n h(\delta^1 b) + \delta(h(b)) & \text{if } b \in B_1, \\ \delta(h(b)) & \text{if } b \in B_0. \end{cases}$$

Using the notion above, Brown and Higgins define an internal hom-functor and tensor product that give to Crs and Crs* the structure of symmetric monoidal closed categories; for a complete description we refer the reader to Brown-Higgins [B-H.1] where these constructions are developed in detail. The crossed complex CRS(B, C) is in dimension 0 the set of all morphisms $B \longrightarrow C$. In dimension $m \geq 1$, it consists of m-fold homotopies $h: B \longrightarrow C$ over morphisms $f: B \longrightarrow C$. Similarly, the category Crs* admits an internal hom-functor CRS*(B,C) if we consider pointed morphisms and homotopies. If A and B are crossed complexes, then $A \otimes B$ is the crossed complex generated by elements $a \otimes b$ in dimension $m + n$, where $a \in A_m$ and $b \in B_n$, satisfying the relations given in [B-H.1]. For the pointed case to define the pointed tensor $A \otimes_* B$, we must add in the new relations

$$\begin{aligned} a \otimes_* * &= 0_* && \text{for } a \in A_k \ (k \geq 2), \\ a \otimes_* * &= 1_* && \text{for } a \in A_1, \\ a \otimes_* * &= * && \text{for } a \in A_0, \end{aligned}$$

similarly $* \otimes_* b = 0_*$, 1_* or $*$.

The tensor product and hom-functor satisfy the following (see [B-H.1]):

Theorem 1.1.

(i) The functor $-\otimes B$ is left adjoint to the functor CRS(B, —) from Crs to Crs.

(ii) For crossed complexes A, B, C, there are natural isomorphisms of crossed complexes

$$(A\otimes B)\otimes C \cong A\otimes (B\otimes C)$$
$$CRS\,(A\otimes B, C) \cong CRS\,(A, CRS(B, C))$$

giving Crs the structure of a monoidal closed category.

(iii) There are similar results for Crs* with the corresponding functors $-\otimes_* B$, and $CRS_*(B, -)$.

Let CCW denote the category of CW-complexes and cellular maps and Π(CCW) the homotopy category obtained using cellular homotopies. For a CW-complex X we have the associated crossed complex πX

$$\longrightarrow \pi_n(X^n, X^{n-1}) \longrightarrow \pi_{n-1}(X^{n-1}, X^{n-2}) \longrightarrow \cdots \longrightarrow \pi_1(X^1, X^0) \underset{\delta^1}{\overset{\delta^0}{\rightrightarrows}} X^0$$

where $\pi_1(X^1, X^0)$ is the fundamental groupoid of X^1 whose objects are the 0-cells of X, and $\pi_n(X^n, X^{n-1})$ is the family $\{\pi_n(X^n, X^{n-1}, p) \mid p \in X^0\}$. This construction, defined by Whitehead [W.1, W.2] for reduced CW-complexes, gives the functors π: CCW $\longrightarrow$ Crs and π: CCW* $\longrightarrow$ Crs*. If we consider the standard n-ball $E^n = e^0 \cup e^{n-1} \cup e^n$ and (n - 1)-sphere $S^{n-1} = e^0 \cup e^{n-1}$, we obtain the crossed complexes

$$C(n) = \pi E^n \qquad n \geq 0,$$
$$S(n-1) = \pi S^{n-1} \quad n \geq 1.$$

The crossed complex C(1), which is also denoted J, has two elements 0 and 1 in dimension 0 and one generator from 0 to 1 in dimension 1. The crossed complex C(0) is also denoted by *.

The crossed complex, J, defines a cylinder functor $J\otimes -$, with end maps,

$$X \sqcup X \xrightarrow{\partial_0 + \partial_1} J \otimes X$$

This defines a homotopy relation on the maps of Crs and induces the corresponding homotopy category $\Pi(\text{Crs})$. If J^+ denotes the crossed complex $J \sqcup *$, we also have the pointed cylinder

$$X \vee X \xrightarrow{\partial_0 + \partial_1} J^+ \otimes_* X$$

and the category $\Pi(\text{Crs}_*)$.

It is easy to check that we also have the induced functors $\pi\colon \Pi(\text{CCW}) \longrightarrow \Pi(\text{Crs})$ and $\pi\colon \Pi(\text{CCW}_*) \longrightarrow \Pi(\text{Crs}_*)$.

Definition 1.2 (c.f. [B-G]). Define the n-th homotopy group of a crossed complex X at a vertex $p \in X_0$ by

$$\Pi_n(X, p) = \Pi(\text{Crs}_*)(\pi S^n, X)$$

where πS^n and X are pointed by e^0 and p, respectively.

Note that $\Pi(\text{Crs}_*)(X, Y) = \Pi_0(\text{CRS}_*(X, Y))$ and for a given crossed complex Z and $p \in Z_0$, $\Pi_0(Z, p)$ is just the p-pointed set of connected components of Z.

Brown and Golasiński have proved that Crs admits a closed model structure taking the following family of fibrations, weak equivalences and cofibrations, see [B-G].

Definition 1.3. A morphism $p\colon E \longrightarrow B$ of crossed complexes will be called a fibration if
(i) $p_1\colon E_1 \longrightarrow B_1$ and $p_0\colon E_0 \longrightarrow B_0$ verify the following property: suppose that $x \in E_0$ and $b \in B_1$ with $p_0 x = \delta^0 b$, then there is $e \in E_1$ such that $\delta^0 e = x$ and $p_1 e = b$.
(ii) For each $n \geq 2$ and $x \in E_0$, the morphism of groups $E_n(x) \longrightarrow B_n(p_0 x)$ is surjective.

Definition 1.4. A morphism $f\colon X \longrightarrow Y$ of crossed complexes is said to be a weak equivalence if for each $n \geq 0$ and $x \in X_0$, $\Pi_n(X, x) \longrightarrow \Pi_n(Y, fx)$ is an isomorphism.

Definition 1.5. A morphism $i\colon A \longrightarrow X$ of crossed complexes is said to be a cofibration if it has the left lifting property with respect to the class of trivial fibrations; that is, morphisms which are fibrations and weak equivalences.

Brown and Golasiński have proved that:

Theorem 1.6. The category Crs of crossed complexes, together with the distinguished classes of weak equivalences, fibrations and cofibrations defined above is a closed model category.

We will denote by Ho(Crs) the category obtained by inverting the weak equivalences in Crs.

2. Pointed and reduced crossed complexes.

To prove that a category admits a closed model structure it suffices to prove the axioms M0, M1, M2, M5 of Quillen, see [Q], and to prove that the classes of cofibrations, fibrations and weak equivalences are closed by retracts in the category of maps of the category. Our initial aim will be to show that replacing weak equivalences in the above by a suitable notion of n-equivalence still leaves one with a closed model structure.

First we will note there is a closed model structure on Crs* induced by the model structure of Crs. Between the categories Crs and Crs* we have the forgetful functor U: Crs* $\longrightarrow$ Crs and the functor $(\ )^+$: Crs $\longrightarrow$ Crs*, where for a crossed complex X we let $X^+ = X \sqcup *$, where $*$ is the crossed complex associated with a one-point space. It is clear that

$$\mathrm{Crs}_*(X^+, Y) \cong \mathrm{Crs}(X, UY)$$

and we also have that

$$\mathrm{CRS}_*(X^+, Y) \cong \mathrm{CRS}(X, UY)$$

In Crs* we can take the following as induced structure:

Definition 2.1. A morphism f of pointed crossed complexes is said to be a cofibration, a fibration or a weak equivalence if Uf is a cofibration, a fibration or a weak equivalence in Crs, respectively.

Now it is easy to check the conditions given at the beginning of this section to obtain:

Theorem 2.2. Taking the definitions above Crs* admits a closed model structure.

An interesting subcategory of both Crs* and Crs is the category of reduced crossed complexes RedCrs* that were introduced by J.H.C. Whitehead [W.1, W.2] in 1949.

Definition 2.3. A (pointed) crossed complex B is said to be reduced if $B_0=\{*\}$.

There is a functor

Re: Crs* $\longrightarrow$ RedCrs*

which associates to a pointed crossed complex C the reduced crossed complex ReC defined by $(\mathrm{Re}C)_0=\{*\}$, $(\mathrm{Re}C)_n=C_n(*)$ for $n\geq 1$. If $f\colon B\longrightarrow C$ is a morphism of pointed crossed complexes, then Ref: ReB $\longrightarrow$ ReC is defined by Ref(b) = f(b).

We will denote by In: RedCrs* $\longrightarrow$ Crs* the inclusion functor, then it is easy to check:

Proposition 2.4. The functor In: RedCrs*$\longrightarrow$Crs* is left adjoint to Re: Crs* $\longrightarrow$ RedCrs*.

The map Re: Crs*(A, B)$\longrightarrow$RedCrs*(A, B) has an extension Re: CRS*(A,B)$\longrightarrow$CRS*(ReA, ReB) defined by Reh(b) = h(b) where $h\colon B\longrightarrow C$ is an n-fold left homotopy. If A is a reduced crossed complex we have that

CRS*(A, B) $\cong$ CRS*(A, ReB)

Any object B of Crs* induces the functor $-\otimes_*$ B: RedCrs* $\longrightarrow$ RedCrs*. If A is reduced, then $(A\otimes_* B)_0=\{*\}$ and if A and B are reduced then, $(A\otimes_* B)_1=\{1_*\}$.

Definition 2.5. For B, C objects in Crs*, we can define REDCRS*(B, C) = Re(CRS*(B, C)).

Proposition 2.6. If A is a reduced crossed complex, there is a natural isomorphism

REDCRS*(A, REDCRS*(B, C) $\cong$ REDCRS* $(A\otimes_* B, C)$

Proof.

REDCRS*(A, REDCRS*(B, C)) $\cong$ Re(CRS*(A, Re(CRS*(B, C)))) $\cong$
$\cong$ Re(CRS*(A, CRS*(B, C))) $\cong$ Re (CRS*$(A\otimes_* B$, C)) $\cong$ REDCRS*$(A\otimes_* B$, C).

Proposition 2.7. Any morphism f: $X \longrightarrow Y$ in RedCrs* can be factorized as $f = pi$ where i and p are morphisms in RedCrs*, i is a cofibration and p is a trivial fibration.

Proof. Firstly, notice that a morphism p: $E \longrightarrow B$ in RedCrs* is a trivial fibration in Crs* if p has the right lifting property with respect the set of morphisms $\{S(n-1) \longrightarrow C(n) | n \geq 1\}$, see proposition 2.2 of [B-G]. We have excluded the case $\emptyset \longrightarrow C(0)$ since E and B are reduced.

Let f: $X \longrightarrow Y$ be a morphism in RedCrs*. Repeating the Quillen's construction [Q] or that of Brown-Golasiński [B-G] with respect the family $\{S(n-1) \longrightarrow C(n) | n \geq 1\}$ we obtain the factorization

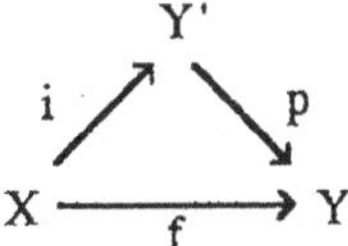

where i is of relative free type (hence is a cofibration), Y' is reduced and by the Quillen's small objects argument, see [Q], and p has the right lifting property with respect to $\{S(n-1) \longrightarrow C(n) | n \geq 1\}$. Therefore p is a trivial fibration.

Corollary 2.8. The category Ho(RedCrs*) obtained by inverting the class of weak equivalences in RedCrs* is isomorphic to the full subcategory Ho(Crs*)|RedCrs* determined by the reduced crossed complexes.

Proof. Since all objects in Crs* are fibrant, Ho(Crs*) admits calculus of right fractions, so any morphism in Ho(Crs*) from X to Y can be expressed by a diagram

$$X \xleftarrow{q} X' \xrightarrow{f} Y$$

where q is a fixed trivial fibration and X' is cofibrant. By the proposition above if X is reduced we can choose q: $X' \longrightarrow X$ with X' reduced too. Then q: $X' \longrightarrow X$ is a weak equivalence of reduced crossed complexes.

Remarks. (1) From the definition of fibration in Crs*, it is easy to check that if p: $E \longrightarrow B$ is a fibration of reduced crossed complexes then $\Pi_1(E, *) \longrightarrow \Pi_1(B, *)$ is a surjection.

(2) The map: $f: * \longrightarrow S(1)$ cannot be factorized as $f = pi$ where i is a weak equivalence and p a fibration in the subcategory RedCrs*. This follows from remark 1 and the fact that $\Pi_1(*, *) \longrightarrow \Pi_1(S(1), *)$ is not surjective.

Notice that the functor Re: Crs* $\longrightarrow$ RedCrs* and In: RedCrs* $\longrightarrow$ Crs* preserve weak equivalences and induce functors Re: Ho(Crs*) $\longrightarrow$ Ho(RedCrs*), In: Ho(RedCrs*) $\longrightarrow$ Ho(Crs*). Let us denote by Ho(Crs*)|(0-connected) the full subcategory of Ho(Crs*) determined by 0-connected crossed complexes.

Proposition 2.9.

(i) In: Ho(RedCrs*) $\longrightarrow$ Ho(Crs*) is left adjoint to Re: Ho(Crs*) $\longrightarrow$ Ho(RedCrs*).

(ii) The functors In: Ho(RedCrs*) $\longrightarrow$ Ho(Crs*)|(0-connected),
Re: Ho(Crs*)|(0-connected) $\longrightarrow$ Ho(RedCrs*) define an equivalence of categories.

Proof. (i) Let X be an object in RedCrs* and Y and object in Crs*, then

$$\mathrm{Ho(Crs*)}\,(\mathrm{In}\,X,\ Y) = \mathrm{Ho(Crs*)}\,(\mathrm{In}\,X',\ Y) = \Pi(\mathrm{Crs*})\,(\mathrm{In}\,X',\ Y),$$

where X' is a reduced cofibrant crossed complex weakly equivalent to X. We also have that

$$\Pi(\mathrm{Crs*})(\mathrm{In}\,X',\ Y) = \Pi_0(\mathrm{CRS*}(\mathrm{In}\,X',\ Y)) = \Pi_0(\mathrm{CRS*}(X',\ \mathrm{Re}\,Y)) = \Pi(\mathrm{Crs*})(X',\ \mathrm{Re}Y)$$
$$= \mathrm{Ho(Crs*)}\,(X',\ \mathrm{Re}\,Y) = \mathrm{Ho(Crs*)}\,(X,\ \mathrm{Re}\,Y) = \mathrm{Ho(RedCrs*)}\,(X,\ \mathrm{Re}\,Y),$$

where the last isomorphism is obtained by applying corollary 2.8.

To prove (ii) it suffices to check that the unit and counit of the adjunction are weak equivalences.

3. Closed model structures for n-types of crossed complexes.

In this section, we will give a sequence of closed model structures on the categories Crs and Crs*, corresponding to various weakened forms of weak equivalence.

Definition 3.1. A morphism $f: X \longrightarrow Y$ of crossed complexes is said to be a weak n-equivalence if $\Pi_k(X, x) \longrightarrow \Pi_k(Y, f_0 x)$ is a isomorphism for each $x \in X_0$ and $0 \leq k \leq n$.

Definition 3.2. A map $p: E \longrightarrow B$ is said to be an n-fibration if it has the right lifting property with respect to $* \longrightarrow S(n+1)$ and $C(k) \longrightarrow J \otimes C(k)$ for every integer k satisfying

$0 \leq k \leq n+1$. A map p is said to be a n-trivial fibration if p is both n-fibration and a weak n-equivalence.

Definition 3.3. A morphism $i: A \longrightarrow X$ of crossed complexes is said to be a n-cofibration if i has the left lifting property with respect to the class of n-trivial fibrations.

Proposition 3.4. Let $p: E \longrightarrow B$ be a morphism of crossed complexes, then the following statements are equivalent

(i) p is an n-trivial fibration.

(ii) p has the right lifting property with respect to $C(n+1) \longrightarrow J \otimes C(n+1)$ and $S(k-1) \longrightarrow C(k)$ for $0 \leq k \leq n+1$.

Proof. Let us recall that Crs$_*$ has a closed model structure, see [B-G]. Therefore for each morphism $p: E \longrightarrow B$ we can consider the exact sequence associated with p

$$\cdots \longrightarrow \Pi_{k+1}(E \longrightarrow B) \longrightarrow \Pi_k(E) \longrightarrow \Pi_k(B) \longrightarrow \cdots \longrightarrow \Pi_0(E) \longrightarrow \Pi_0(B).$$

To prove that (i) $\longrightarrow$ (ii), we have that p is a weak n-equivalence, then $\Pi_k(E \longrightarrow B)$ is trivial for $1 \leq k \leq n$. Since $\tilde{p}$ has the RLP with respect to $* \longrightarrow S(n+1)$ we also have that $\Pi_{n+1}(E \longrightarrow B) = 0$. Now we use the fact that p has the RLP property with respect to $C(k) \longrightarrow J \otimes C(k)$ for $1 \leq k \leq n+1$ to obtain that p has the RLP with respect to $S(k-1) \longrightarrow C(k)$ for $1 \leq k \leq n+1$. Since $\Pi_0(E) \longrightarrow \Pi_0(B)$ is a bijection and p has the RLP with respect to $C(0) \longrightarrow J \otimes C(0)$, we also have that p has the RLP with respect to $S(k-1) \longrightarrow C(k)$ for $k=0$.

Conversely, because $C(k) \longrightarrow J \otimes C(k)$ is a morphism of free relative type, see [B-G], and p has the RLP with respect to $S(k-1) \longrightarrow C(k)$ for $0 \leq k \leq n+1$, then p has the RLP with respect to $C(k) \longrightarrow J \otimes C(k)$, $0 \leq k \leq n$. By hypothesis the same ocurs for $k = n+1$. It is clear that p is a weak n-equivalence, so we only need to prove that p has the RLP with respect to $* \longrightarrow S(n+1)$. Taking into account that $\Pi_{n+1}(E) \longrightarrow \Pi_{n+1}(B)$ is surjective and that p has the RLP with respect to $C(k) \longrightarrow J \otimes C(k)$ for $0 \leq k \leq n+1$, we can conclude that p has the RLP with respect to $* \longrightarrow S(n+1)$.

Proposition 3.5. Any morphism $f: X \longrightarrow Y$ of crossed complexes can be factorised as $f = pi$ where i is an n-cofibration and p is an n-trivial fibration.

Proof. We repeat Quillen's argument of [Q; 3.3] giving a diagram

$$\begin{array}{ccccc} X & \xrightarrow{i_0} & Z^0 & \xrightarrow{i_1} & Z^1 \longrightarrow \cdots \\ & {\scriptstyle f}\searrow & \downarrow{\scriptstyle p_0} & \swarrow{\scriptstyle p_1} & \\ & & Y & & \end{array}$$

where $Z^{-1}=X$ and $p_{-1}=f$ and having obtained Z^{n-1}, taking all diagrams of the form

$$\begin{array}{ccc} S(k-1) & \xrightarrow{u_\lambda} & Z^{n-1} \\ \downarrow & & \downarrow{\scriptstyle p_{n-1}} \\ C(k) & \xrightarrow[v_\lambda]{} & Y \end{array} \qquad \begin{array}{ccc} C(n+1) & \xrightarrow{u_\mu} & Z^{n-1} \\ \downarrow & & \downarrow \\ J\otimes C(n+1) & \xrightarrow{v_\mu} & Y \end{array}$$

for $0 \leq k \leq n+1$, i_n: $Z^{n-1} \longrightarrow Z^n$ is defined by the co-cartesian diagram

$$\begin{array}{ccc} (\coprod_\lambda S(k-1)) \sqcup (\coprod_\mu C(n+1)) & \longrightarrow & Z^{n-1} \\ \downarrow & & \downarrow{\scriptstyle i_n} \\ (\coprod_\lambda S(k)) \sqcup (\coprod_\mu J\otimes C(n+1)) & \longrightarrow & Z^n \end{array}$$

Define $Z = \operatorname{colim} Z^n$ and $p = \operatorname{colim} p_n$. Using the small object argument, see [Q], we prove the p has the RLP with respect $C(n+1) \longrightarrow J\otimes C(n+1)$ and $S(k-1) \longrightarrow C(k)$ for $0 \leq k \leq n+1$. Applying the previous proposition we have that p is an n-trivial fibration. To prove that i is an n-cofibration we again use that proposition.

Proposition 3.6. Any morphism $f: X \longrightarrow Y$ of crossed complexes can be factorised as $f=qj$ where j is a weak n-equivalence which has the LLP with respect to n-fibrations (therefore j is also an n-cofibration) and q is an n-fibration.

Proof. We use the same type of construction as in the above proposition but in this case taking diagrams of the form

$$\begin{array}{ccc} * & \xrightarrow{u_\lambda} & Z^{n-1} \\ \downarrow & & \downarrow{\scriptstyle q_{n-1}} \\ S(n+1) & \xrightarrow[v_\lambda]{} & Y \end{array} \qquad \begin{array}{ccc} C(k) & \xrightarrow{u_\mu} & Z^{n-1} \\ \downarrow & & \downarrow{\scriptstyle q_{n-1}} \\ J\otimes C(k) & \xrightarrow{v_\mu} & Y \end{array}$$

for $0 \leq k \leq n+1$, and defining j_n: $Z^{n-1} \longrightarrow Z^n$ by the co-cartesian diagram

$$\begin{array}{ccc} (\coprod_{\lambda} *) \sqcup (\coprod_{\mu} C(k)) & \longrightarrow & Z^{n-1} \\ \downarrow & & \downarrow j_n \\ (\coprod_{\lambda} S(n+1)) \sqcup (\coprod_{\mu} J \otimes C(k)) & \longrightarrow & Z^n \end{array}$$

It is not hard to see that the corresponding j: $X \longrightarrow Z$ is weak n-equivalence ($\pi_n(S^{n+1}) = 0$) and that j has the LLP with respect to n-fibrations. Again by the small object argument it follows that $q = \operatorname{colim} q_n$ has the RLP with respect to $* \longrightarrow S(n+1)$ and $C(k) \longrightarrow J \otimes C(k)$, $0 \leq k \leq n+1$. Therefore q is an n-fibration.

Proposition 3.7. The classes of weak n-equivalences, n-fibrations and n-cofibrations are closed by retracts in the category of morphisms of crossed complexes.

Proof. This is routine.

Proposition 3.8. Let i: $A \longrightarrow X$ be an n-trivial cofibration; that is, a weak n-equivalence and an n-cofibration. Then i has the LLP with respect to n-fibrations.

Proof. By proposition 3.6, we can factorise i=qj where j is a weak n-equivalence which has the LLP with respect n-fibrations and q is an n-fibration since i is a weak n-equivalence we have that q is also a weak n-equivalence. Now in the diagram

$$\begin{array}{ccc} A & \xrightarrow{j} & Z \\ i \downarrow & & \downarrow q \\ X & \xrightarrow[id]{} & X \end{array}$$

we have that i is an n-cofibration and q an n-trivial fibration. By definition of n-cofibration there is a map $s: X \longrightarrow Z$ such that $qs = id_X$ and $si = j$. Therefore i is a rectract of j in the category of morphisms of crossed complexes. Since j has the LLP with respect to n-fibrations and i is a retract of j, it follows that i has the LLP with respect to n-fibrations.

To prove that a category admits a closed model structure if suffices to prove that the classes of cofibrations, fibrations and weak equivalences is closed by retracts in the category of morphisms of the category and to prove the axioms M0, M1, M2 and M5 of [Q]. Using the propositions of this section we thus have:

Theorem 3.9. The category Crs of crossed complexes together with the distinguished classes of weak n-equivalences, n-fibrations and n-cofibrations defined above is a closed model category.

Using the forgetful functor U: Crs* $\longrightarrow$ Crs, we can say that a morphism f of pointed crossed complexes is a weak n-equivalence, an n-fibration or an n-cofibration if Uf is the same in Crs. With these definitions we also have.

Corollary 3.10. For each integer $n \geq 0$, Crs* admits a closed model category for the classes of weak n-equivalences, etc as defined above.

The categories obtained from Crs or Crs* by inverting weak n-equivalences will be denoted by Ho_n(Crs), Ho_n(Crs*). It is interesting to note that in both cases all the objects are n-fibrant and if X is a crossed complex of free type such that dim $X \leq n+1$, then X is n-cofibrant. We shall show later that a crossed complex n-type (i.e. isomorphism class in Ho_n(Crs)) can be represented (not uniquely) by a crossed complex with 0 in dimension >n.

For reduced crossed complexes as in the section above we have that

$Ho_n(\mathrm{RedCrs*}) \cong Ho_n(\mathrm{Crs*}) \mid \mathrm{RedCrs*}$.

4. Categories of n-types for pro-crossed complexes.

Given a category C, we can consider the category proC whose objects are functors $X: \Lambda \longrightarrow C$ where Λ is a small left filtering category. This category was first defined by Grothendieck in [Gr] and detailed descriptions can be found in [A-M] or [E-H].

Let us recall that for each small left filtering category Λ, there is a functor $F: C^{\Lambda} \longrightarrow \mathrm{proC}$, where C^{Λ} is the category of covariant functors $\Lambda \longrightarrow C$ and natural transformations. A morphism of the form $FX \xrightarrow{Ff} FY$ is said to be a level morphism of proC. For a given closed model structure on a category C satisfying the condition N [E-H, p. 45], Edwards and Hastings [E-H, p. 72] have proved that proC admits a closed model structure if one takes the following definitions:

Definition 4.1. A map $f: X \longrightarrow Y$ in C^{Λ} is a cofibration in C^{Λ} if for all $\lambda \in \Lambda$ the maps f_{λ} are cofibrations in C. A map f is a weak equivalence in C^{Λ} if each f_{λ} is a weak equivalence in C. A map f is a fibration if it has the RLP with respect all the maps i which are both cofibrations and weak equivalences.

Definition 4.2. A map $f: X \longrightarrow Y$ in proC is a strong cofibration if f is the image of a level cofibration (in some C^{Λ}). Similar definitions will be assumed to have been made for strong fibration, strong trivial cofibration and strong trivial fibration. A map in proC is a cofibration if it is a retract in Maps(proC) of a strong cofibration. Fibrations, trivial cofibrations and trivial fibrations are defined similarly. The image of a level weak equivalence is said to be a strong equivalence. A map f in proC is a weak equivalence if $f = pi$ where p is a trivial fibration and i is a trivial cofibration.

In section 3, we have given for each integer $n \geq 0$ a closed model structure to the category of crossed complexes Crs (similarly Crs∗). We extend this family of structures taking for $n = +\infty$ the closed model structure given by Brown and Golasiński in [B-G]. Now using the results of Edwards and Hastings, for each $n \geq 0$, a new closed model structure is obtained in the categories proCrs, proCrs∗, (proCrs, Crs), (proCrs∗, Crs∗), towCrs, ... The corresponding categories obtained by inverting weak n-equivalences will be denoted by Ho_n(proCrs), Ho_n(proCrs∗), etc. For the case $n = +\infty$ we also use the notation $Ho_{\infty} = Ho$. The category Ho_n(proCrs) is called the category of n-types of pro-crossed complexes and Ho_n(proCrs∗) is the category of n-types of pro-crossed pointed complexes.

We will also use the category of towers of C, towC, which is the full subcategory of proC determined by proobjects indexed by the natural numbers. As above towC has a closed model structure induced by that of C.

The following elementary result will be applied to induce functors between homotopy categories, for a closed model category C, we consider in proC the closed model structure given by Edwards and Hastings.

Lema 4.3. If $F: \text{proC} \longrightarrow D$ is a functor which sends strong equivalences into isomorphisms, then F sends weak equivalences into isomorphisms.

Proof. A map f in proC is a weak equivalence if $f = pi$ where p is a trivial fibration and i is a trivial cofibration. Then p is a retract of some strong trivial fibration q (similarly for i). It suffices to see that F sends strong trivial fibrations and strong trivial cofibrations into isomorphisms, but both classes of maps are contained in the class of strong equivalences. Taking into account that F preserve retracts and commutes with composition it follows that Ff is also an isomorphism.

5. The skeleton, coskeleton and truncation functors.

Let us consider the (n+1)-skeleton functor $sk_{n+1}\colon Crs \longrightarrow Crs$ which associates to a crossed complex X the crossed complex $sk_{n+1} X$ defined by

$$(sk_{n+1} X)_q = \begin{cases} X_q & q \le n+1, \\ 0 & q > n+1, \end{cases}$$

and for a morphism f of crossed complexes the morphism $sk_{n+1}f$ is defined by

$$(sk_{n+1}f)_q = \begin{cases} f_q & q \le n+1, \\ 0 & q > n+1. \end{cases}$$

Associated with the functor sk_{n+1} we have the natural transformation $sk_{n+1}X \longrightarrow X$ which is a weak n-equivalence. If $Crs|(\dim \le n+1)$ denotes the full subcategory of Crs determined by crossed complexes of dimension $\le n+1$, we also have the functors

$$sk_{n+1}\colon Crs \longrightarrow Crs|(\dim \le n+1)$$

$$In_{n+1}\colon Crs|(\dim \le n+1) \longrightarrow Crs$$

that preserve weak n-equivalences and induce an equivalence of categories

$$Ho_n(Crs) \underset{In_{n+1}}{\overset{sk_{n+1}}{\rightleftarrows}} Ho_n(Crs|(\dim \le n+1))$$

Remark. It is easy to check that $Ho_n(Crs|(\dim \le n+1))$ and $Ho_n(Crs)|(\dim \le n+1)$ are isomorphic.

The above functors extend to new functors

$$sk_{n+1}\colon proCrs \longrightarrow proCrs|(\dim \le n+1)$$

$$In_{n+1}\colon proCrs|(\dim \le n+1) \longrightarrow proCrs$$

It is clear that sk_{n+1}, In_{n+1} preserve strong n-equivalences, so that we also have an equivalence of categories

$$Ho_n(\text{proCrs}) \underset{In_{n+1}}{\overset{sk_{n+1}}{\rightleftarrows}} Ho_n\ (\text{proCrs} \mid (\dim \leq n+1))$$

Consider next the (n+1)-coskeleton functor $cosk_{n+1}$: Crs $\longrightarrow$ Crs defined by

$$(cosk_{n+1} X)_q = \begin{cases} X_q & q \leq n+1, \\ \text{Ker } \delta_{n+1} & q = n+2, \\ 0 & q > n+2. \end{cases}$$

$$(cosk_{n+1} f)_q = \begin{cases} f_q & q \leq n+1, \\ f_{n+1} \mid \text{Ker } \delta_{n+1} & q = n+2, \\ 0 & q > n+2. \end{cases}$$

With these definitions we have that:

Proposition 5.1.

(i) sk_{n+1}: Crs $\longrightarrow$ Crs is left adjoint to $cosk_{n+1}$: Crs $\longrightarrow$ Crs.
(ii) sk_{n+1}: proCrs $\longrightarrow$ proCrs is left adjoint to $cosk_{n+1}$: proCrs $\longrightarrow$ proCrs.

It is interesting to note that classes of n-trivial fibrations satisfy

(∞-trivial fibrations) $\subset \ldots \subset$ ((n+1)-trivial fibrations) $\subset$ (n-trivial fibrations) $\subset \ldots$,

therefore the classes of n-cofibrations verify:

$\ldots \subset$ (n-cofibrations) $\subset$ ((n+1)-cofibrations) $\subset \ldots \subset$ (∞-cofibrations).

We can use these relations to prove:

Proposition 5.2.

(i) sk_{n+1}: Crs $\longrightarrow$ Crs sends cofibrations into n-cofibrations, weak equivalences into weak n-equivalences and preserves n-cofibrations and weak n-equivalences.
(ii) The same properties are satisfied by sk_{n+1}: proCrs $\longrightarrow$ proCrs.
(iii) $cosk_{n+1}$: Crs $\longrightarrow$ Crs sends n-fibrations into fibrations, weak n-equivalences into weak equivalences and preserves n-fibrations and weak n-equivalences.
(iv) Similar properties are satisfied by $cosk_{n+1}$: proCrs $\longrightarrow$ proCrs.

Proof. Let i: $A \longrightarrow X$ be a cofibration in Crs. By Proposition 2.7 of [B-G], i is a retract of a cofibration j: $A \longrightarrow Z$ such that $A \longrightarrow Z$ is of relative free type. It is clear that $sk_{n+1} A \longrightarrow sk_{n+1} Z$ is also of relative free type, because dim $sk_{n+1} Z \leq n+1$, so that $sk_{n+1} A \longrightarrow sk_{n+1} Z$ is an n-cofibration. Since i is a retract of j, then sk_{n+1} i is a retract of sk_{n+1} j. Because sk_{n+1} j is an n-cofibration, we have that sk_{n+1} i is also an n-cofibration. Taking into account that an n-cofibration is also a cofibration we have that sk_{n+1} preserves n-cofibrations. From the definition of $\Pi_k(X, *)$ it follows easily that sk_{n+1} preserves weak n-equivalences.

To prove (ii), take a cofibration i in proCrs. By definition, i is a retract of a strong (level) cofibration j. Applying (i), we have that sk_{n+1} j is a strong (level) n-cofibration and sk_{n+1} i is a retract of sk_{n+1} j, then sk_{n+1} i is an n-cofibration of proCrs. Similarly, it is proved that sk_{n+1} preserves n-cofibrations. To see that sk_{n+1} preserves weak n-equivalences, by Lemma 4.3, it suffices to check that sk_{n+1} preserves strong n-equivalences and this follows from (i). Similarly for the case of weak equivalences.

To prove (iii) and (iv) we can use the properties of cofibrations, fibrations and weak equivalences in a closed model category (axiom M6 [Q]), and the fact that sk_{n+1} is left adjoint to $cosk_{n+1}$.

Now, recall theorem 3 of chI.§4 of [Q]. Suppose that C and C' are model categories and F: $C \longrightarrow C'$ is left adjoint to G: $C' \longrightarrow C$. If F preserves cofibrations and weak equivalences and G preserves fibrations and weak equivalences, then the induced functor F: Ho $C \longrightarrow$ Ho C' is left adjoint to the induced functor G: Ho $C' \longrightarrow$ Ho C. By using this theorem and the proposition above we have:

Corollary 5.3.

(i) sk_{n+1}: Ho(Crs) $\longrightarrow$ Ho_n(Crs) is left adjoint to $cosk_{n+1}$: Ho_n(Crs) $\longrightarrow$ Ho(Crs).

(ii) sk_{n+1}: Ho_n(Crs) $\longrightarrow$ Ho_n(Crs) is left adjoint to $cosk_{n+1}$: Ho_n(Crs) $\longrightarrow$ Ho_n(Crs).

(iii) sk_{n+1}: Ho(proCrs) $\longrightarrow$ Ho_n(proCrs) is left adjoint to
$cosk_{n+1}$: Ho_n(proCrs) $\longrightarrow$ Ho(proCrs).

(iv) sk_{n+1}: Ho_n(proCrs) $\longrightarrow$ Ho_n(proCrs) is left adjoint to
$cosk_{n+1}$: Ho_n(proCrs) $\longrightarrow$ Ho_n(proCrs).

A crossed complex X is said to (n + 1)-coconnected if $\Pi_k(X, x) = 0$ for each $x \in X_0$ and $k \geq n+1$. Let us denote by Crs|((n + 1)-cc), Ho(Crs)|((n + 1)-cc) the full categories of Crs, Ho(Crs), respectively, determined by (n + 1)-coconnected crossed complexes. Using the functors of the corollary above we have the following results:

Corollary 5.4.

(i) The functors sk_{n+1}: Ho(Crs)|((n + 1)-cc) $\longrightarrow$ Ho_n(Crs) $cosk_{n+1}$: Ho_n(Crs) $\longrightarrow$ Ho(Crs)|((n + 1)-cc) give an equivalence of categories.

(ii) The functors sk_{n+1}: Ho(proCrs)|(pro((n + 1)-cc)) $\longrightarrow$ Ho_n(proCrs) and $cosk_{n+1}$: Ho_n(proCrs) $\longrightarrow$ Ho(proCrs)|(pro((n + 1)-cc)) give an equivalence of categories.

Remarks. (1) The importance of this corollary is that each crossed complex n-type is determined by a crossed complex with zeros in dimensions greater than n+2 (see also below 5.5)

(2) We can also define notions of n-fibrations, weak n-equivalences and n-cofibrations in the categories SS, Top, proSS and proTop. The properties of the skeleton and coskeleton functors are well known in these categories, see [G-Z, E-H]. As in the case of crossed and pro-crossed complexes we also have that sk_{n+1}: Ho($\mathfrak{C}$) $\longrightarrow$ $Ho_n(\mathfrak{C})$ is left adjoint to $cosk_{n+1}$: $Ho_n(\mathfrak{C})$ $\longrightarrow$ Ho($\mathfrak{C}$) when $\mathfrak{C}$ is one of the categories SS, Top, proSS and proTop. These functors induce equivalences of categories between the following pair of categories: Ho(proSS)|(pro((n + 1)-cc)), Ho_n(proSS); Ho(proTop)|(pro((n + 1)-cc)), Ho_n(proTop); Ho(SS)|((n + 1)-cc), Ho_n(SS) and Ho(Top)|(n + 1)-cc), Ho_n(Top).

If a crossed complex X satisfies $\dim X \leq n+2$, $\Pi_{n+1}(X, x) = 0$ for all $x \in X_0$ and $\delta_{n+2}: X_{n+2} \longrightarrow X_{n+1}$ is a monomorphism, then X is (n + 1)-coconnected and it is said to be a canonical (n + 1)-coconnected crossed complex ((n + 1)-ccc). Because $cosk_{n+1}X$ is an (n + 1)-ccc crossed complex we also have that:

Corollary 5.5. By considering the functors of corollary above Ho(Crs)|((n + 1)-ccc) is equivalent to Ho_n(Crs) and Ho(proCrs)|(pro((n+1)-ccc)) is equivalent to Ho_n(proCrs).

We also will need to study the n-truncation functor tr_n: Crs $\longrightarrow$ Crs that can be defined by

$$(\mathrm{tr}_n X)_q = \begin{cases} X_q & q<n \\ X_n/\mathrm{Im}\,\delta_{n+1} & q=n \\ 0 & q>n \end{cases}$$

and on maps

$$(\mathrm{tr}_n f)_q = \begin{cases} f_q & q<n \\ \bar{f}_n & q=n \\ 0 & q>n \end{cases}$$

where $\bar{f}_n$ is the map induced by f_n.

Notice that we have a natural transformation $P: X \longrightarrow \mathrm{tr}_n X$ where $P_q = \mathrm{id}$ for $q<n$, $P_n: X_n \longrightarrow X_n/\mathrm{Im}\,\delta_{n+1}$ is the quotient map and P_q is trivial for $q>n$. It is clear that P is a weak n-equivalence.

Proposition 5.6.

(i) The functor tr_n: Crs $\longrightarrow$ Crs is left adjoint to sk_n: Crs $\longrightarrow$ Crs.
(ii) The functor tr_n: proCrs $\longrightarrow$ Crs if left adjoint to sk_n: proCrs $\longrightarrow$ proCrs.
(iii) tr_n: Crs $\longrightarrow$ Crs$|(\dim \leq n)$ is left adjoint to the inclusion
In $= \mathrm{sk}_n$: Crs$|(\dim \leq n) \longrightarrow$ Crs.
(iv) tr_n: proCrs $\longrightarrow$ pro(Crs$|(\dim \leq n)$) is left adjoint to the inclusion
In $= \mathrm{sk}_n$: pro(Crs$|(\dim \leq n)$) $\longrightarrow$ proCrs.

If f is a weak n-equivalence of crossed complexes, then $\mathrm{tr}_n f$ is a weak equivalence and if $f: X \longrightarrow Y$ is a weak equivalence, $\dim X \leq n$ and $\dim Y \leq n$, then $\mathrm{tr}_n g = g = \mathrm{In}\, g$ is a weak n-equivalence. Let us denote by Ho(proCrs$|(\dim \leq n)$) and Ho(Crs$|(\dim \leq n)$) the categories obtained by inverting the weak equivalences of the categories Crs$|(\dim \leq n)$ and proCrs$|(\dim \leq n)$, respectively. As usual, Ho(Crs)$|(\dim \leq n)$, Ho(proCrs)$|(\dim \leq n)$ will denote full subcategories of Ho(Crs) and Ho(proCrs), respectively. With this notation we have:

Proposition 5.7. The following pairs of functors define equivalence of categories

(i) tr_n: $Ho_n(Crs) \longrightarrow Ho(Crs|(\dim \leq n))$,
$In = sk_n$: $Ho(Crs|(\dim \leq n)) \longrightarrow Ho_n(Crs)$.

(ii) tr_n: $Ho_n(Crs) \longrightarrow Ho(Crs)|(\dim \leq n)$,
sk_{n+1}: $Ho(Crs)|(\dim \leq n) \longrightarrow Ho_n(Crs)$.

(iii) tr_n: $Ho_n(proCrs) \longrightarrow Ho(proCrs|(\dim \leq n))$,
$In=sk_n$: $Ho(proCrs|(\dim \leq n)) \longrightarrow Ho_n(proCrs)$.

(iv) tr_n: $Ho_n(proCrs) \longrightarrow Ho(proCrs)|(\dim \leq n)$,
sk_{n+1}: $Ho(proCrs)|(\dim \leq n) \longrightarrow Ho_n(proCrs)$.

6. The functors B and πR for crossed and pro-crossed complexes.

Let SS denote the category of simplicial sets and consider the realisation functor R: SS ⟶ Top and the singular functor S: Top ⟶ SS that give the standard equivalence between the categories Ho(SS), Ho(Top). The weak equivalences of Top are defined by the maps that induce isomorphisms on the homotopy groups. The closed model structure of Top given by weak equivalences and Serre fibrations extends to a closed model structure on proTop by using the Edwards-Hastings [E-H] method. The closed model structure of SS has as weak equivalences the simplicial maps that induce isomorphism on the homotopy groups and a map f in SS is a fibration if f has the RLP with respect to $V(n,k) \longrightarrow \Delta[n]$, where $V(n,k)$ denote the simplicial subset of $\Delta[n]$ which is the union of the images of the i-faces, $0 \leq i \leq n$, $i \neq k$. It is easy to check that R: proSS ⟶ proTop is left adjoint to S: proTop ⟶ proSS, that R preserves cofibrations and weak equivalences and S preserves weak equivalences and fibrations. Therefore applying theorem 3 of chI.§4 of [Q], we have that R: Ho(proSS) ⟶ Ho(proTop) and S: Ho(proTop) ⟶ Ho(proSS) define an equivalence of categories.

Denote by $Ho_{Strøm}$(proTop) the category obtained when we consider the Strøm structure of Top and the induced closed model structure on proTop. The closed model structure of Top given by Strøm [S] has as weak equivalences the homotopy equivalences and as fibrations the Hurewicz fibrations; that is, the maps p that have the homotopy lifting property for any space. It is not difficult to see that RS: Ho(proTop) ⟶ $Ho_{Strøm}$(proTop) is left adjoint to the "inclusion" functor $Ho_{Strøm}$(proTop) ⟶ Ho(proTop) and induces an equivalence between the categories Ho(proTop)|proCW and $Ho_{Strøm}$(proTop)|proCW.

Let CCW be the category of CW-complexes and cellular maps and $\Pi(CCW)$ the quotient category obtained by considering cellular homotopies. We can consider the Whitehead functor π: CCW $\longrightarrow$ Crs defined by the crossed complex.

$$\cdots \longrightarrow \pi_n(X^n, X^{n-1}) \longrightarrow \pi_{n-1}(X^{n-1}, X^{n-2}) \longrightarrow \cdots \longrightarrow \pi_1(X^1, X^0) \underset{\delta^1}{\overset{\delta^0}{\rightrightarrows}} \pi_0(X^0)$$

where X is a CW-complex, X^k the k-skeleton, $\pi_1(X^1, X^0)$ the fundamental groupoid on vertex set X^0 and for $n > 1$, $\pi_n(X^n, X^{n-1})$ denotes the family $\{\pi_n(X^n, X^{n-1}, p) | p \in X^0\}$.

A cellular homotopy $F: I \times X \longrightarrow Y$ induces a 1-fold homotopy $J \otimes \pi X \longrightarrow \pi Y$, therefore we have the induced functor π: Π(CCW) $\longrightarrow$ Ho(Crs). We can translate the functor π to SS and Top by considering the functors

$$SS \xrightarrow{R} CCW \xrightarrow{\pi} Crs$$

$$Top \xrightarrow{S} SS \xrightarrow{R} CCW \xrightarrow{\pi} Crs$$

If f is a weak equivalence in SS, then Rf is a homotopy equivalence. Since πRf is a homotopy equivalence we also have that πRf is a weak equivalence in Crs. Therefore we also have the functors πR: Ho(SS) $\longrightarrow$ Ho(Crs) and πRS: Ho(Top) $\longrightarrow$ Ho(Crs).

We next consider a simplicial version of the classification functor B that was first studied by Brown and Higgins [B-H.2]. They define B by using the realisation of the cubical nerve of a crossed complex.

Denote Δ the category whose objects are ordered sets $[n] = \{0, \ldots, n\}, n \geq 0$, and whose morphisms are the ordered-preserving maps between them. Consider the functors

$$\Delta \xrightarrow{\Delta} SS \xrightarrow{R} CCW \xrightarrow{\pi} Crs$$

where $\Delta[n]$ is the simplicial set $\Delta(-, [n])$. For each crossed complex X, we have the contravariant functor Crs(–, X): Crs $\longrightarrow$ Set. Then we can define the functor B: Crs $\longrightarrow$ SS by

$$(BX)_q = Crs(\pi R \Delta[q], X)$$

The above functors extend to the corresponding procategories giving

πR: proSS $\longrightarrow$ proCrs
πRS: proTop $\longrightarrow$ proCrs

$B: proCrs \longrightarrow proSS$
$RB: proCrs \longrightarrow proTop$

Proposition 6.1.

(i) $\pi R: SS \longrightarrow Crs$ is left adjoint to $B: Crs \longrightarrow SS$.
(ii) $\pi R: proSS \longrightarrow proCrs$ is left adjoint to $B: proCrs \longrightarrow proSS$.

Proof. Let X be a simplicial set and G a crossed complex. If $f \in Crs(\pi RX, G)$ then define $f^b \in SS(X, BG)$ by $f^b x = f\, \pi R\bar{x}$ where $x \in X_p$ and $\bar{x}: \Delta[p] \longrightarrow X$ is the unique simplicial map such that $\bar{x}(i_p) = x$ and i_p is the identify of $[p]=\{0,\dots,p\}$. If $g \in SS(X, BG)$ taking into account that πRX is a crossed complex of free type it suffices to define the composed maps, $g^{\#} . \pi R\bar{x} = gx$ (c.f.[B-G]). It is easy to check that $(f^b)^{\#} = f$ and $(g^{\#})^b = g$. By the definition of the hom-set the result extends to the procategories.

Lemma 6.2.

(i) The functors $\pi R: SS \longrightarrow Crs$ and $\pi R: proSS \longrightarrow proCrs$ preserve cofibrations and weak equivalences.
(ii) The functors $B: Crs \longrightarrow SS$ and $B: proCrs \longrightarrow proSS$ preserve fibrations and weak equivalences.

Proof. If f is a weak equivalence in SS, then Rf is a homotopy equivalence in Top. As πRf is a homotopy equivalence, we have that it is a weak equivalence. If $i: A \longrightarrow X$ is an inclusion map in SS, then $\pi Ri: \pi RA \longrightarrow \pi RX$ is a morphism of crossed complexes of relative free type. Therefore πRi is also a cofibration, see [B-G].

The functor $B: Crs \longrightarrow SS$ has the following property if $f: G \longrightarrow G'$ is a map of crossed complexes then there is a commutative diagram

$$\begin{array}{ccc} \Pi_q(G, p) & \longrightarrow & \Pi_q(G', fp) \\ \downarrow & & \downarrow \\ \pi_q(BG, p) & \longrightarrow & \pi_q(BG', fp) \end{array}$$

such that the vertical maps are isomorphisms, see [B-H.2]. Therefore if $f: G \longrightarrow G'$ is a weak equivalence, then $Bf: BG \longrightarrow BG'$ is also a weak equivalence. Since πR is left adjoint to B and πR preserves trivial cofibrations it follows that B preserves fibrations.

By considering the definition of the closed model structures in the procategories it is easy to see that the corresponding properties are also satisfied by the induced profunctors.

As a consequence of this lemma we have:

Proposition 6.3.
(i) πR: Ho(SS) $\longrightarrow$ Ho(Crs) is left adjoint to B: Ho(Crs) $\longrightarrow$ Ho(SS).
(ii) πR: Ho(proSS) $\longrightarrow$ Ho(proCrs) is left adjoint to
B: Ho(proCrs) $\longrightarrow$ Ho(proSS).
(iii) πRS: Ho(Top) $\longrightarrow$ Ho(Crs) is left adjoint to RB: Ho(Crs) $\longrightarrow$ Ho(Top).
(iv) πRS: Ho(proTop) $\longrightarrow$ Ho(proCrs) is left adjoint to
RB: Ho(proCrs) $\longrightarrow$ Ho(proTop).

Proof. These are a consequence of theorem 3 of [Q; chI.§4] and the fact that R: Ho(SS) $\longrightarrow$ Ho(Top) and S: Ho(Top) $\longrightarrow$ Ho(SS) define an equivalence of categories, similarly in the case of procategories.

Remarks.
(1) The functors πR: Ho(SS) $\longrightarrow$ Ho(Crs) and πR: Ho(proSS) $\longrightarrow$ Ho(proCrs) commute with suspension functors and preserve cofibration sequences.
(2) The functors B: Ho(SS) $\longrightarrow$ Ho(Crs) and B: Ho(proCrs) $\longrightarrow$ Ho(proSS) commute with loop functors and preserve fibration sequences.

The first definition of J_n-complex was given by Whitehead [W1, W2] to study the realisability problem for proper maps between homotopy systems. Using this condition Whitehead proved that if K is a reduced CW-complex and L is a reduced J_n-complex, then any map $f:\pi K \longrightarrow \pi L$ can be realised by a continuous map $\phi:K \longrightarrow L$. A CW-complex was said to be a J_n-complex if $\pi_r(X^r) \longrightarrow \pi_r(X^r,X^{r-1})$ is a monomorphism for $1 \leq r \leq n$, and for the case $n=+\infty$ X was said to be a J-complex. A CW-complex L is a J_n-complex if and only if the Hurewicz homomorphism of the universal covering $\tilde{L}$ of L is an isomorphism in dimension $\leq n$ and an epimorphism in dimension n+1. In the next section we will define the notion of

$\mathfrak{J}_n$-complex by removing the condition of being an epimorphism in dimension n+1. Now we define J-complex by using an equivalent condition, see the remark of corollary 6.7.

Definition 6.4. A simplicial set X is said to be a J-simplicial set if the natural transformation $X \longrightarrow B\pi RX$ is a weak equivalence. A topological space Y is said to be a J-space if $Y \longrightarrow RB\pi RSY$ is a weak equivalence or equivalently SY is a J-simplicial set. A pro-simplicial set X is said to be a strong J-pro-simplicial set if the transformation $X \longrightarrow B\pi RX$ is a strong equivalence and X is said to be a J-pro-simplicial set if $X \longrightarrow B\pi RX$ is a weak equivalence in proSS. A prospace Y is said to be a (strong) J-prospace if SY is a (strong) J-pro-simplicial set.

Definition 6.5. A crossed complex G is said to be a J-crossed complex if the natural transformation $\pi RBG \longrightarrow G$ is a weak equivalence. A pro-crossed complex H is said to be a (strong) J-crossed complex if the natural transformation $\pi RBH \longrightarrow H$ is a weak equivalence (a strong equivalence).

Recall the following properties of adjoint functors. Assume that $F:\mathcal{A} \longrightarrow \mathcal{B}$ is left adjoint $G:\mathcal{B} \longrightarrow \mathcal{A}$ and denote by $f \to f^b$, $g^\# \leftarrow g$ the corresponding maps of the adjuntion $\mathcal{B}(FA,B) \cong \mathcal{A}(A,GB)$. The unit and counit of the adjunction are given by $id^b_{FA}:A \longrightarrow GFA$ and $id^\#_{GB}:FGB \longrightarrow B$, respectively. The unit and counit satisfy the relations:

$$G(id^\#_{GB})id^b_{FGB}=id_{GB}$$

$$id^\#_{GFA}G(id^b_{FA})=id_{FA}$$

where A is an object in $\mathcal{A}$ and B is an object in $\mathcal{B}$.

Proposition 6.6.

(i) If X is a J-simplicial set, then πRX is a J-crossed complex.

(ii) If X is a (strong) J-pro-simplicial set, then πRX is a (strong) J-pro-crossed complex.

(iii) G is a J-crossed complex if and only if BG is a J-simplicial set.

(iv) G is a strong J-pro-crossed complex if and only if BG is a strong J-pro-simplicial set.

(v) If G is a J-pro-crossed complex, then BG is a J-pro-simplicial set.

Proof. We have the relations

$$id^\#_{B\pi RX}\, \pi R(id^b_{\pi RX})=id_{\pi RX}$$

$B(id^{\#}_{BG})\, id^{b}_{\pi RBG} = id_{BG}$.

To prove (i), suppose that X is a J-simplicial set, then $id^{b}_{\pi RX}$ is a weak equivalence. By lemma 6.2, $\pi R(id^{b}_{\pi RX})$ is also a weak equivalence. Now taking into account axiom M5 of the closed model structure, we have that $id^{\#}_{B\pi RX}$ is a weak equivalence. Therefore πRX is a J-crossed complex.

We also have to prove that if BG is a J-simplicial set, then G is a J-crossed complex. The rest of the proposition can be proved by using similar arguments. If BG is a J-simplicial set, then $id^{b}_{\pi RBG}$ is a weak equivalenece. Therefore $B(id^{\#}_{BG})$ is also a weak equivalence. From the commutativity of the diagram of the proof of lemma 6.2, it follows that $\Pi_q(id^{\#}_{BG})$ is a isomorphism for $q \geq 0$. Therfore G is a J-crossed complex.

Corollary 6.7. The following pairs of functors define equivalences of categories

(i) πR: Ho(SS)|(J) $\longrightarrow$ Ho(Crs)|(J) and
B: Ho(Crs)|(J) $\longrightarrow$ Ho(SS)|(J).

(ii) πR: Ho(proSS)|(J) $\longrightarrow$ Ho(proCrs)|(J) and
B: Ho(proCrs)|(J) $\longrightarrow$ Ho(proSS)|(J).

(iii) πRS: Ho(Top)|(J) $\longrightarrow$ Ho(Crs)|(J) and
RB: Ho(Crs)|(J) $\longrightarrow$ Ho(Top)|(J).

(iv) πRS: Ho(proTop)|(J) $\longrightarrow$ Ho(proCrs)|(J) and
RB: Ho(proCrs)|(J) $\longrightarrow$ Ho(proTop)|(J).

Remarks.(1) It is well known that the following sequence is exact

$$\cdots \longrightarrow \mathrm{Img}(\pi_{n+1}\, sk_n X \longrightarrow \pi_{n+1} sk_{n+1} X) \longrightarrow \pi_{n+1}(X) \longrightarrow \pi_{n+1}(B\pi RX) \longrightarrow$$
$$\longrightarrow \mathrm{Img}(\pi_n\, sk_{n-1} X \longrightarrow \pi_n\, sk_n X) \longrightarrow \cdots$$

for any simplicial set X, see [C-C, B-H.2]. Therefore X is a J-simplicial set if and only if $\pi_{n+1}(sk_n X, *) \longrightarrow \pi_{n+1}(sk_{n+1} X, *)$ is trivial for any choice of base point and $n \geq 1$.

(2) If X is a CW-complex, then the condition above is equivalent to saying that $\pi_{n+1}(X^n, *) \longrightarrow \pi_{n+1}(X^{n+1}, *)$ is trivial for any choice of base point and $n \geq 1$.

(3) A pro-simplicial set $X = \{X_\lambda | \lambda \in \Lambda\}$ is a strong J-pro-simplicial set if and only if each X_λ is a J-simplicial set.

Examples. (a) Let $(\mathbb{Z},2)$ be the crossed complex that is $\mathbb{Z}$ in dimension 2 and it is trivial in other dimensions. It is clear that $RB(\mathbb{Z},2)$ is an Eilenberg-Mac Lane space $K(\mathbb{Z},2)$. But we can take as $K(\mathbb{Z},2)$ the projective space, $\mathbb{C}P^{\infty}=e^0\cup e^2\cup e^4...$, which has non trivial homology in even dimensions. Then the Hurewicz homomorphism of the universal covering of $RB(\mathbb{Z},2)$ is not an isomorphism in dimensions $=4, 6, 8, \ldots$. Therefore $RB(\mathbb{Z},2)$ is not a J-space. Applying proposition 6.6, we also have that $(\mathbb{Z},2)$ is not a J-crossed complex.

(b) Let $(G,1)$ be the crossed complex having G in dimension 1, then $RB(G,1)$ is a Eilenberg-Mac Lane space $K(G,1)$. It is clear that $\Pi_q(\pi RB(G,1))\cong\Pi_q(G,1)$ for $q=0,1$. For $q>1$, we have that $\Pi_q(\pi RB(G,1))\cong H_q(\text{Univ. Cov. of } RB(G,1))$, but the universal covering of $RB(G,1)$ is a contractible space, then $\Pi_q(\pi RB(G,1))\cong 0\cong\Pi_q(G,1)$ for $q>1$. This proves that $(G,1)$ is a J-crossed complex.

Let us denote by Kan*, the category of pointed Kan simplicial sets. A simplicial set X is said to be reduced if $X_0=\{*\}$. It is easy to define a functor Re: Kan* $\longrightarrow$ RedKan* that induces a functor Re: Ho(SS*) $\longrightarrow$ Ho(SS*)|(RedKan*). If we also consider the inclusion functor In: Ho(SS*)|(RedKan*) $\longrightarrow$ Ho(SS*) we obtain an equivalence of categories given by

Re: Ho(SS*)|(0-connected) $\longrightarrow$ Ho(SS*)|(RedKan*) and

In: Ho(SS*)|(RedKan*) $\longrightarrow$ Ho(SS*)|(0-connected).

Now it is easy to check that the following squares are commutative (up to natural equivalence).

$$\begin{array}{ccc} \mathrm{Kan}_* & \xrightarrow{\mathrm{Re}} & \mathrm{RedKan}_* \\ {\scriptstyle \pi R}\downarrow & & \downarrow{\scriptstyle \pi R} \\ \mathrm{Crs}_* & \xrightarrow[\mathrm{Re}]{} & \mathrm{RedCrs}_* \end{array} \qquad\qquad \begin{array}{ccc} \mathrm{Kan}_* & \xrightarrow{\mathrm{Re}} & \mathrm{RedKan}_* \\ {\scriptstyle B}\uparrow & & \uparrow{\scriptstyle B} \\ \mathrm{Crs}_* & \xrightarrow[\mathrm{Re}]{} & \mathrm{RedCrs}_* \end{array}$$

Therefore we can state the following results:

Corollary 6.8. (i) The following

$$\begin{array}{ccc} \mathrm{Ho(proSS_*)|(J, pro(0\text{-}connected))} & \underset{\mathrm{In}}{\overset{\mathrm{Re}}{\leftrightarrows}} & \mathrm{Ho(proSS_*)|(J, pro(RedKan_*))} \\ B \uparrow \downarrow \pi R & & B \uparrow \downarrow \pi R \\ \mathrm{Ho(proCrs_*)|(J, pro(0\text{-}connected))} & \underset{\mathrm{In}}{\overset{\mathrm{Re}}{\leftrightarrows}} & \mathrm{Ho(proCrs_*)|(J, pro(RedCrs_*))} \end{array}$$

is a commutative diagram of equivalences of categories.

(ii) There is a similar diagram by replacing proSS* by proTop*. Similar results hold by considering spaces, simplicial sets or crossed complexes instead of the corresponding proobjects, in this case we also have:
(iii) The category Ho(Crs*)|(J, RedCrs*) is equivalent to Ho(J, RedCrs*) the category obtained by inverting weak equivalences between J-reduced-crossed complexes.

Remark. If X is a simplicial set, ReX is obtained by reducing some simplicial set X' with the Kan extension property and weakly equivalent to X.

As a consequence of the results above, we can translate the problem of computing the homotopy hom-set between J-prospaces to computing the hom-set in the homotopy categories of pro-crossed complexes. We can also use this procedure in the following cases:

Corollary 6.9.
(i) If X is a space and Y a J-space then Ho(Top)(X, Y) ≅ Ho(Crs) (πRSX, πRSY).
(ii) If X is a prospace and Y a J-prospace, then
Ho(proTop) (X, Y) ≅ Ho(proCrs) (πRSX, πRSY).
(iii) If X and Y are objects in pro(CCW), and Y is a J-prospace, then
Ho(proTop) (X, Y) ≅ Ho(proCrs) (πX, πY).
(iv) If G is a J-crossed complex and H a crossed complex, then
Ho(Crs)(G,H) ≅ Ho(SS)(BG, BH)
(v) If G is a J-pro-crossed complex and H a pro-crossed complex, then
Ho(proCrs)(G,H) ≅ Ho(proSS)(BG, BH).

Proof. For (i) let Y be a J-space, then $Y \longrightarrow RB\pi RSY$ is a weak equivalence. Therefore $Ho(Top)(X, Y) \cong Ho(Top)(X, RB\pi RS\,Y) \cong Ho(Top)(\pi RSX, \pi RS\,Y)$; similarly it is for (ii).

To prove (iii), consider the functor $s: CCW \longrightarrow SS$ defined by $sX = \{x \in SX | x \text{ is cellular}\}$. It is well known that the inclusion $sX \longrightarrow SX$ is a weak equivalence, then we have the diagram

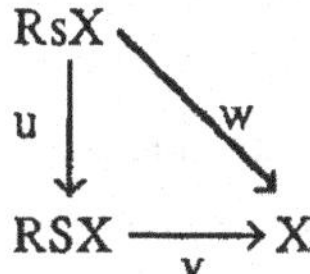

where u, v, w are homotopy equivalences and u, w are also cellular. Thus $\pi RsX \longrightarrow \pi RSX$ and $\pi RsX \longrightarrow \pi X$ are homotopy equivalences in Crs. Therefore πX is isomorphic to πRSX in Ho(Crs). Now we can prove (iii) for prospaces by using similar arguments.

The proof of (iv) and (v) is analogous to (i) and (ii), respectively.

Remarks. (1) Result (i) of the above is essentially the result of Whitehead on realisability of maps with codomain the homotopy system of a J-space. We thus can consider the other four results as generalisations and/or extensions of that result.
(2) The equivalence of categories and results of this section can be stated in the pointed case and for the categories of the form (proSS, SS),.(proTop,.Top), (proCrs, Crs), (towSS, SS), etc. For see applications of these categories, see [H-P.1], [H-P.2].

7. The n-type of a $\mathcal{J}_n$-prospace.

In this section we consider the notions of $\mathcal{J}_n$-prospace and $\mathcal{J}_n$-pro-crossed complex and prove that the category of n-types of $\mathcal{J}_n$-prospaces is equivalent to the category of n-types of $\mathcal{J}_n$-pro-crossed complexes. By using the properties of the skeleton, coskeleton and truncation functors we give other algebraic models for the n-types of $\mathcal{J}_n$-prospaces.

Lemma 7.1.

(i) The functors $\pi R:SS \longrightarrow Crs$ and $\pi R:proSS \longrightarrow proCrs$ preserve n-cofibrations and weak n-equivalences.

(ii) The functors $B:Crs \longrightarrow SS$, and $B:proCrs \longrightarrow proSS$, preserve n-fibrations and weak n-equivalences.

Proof. Let $V(p,k)$ denote the simplicial subset of $\Delta[p]$ which is the union of the images of the i-faces, $0\leq i \leq p$, $i\neq k$, and let $\dot{\Delta}[n+1]$ denote the boundary of $\Delta[n+1]$ and $\Delta[n+1]/\dot{\Delta}[n+1]$ the (n+1)-sphere. A map f in SS is a n-fibration if f has the RLP with respect to $V(p,k) \longrightarrow \Delta[p]$, $0\leq p\leq n+2$, $0\leq k\leq p$, and $* \longrightarrow \Delta[n+1]/\dot{\Delta}[n+1]$. By the properties of the functor π it is not difficult to check that $\pi RV(p,k) \longrightarrow \pi R\Delta[p]$, $0\leq p\leq n+2$, $0\leq k\leq p$, and $* \longrightarrow \pi R(\Delta[n+1]/\dot{\Delta}[n+1])$ have the LLP with respect to nfibrations.

Now if $q:G \longrightarrow H$ is an n-fibration of crossed complexes, because $\pi R:SS \longrightarrow Crs$ is left adjoint to $B:Crs \longrightarrow SS$ it follows that Bq has the RLP with respect to $V(p,k) \longrightarrow \Delta[p]$, $0\leq p\leq n+2$, $0\leq k\leq p$, and $* \longrightarrow \Delta[n+1]/\dot{\Delta}[n+1]$. Therefore Bq is also an n-fibration.

Assume that f is a weak n-equivalence in Crs. By using the commutative diagram of lemma 6.2 we obtain thaat Bf is a weak n-equivalence in SS.

For (i), suppose that g is a weak n-equivalence in SS, then Rf is a n-homotopy equivalence. Since πRf is a n-homotopy equivalence we also have that πRf is a weak n-equivalence in Crs. Now taking into account that πR is left adjoint to B it follows that πR preserves n-cofibrations.

Using arguments similar to those of proposition 5.2 (ii), these results extend to the corresponding procategories.

Now, we can apply again theorem 3 of [Q; chI.§4] to obtain:

Proposition 7.2.

(i) $\pi R: Ho_n(SS) \longrightarrow Ho_n(Crs)$ is left adjoint to $B: Ho_n(Crs) \longrightarrow Ho_n(SS)$.

(ii) $\pi R: Ho_n(proSS) \longrightarrow Ho_n(proCrs)$ is left adjoint to
$B: Ho_n(proCrs) \longrightarrow Ho_n(proSS)$.

(iii) $\pi RS: Ho_n(Top) \longrightarrow Ho_n(Crs)$ is left adjoint to $RB: Ho_n(Crs) \longrightarrow Ho_n(Top)$.

(iv) $\pi RS: Ho_n(proTop) \longrightarrow Ho_n proCrs)$ is left adjoint to
$RB: Ho_n(proCrs) \longrightarrow Ho_n(proTop)$.

Definition 7.3. A simplicial set X is said to be a $\mathfrak{J}_n$-simplicial set if the natural transformation $X \longrightarrow B\pi RX$ is a weak n-equivalence. A topological space X is said to be a $\mathfrak{J}_n$-space if SX is a $\mathfrak{J}_n$-simplicial set. As in definition 6.4, we define similarly strong $\mathfrak{J}_n$-prospace and $\mathfrak{J}_n$-prospace.

Definition 7.4. A crossed complex G is said to be a $\mathcal{J}_n$-crossed complex if the natural transformation $\pi RBG \longrightarrow G$ is a weak n-equivalence. A pro-crossed complex H is said to be a (strong) $\mathcal{J}_n$-crossed complex if the natural transformation $\pi RBH \longrightarrow H$ is a weak n-equivalence (a strong equivalence).

By using arguments similar to those of proposition 6.6 we also have that:

Proposition 7.5.

(i) If X is a $\mathcal{J}_n$-simplicial set, then πRX is a $\mathcal{J}_n$-crossed complex.

(ii) If X is a (strong) $\mathcal{J}_n$-pro-simplicial set, then πRX is a (strong) $\mathcal{J}_n$-pro-crossed complex.

(iii) G is a $\mathcal{J}_n$-crossed complex if and only if BG is a $\mathcal{J}_n$-simplicial set.

(iv) G is a strong $\mathcal{J}_n$-pro-crossed complex if and only if BG is a strong $\mathcal{J}_n$-pro-simplicial set.

(v) If G is a $\mathcal{J}_n$-pro-crossed complex, then BG is a $\mathcal{J}_n$-pro-simplicial set.

Remarks.

(1) A simplicial set X is a $\mathcal{J}_n$-simplicial set if and only if $\pi_k(sk_{k-1} X, *) \longrightarrow \pi_k(sk_k X, *)$, $1 < k \leq n-1$, and $\pi_n(sk_{n-1} X, *) \longrightarrow \pi_n(sk_{n+1} X, *)$ are trivial for any choice of base point, see [C-C]. A CW-complex X is a $\mathcal{J}_n$-space if and only if $\pi_k(X^{k-1}, *) \longrightarrow \pi_k(X^k, *)$, $1 < k \leq n-1$, and $\pi_n(X^{n-1}, *) \longrightarrow \pi_n(X^{n+1}, *)$ are trivial for any choice of base point. The notion of $\mathcal{J}_n$-space is exactly the condition given in [C-C] to prove that the hypercrossed complex $\mu(X)$ associated with a CW-complex is quasi-isomorphic to the crossed complex .

(2) Notice that a J_n-space in the sense of Whitehead is always a $\mathcal{J}_n$-space. However, $K(\mathbb{Z},2)$ is a $\mathcal{J}_3$-space which is not a J_3-space. If G is a group (progroup), then $K(G,n)$ and $K(G,n-1)$ are $\mathcal{J}_n$-spaces ($\mathcal{J}_n$-prospaces). A reduced CW-complex X with $\pi_i(X)=0$ for $1< i <n$ is a $\mathcal{J}_n$-space, if $\pi_i(X)=0$ for $1< i <n-1$ and $\pi_n(X)=0$, then X is also a $\mathcal{J}_n$-space. Any crossed (pro-crossed) complex G of dimension ≤ 3 is a $\mathcal{J}_3$-crossed ($\mathcal{J}_3$-procrossed) complex, hence BG is a $\mathcal{J}_3$-simplicial set ($\mathcal{J}_3$-pro-simplicial set).

(3) There is a CW-complex X such that the associated crossed complex πX is a $\mathcal{J}_n$-crossed complex but X is not a $\mathcal{J}_n$-complex. For example $S^2 = e^0 \cup e^2$ is not a $\mathcal{J}_3$-complex and πS^2 is a $\mathcal{J}_3$-crossed complex.

Corollary 7.6. The restriction of the functors of proposition 7.2 to $\mathcal{J}_n$-objects or $\mathcal{J}_n$-pro-objects induce equivalences of categories.

Recall that in the category of simplicial sets we have the functors $sk_{n+1}:SS \longrightarrow SS$ and $cosk_{n+1}:SS \longrightarrow SS$. If X is a simplicial set, $(sk_{n+1}X)_q=X_q$ for $q\leq n+1$ and for $q>n+1$ it only has degenerate simplexes. The functor $cosk_{n+1}$ is defined by $(cosk_{n+1}X)_q=SS(sk_{n+1}\Delta[q],X)$.

Lemma 7.7.

(i) The following diagrams are commutative (up to natural isomorphism).

$$\begin{array}{ccc} SS & \xrightarrow{sk_{n+1}} & SS \\ \pi R \downarrow & (1) & \downarrow \pi R \\ Crs & \xrightarrow{sk_{n+1}} & Crs \end{array} \qquad\qquad \begin{array}{ccc} SS & \xrightarrow{cosk_{n+1}} & SS \\ B \uparrow & (2) & \uparrow B \\ Crs & \xrightarrow{cosk_{n+1}} & Crs \end{array}$$

(ii) The same results hold replacing SS, and Crs by proSS and proCrs, respectively.

Proof. It is easy to check that (1) commutes and an immediate consequence of the adjointness of πR and B is the commutativity of (2).

Theorem 7.8. The diagram

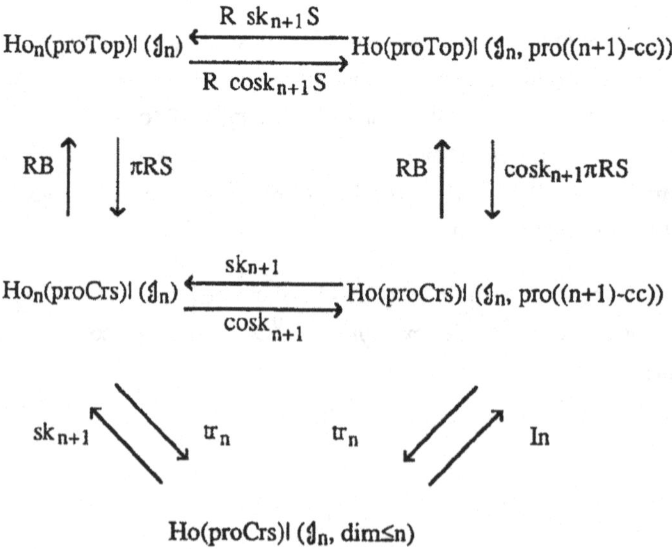

has the following properties:

(i) Each pair of opposite arrows ($\leftrightarrows$) gives an equivalence of categories.

(ii) Up to natural isomorphism, it is a commutative diagram of equivalences of categories.

(iii) There are similar results, replacing the category proTop by proSS, Top, SS and the corresponding pointed cases.

Proof. In the following diagram

$$\begin{array}{ccc}
Ho_n(proSS) & \underset{cosk_{n+1}}{\overset{sk_{n+1}}{\leftrightarrows}} & Ho(proSS)|\,(\,pro((n+1)\text{-}cc)) \\
RB \uparrow \;\; \downarrow \pi RS & & \\
Ho_n(proCrs) & \underset{cosk_{n+1}}{\overset{sk_{n+1}}{\leftrightarrows}} & Ho(proCrs)|\,(\,pro((n+1)\text{-}cc)),
\end{array}$$

by corollary 5.4 (iii) we have that sk_{n+1} and $cosk_{n+1}$ define an equivalence of categories for pro-crossed complexes. Taking into account the remark of this corollary, the same results follows for pro-simplicial sets. Proposition 7.2 proves that πR is left adjoint to B in diagram above and by corollary 7.6 the restriction of these functors to $\mathcal{J}_n$-proobjects define an equivalence of categories.

Since $X \longrightarrow cosk_{n+1} X$, $sk_{n+1}Y \longrightarrow Y$ are weak n-equivalences we have that if X or Y is a $\mathcal{J}_n$-proobject, then $cosk_{n+1} X$ or $sk_{n+1}Y$ is also a $\mathcal{J}_n$-proobject, respectively. Thus the restrictions of sk_{n+1} and $cosk_{n+1}$ to $\mathcal{J}_n$-proobjects in the diagrams above give equivalences of categories.

The commutativity of the square of the threorem follows from

$cosk_{n+1}\pi R sk_{n+1} \cong cosk_{n+1} sk_{n+1} \pi R \cong cosk_{n+1} \pi R$

$cosk_{n+1} B\, sk_{n+1} \cong B\, cosk_{n+1} sk_{n+1} \cong B\, cosk_{n+1} \cong B$

where we have used the lemma 7.7 and the fact that if X is an object in proCrs|(pro(n+1)-cc)), $X \longrightarrow cosk_{n+1} X$ is an strong equivalence.

Now we consider the part of the diagram with the truncation functors. By proposition 5.7 we have that

$tr_n\colon Ho_n(proCrs) \longrightarrow Ho(proCrs)|(\dim \leq n),$

sk_{n+1}: Ho(proCrs)|(dim ≤ n) ⟶ Ho_n(proCrs)

give an equivalence of categories. Since X ⟶ $tr_n X$, $sk_{n+1}Y$ ⟶ Y are weak n-equivalences we have that if X or Y is a $\mathcal{J}_n$-pro-crossed complex, then $tr_n X$ or $sk_{n+1}Y$ is also a $\mathcal{J}_n$-pro-crossed complex, respectively. Finally, if X is an object in Ho(proCrs)|($\mathcal{J}_n$,pro(n+1)-cc), then X ⟶ $tr_n X$ is a weak equivalence and if Y is an object in Ho(proCrs)|($\mathcal{J}_n$,dim≤n), InY=Y=tr_nInY. Thus tr_n and In give an equivalence of categories in the diagram of the theorem.

Remark. In proposition 5.7, we have seen that the category Ho(proCrs)|(dim ≤ n) defined as the full subcategory of Ho(proCrs) determined by pro-crossed complexes of dimension ≤ n is equivalent to Ho(proCrs|(dim ≤ n)) the category obtained by inverting the weak n-equivalences of the category proCrs|(dim ≤ n).

As in Corollary 6.9, we can translate the problem of computing the hom-set in the following cases:

Corollary 7.9.

Let X and Y be CW-complexes

(i) $Ho_n(Crs)(\pi X,\pi Y) \cong Ho(Crs)(\pi X, cosk_{n+1}\pi Y) \cong Ho(Crs)(cosk_{n+1}\pi X, cosk_{n+1}\pi Y)$

$Ho_n(Crs)(\pi X,\pi Y) \cong Ho(Crs)(\pi X, tr_n\pi Y) \cong Ho(Crs)(tr_n\pi X, tr_n\pi Y)$

(ii) If Y is a $\mathcal{J}_n$-space, then $Ho_n(Top)(X, Y) \cong Ho_n(Crs)(\pi X, \pi Y)$.

(iii) If dim X≤n and Y is a $\mathcal{J}_n$-space, then $Ho(Top)(X, Y) \cong Ho(Crs)(\pi X, \pi Y)$.

(iv) If dim X≤m, n<m≤+∞, Y Y is a $\mathcal{J}_n$-space and $\pi_k(Y)=0$ for n< k ≤m, then

$Ho(Top)(X, Y) \cong Ho_n(Crs)(\pi X, \pi Y)$.

Let G, H be crossed complexes

(v) $Ho_n(SS)(BG,BH) \cong Ho(SS)(BG, cosk_{n+1}BH) \cong Ho(SS)(cosk_{n+1}BG, cosk_{n+1}BH)$

(vi) If G is a $\mathcal{J}_n$-crossed complex, then $Ho_n(Crs)(G,H) \cong Ho_n(SS)(BG, BH)$

(vii) If G is a $\mathcal{J}_n$-crossed complex and H is (n+1)-coconnected, then

$Ho(Crs)(G,H) \cong Ho_n(SS)(BG, BH) \cong Ho(SS)(BG, BH)$.

Remarks. (1) Notice that Corollary 7.9(iii) is the Whitehead' theorem on realisability of morphisms for J_n-complexes, however our condion $\mathcal{J}_n$ is weaker than condition J_n.

(2) The results (i), (ii), (v), (vi), (vii) extend to the corresponding procategories, but the following example proves that (iii) and (iv) does not extend. Consider the Eilenberg-Mac Lane tow-space K(G,2) associated to the tower of groups $G=\{\dots \xrightarrow{2} \mathbf{Z} \xrightarrow{2} \mathbf{Z} \xrightarrow{2} \mathbf{Z}\}$, it is easy to see that $Ho(towTop)(S^1, K(G,2)) \cong \lim^1 G \neq 0$ and $Ho_1(towCrs)(\pi S^1, \pi K(G,2)) \cong 0$.

It is easy to check that any space or prospace satisfies the properties $\mathfrak{J}_0$, $\mathfrak{J}_1$ and $\mathfrak{J}_2$. Therefore we have the following results.

Corollary 7.10.

(i) Ho_0(proTop∗) is equivalent to the category of pro-pointed sets, proSet∗.

(ii) Ho_1(proTop∗)|(pro(0-connected)) is equivalent to the category of progroups, proGps.

(iii) Ho_2(proTop∗)|(pro(0-connected)) is equivalent to the category Ho_2(proCrs∗)|(proRedCrs∗, dim≤2).

Using the characterization of weak equivalences between towers of reduced crossed complexes of dimension ≤2 given by the tower version of the Whitehead theorem given by Edwards-Hastings [E-H; p. 193], we also have:

Corollary 7.11. The category Ho_2(towTop∗)|(pro(0-connected)) is equivalent to the category Ho(tow-crossed modules) obtained by inverting morphisms $G \longrightarrow G'$ of tow-crossed modules that induce isomorphisms $\Pi_1(G) \longrightarrow \Pi_1(G')$ and $\Pi_2(G) \longrightarrow \Pi_2(G')$ of towers of groups.

Remarks. (1) Two pro-pointed spaces X and Y have the same 0-type if and only if the pro-pointed sets $\pi_0(X)$ and $\pi_0(Y)$ are isomorphic. If X and Y are objects in pro(Top∗|(0-connected)) then X and Y have the same 1-type if and only if the fundamental progroups of X and Y are isomorphic. In last case, X and Y have the same 2-type if and only if the pro-crossed modules $\pi_2(SX, sk_1SX, *) \longrightarrow \pi_1(sk_1SX, *)$ and $\pi_2(SY, sk_1SY, *) \longrightarrow \pi_1(sk_1SY, *)$ are isomorphic in the category Ho_2(proCrs∗)|(proRedCrs∗, dim≤2). If X and Y are objects in tow(Top∗|(0-connected)) it suffices to see that the corresponding tow-crossed modules are isomorphic in Ho(tow-crossed modules). In the case that X, Y are objects in tow(CCW∗|(0-connected)) we can replace these tow-crossed modules by the equivalent ones $\pi_2(X, X^1, *) \longrightarrow \pi_1(X^1, *)$, etc

(2) Using the representation of elements of $H^3(G;H)$ given by Mac Lane, Huebschmann, Holt and Lue, see [M], as crossed modules, C, such that $\Pi_1(C) \cong G$, $\Pi_2(C) \cong H$, we have that Corollary 7.11 is the extension of the Mac Lane-Whitehead Theorem to the category of tower of spaces.

REFERENCES

[A-M] M. Artin and B. Mazur, "Étale Homotopy", Lect. Notes in Math., nº 100, Springer, 1969.

[E-H] D. Edwards and H. Hastings, "Čech and Steenrod homotopy theories with applications to Geometric Topology", Lect. Notes Math., 542, Springer, 1976.

[Ba.1] H.J. Baues, "Combinatorial Homotopy and 4-dimensional Complexes", deGruyter Expositions in Mathematics Vol. 2, deGruyter Berlin-New York, 1991.

[Ba.2] H.J. Baues, "Algebraic Homotopy", Cambridge Studies in advanced mathematics 15, Cambridge University Press, 1989.

[B-G] R. Brown and M. Golasiński, "The closed model category of crossed complexes", Cah. Top. Geom. Diff. Cat. 30 (1989) 61-82.

[B-H.1] R. Brown and P.J. Higgins, "Tensor products and homotopies for ω-groupoids and crossed complexes", J. Pure Appl. Algebra 47(1987) 1-33.

[B-H.2] R. Brown and P.J. Higgins, "The classifying space of a crossed module", U.C.N.W. Maths. Preprint 90.20 (36p).

[B-H.3] R. Brown and P.J. Higgins, "Crossed complexes and chain complexes with operators", Math. Proc. Camb. Phil. Soc. 107 (1990) 33-57.

[C-C] P. Carrasco and A.M. Cegarra, "Group theoretic models for homotopy types", J. Pure Appl. Algebra, (1989).

[G-Z] P. Gabriel and M. Zisman, "Calculus of fractions and Homotopy Theory", Springer, Berlin (1966).

[Gr] A. Grothendieck, "Technique De Descente Et Théorèmes d'Existence En Géométrie Algèbrique I-IV", Seminar Bourbaki, Exp. 190, 195, 212, 221, 1959-60, 1960-61.

[H-P.1] L.J. Hernández and T. Porter, "Proper pointed maps from $\mathbb{R}^{n+1}$ to a σ-compact space", Math. Proc. Camb. Phil. Soc. 103 (1988) 462-475.

[H-P.2] L.J. Hernández and T. Porter, "Global analogues of the Brown-Crossman proper homotopy groups", Math. Proc. Camb. Phil. Soc. 104 (1988) 483-496.

[H-P.3] L.J. Hernández and T. Porter, "An embedding theorem for proper n-types at infinity", preprint (1990).

[M] S. Mac Lane, "Historical note", J. Algebra 60 (1979) 319-320.

[M-W] S. Mac Lane and J.H.C. Whitehead, "On the 3-type of a complex", Proc., Nat., Acad., Sci., Washington, 36, (1950) 41-48.

[P] T. Porter, "Čech homotopy I", J. London Math. Soc. 6 (1963) 429-436.

[Q] D. Quillen, "Homotopical Algebra", Lect. Notes in Math, 43, Springer, 1967.

[S] A. Strøm,"The homotopy category is a homotopy category", Arch. Math. 23 (1973) 435-441.

[W.1] J.H.C. Whitehead, "Combinatorial homotopy. I", Bull. Amer. Math. Soc., 55(1949) 213-245.

[W.2] J.H.C. Whitehead, "Combinatorial homotopy. II", Bull. Amer. Math. Soc., 55(1949) 453-496.

L.J. Hernández: Departamento de Matemáticas
Universidad de Zaragoza
50009 Zaragoza
Spain

T. Porter: School of Mathematics
University College of North Wales
Dean Street, Bangor
Gwynedd LL57 1UT,
U.K.

1990 Barcelona Conference
on Algebraic Topology.

MORAVA K–THEORIES OF CLASSIFYING SPACES AND GENERALIZED CHARACTERS FOR FINITE GROUPS

MICHAEL J. HOPKINS[1], NICHOLAS J. KUHN [2]
AND
DOUGLAS C. RAVENEL[3]

This paper is intended to be an informal introduction to [HKR], where we study the Morava K–theories (which will be partially defined below) of the classifying space of a finite group G and related matters. It will be long on exposition and short on proofs, in the spirit of the third author's lecture at the conference.

In Section 1 we will recall Atiyah's theorem relating the complex K–theory of BG to the complex representation ring $R(G)$. We will also define classical characters on G. In Section 2 we will introduce the Morava K–theories $K(n)^*$ and the related theories $E(n)^*$. In Section 3 we will state our main results and conjectures. Most of the former are generalizations of the classical results stated in Section 1.

In the remaining five sections we will outline the proofs of our results. Section 4 is purely group–theoretic, i.e., it makes no use of any topology. In it we will prove our formula for the number $\chi_{n,p}(G)$ of conjugacy classes of commuting n–tuples of elements of prime power order in a finite group G. We have discussed this material with several prominent group theorists, but we have yet to find one who admits to ever having considered this question. In Section 5 we equate this number with the Euler characteristic of $K(n)^*(BG)$. Our generalized characters are functions on the set of such conjugacy classes with values in certain p–adic fields.

In Section 6 we recall the Lubin–Tate construction from local algebraic number theory. It uses formal group laws to construct abelian extensions of finite extensions of the field of p–adic numbers. We need it in Section 7 where we describe the connection between $E(n)^*(BG)$ and our generalized characters.

In Section 8, we prove a theorem about wreath products. A corollary of this is that our main conjecture (3.5) about $K(n)^*(BG)$ holds for all the symmetric groups.

[1]Partially supported by the National Science Foundation
[2]Partially supported by the NSF, the Sloan Foundation and the SERC
[3]Partially supported by the NSF

Contents

1 Atiyah's theorem and other classical results

Atiyah determined the complex K–theory of the classifying space BG in [Ati61]. A complex representation ρ of a finite group G is a homomorphism $G \to U$ (where U denotes the stable unitary group) and hence gives a map of classifying spaces $BG \to BU$, which defines a vector bundle over BG and an element in $K^*(BG)$. A virtual representation leads to a virtual vector bundle, and we still get an element in $K^*(BG)$. It follows that there is a natural ring homomorphism

$$R(G) \longrightarrow K^*(BG),$$

where $R(G)$ denotes the complex representation ring of G.

This map is not an isomorphism, but it is close to being one. The representation ring $R(G)$ has an augmentation ideal I, the ideal of all virtual representations of degree 0. Let $R(G)\hat{}$ denote the I–adic completion of $R(G)$.

Theorem 1.1 (Atiyah) *For a finite group G,*

$$K^i(BG) \cong \begin{cases} R(G)\hat{} & \text{if } i \text{ is even} \\ 0 & \text{if } i \text{ is odd.} \end{cases}$$

When G is a p-group, the I–adic completion is roughly the same as the p–adic completion. More precisely, we have

$$R(G)\hat{} \cong \mathbf{Z} \oplus (\mathbf{Z}_p \otimes I),$$

where $\mathbf{Z}_p$ denotes the p–adic integers. For more general finite groups G, I–adic completion is more drastic. The map $R(G) \to R(G)\hat{}$ has a nontrivial kernel, as can be seen by studying the case $G = \mathbf{Z}/(6)$.

In view of 1.1, it is useful to recall some classical facts about the representation ring $R(G)$. A good reference is Serre's book [Ser67]. Additively, $R(G)$ is a free abelian group whose rank is the number of conjugacy classes of elements in G.

The mod p analog of 1.1 is the following, which was proved in [Kuh87].

Proposition 1.2 *For a finite group G and a prime number p,*

$$K^i(BG;\mathbf{Z}/(p)) \cong \begin{cases} R(G)\hat{} \otimes \mathbf{Z}/(p) & \text{if } i \text{ is even} \\ 0 & \text{if } i \text{ is odd,} \end{cases}$$

and the rank of this vector space is the number of conjugacy classes of elements whose order is some power of the prime p.

A representation ρ induces a complex valued function χ, called a *character*, on the set of conjugacy classes as follows. Pick a basis for the complex vector space V on which G acts via ρ. Then for $G \in G$ we define $\chi(g)$ to be the trace of the matrix $\rho(g)$. This turns out to be invariant under conjugation and to be independent of the choice of basis. It leads to a ring isomorphism

$$R(G) \otimes \mathbf{C} \xrightarrow{\chi_G} \mathrm{Cl}(G), \tag{1.3}$$

where $\mathrm{Cl}(G)$ denotes the ring of complex valued functions on the conjugacy classes of elements in G.

It should be noted here that this isomorphism does *not* have an analog in characteristic p, namely $K^*(BG;\mathbf{Z}/(p)) \otimes \bar{\mathbf{F}}_p$ is not naturally isomorphic to the ring of $\bar{\mathbf{F}}_p$–valued functions on conjugacy classes of prime power order, where $\bar{\mathbf{F}}_p$ denote the algebraic closure of the field $\mathbf{Z}/(p)$.

We will illustrate this by studying the representation ring for the quaternion group Q_8. Using the usual notation for quaternions, its elements are ± 1, $\pm i$, $\pm j$, and $\pm k$. There are five conjugacy classes, namely $\{1\}$, $\{-1\}$, $\{\pm i\}$, $\{\pm j\}$, and $\{\pm k\}$. The center is the subgroup of order 2 generated by -1, and there is a group extension

$$1 \longrightarrow \mathbf{Z}/(2) \longrightarrow Q_8 \longrightarrow \mathbf{Z}/(2) \oplus \mathbf{Z}/(2) \longrightarrow 1$$

There are four irreducible representations of Q_8 of degree 1 induced by those on the quotient group. We denote these by 1, α, β and γ. There is a fifth irreducible representation of degree 2, which we denote by σ, obtained by the usual action of Q_8 on the space $\mathbf{H} = \mathbf{C}^2$ of quaternions.

The corresponding characters are displayed in Table 1.4. This is a *character table* for Q_8. There is a column for each irreducible representation and a row for each conjugacy class. The numbers shown are that various values of $\chi(g)$.

From this table we can read off the multiplicative structure of $R(Q_8)$. We see that $\alpha\beta\gamma = \alpha^2 = \beta^2 = \gamma^2 = 1$, $\sigma\alpha = \sigma\beta = \sigma\gamma = \sigma$, and $\sigma^2 = 1 + \alpha + \beta + \gamma$.

It follows that there is a unique homomorphism from $R(Q_8)$ to a field k of characteristic 2, and it sends α, β and γ to 1 and σ to 0. On the other hand, the ring of k–valued functions on the conjugacy classes of Q_8 admits at least five such homomorphisms, namely evaluation on each of the five conjugacy classes.

Table 1.4: Character table for Q_8

	1	α	β	γ	σ
1	1	1	1	1	2
-1	1	1	1	1	−2
$\pm i$	1	1	−1	−1	0
$\pm j$	1	−1	1	−1	0
$\pm k$	1	−1	−1	1	0

In view of (1.3), one can ask for a characterization of $R(G)$ as a subring of $\mathrm{Cl}(G)$. This seems to be a very delicate business, but one can recover its rational form $R(G)\otimes\mathbf{Q}$ in the following way.

First, one does not need all of $\mathbf{C}$ to get the isomorphism (1.3). The trace of $\rho(g)$ is a linear combination of eigenvalues of g, i.e., it lies in a cyclotomic extension of $\mathbf{Q}$. To fix notation let L denote the union of all such extensions, i.e., the maximal abelian extension of $\mathbf{Q}$. Then (1.3) can be replaced by an isomorphism

$$R(G)\otimes L \longrightarrow \mathrm{Cl}(G,L),$$

where $\mathrm{Cl}(G,L)$ denotes the ring of L-valued class functions.

Now the Galois group $\mathrm{Gal}(L\colon\mathbf{Q})$ is known to be isomorphic to the group of units in the profinite integers. Each automorphism of L raises each root of unity to a certain power, depending on the order of the root. This group also acts on the set of conjugacy classes in G in a similar way. Elements in $\mathrm{Cl}(G,L)$ coming from representations are *equivariant* with respect to these two Galois actions.

Proposition 1.5 *The subring of* $\mathrm{Cl}(G,L)$ *consisting of Galois equivariant class functions is isomorphic to* $R(G)\otimes\mathbf{Q}$.

Finally, we must mention a theorem of Artin about $R(G)$. Given a subgroup H of G, one has a restriction homomorphism $R(G)\to R(H)$. Given a category $\mathcal{C}$ of subgroups (such as that of all abelian subgroups or all cyclic subgroups) one has a map

$$R(G) \longrightarrow \lim_{\substack{\longleftarrow \\ \mathcal{C}}} R(H) \tag{1.6}$$

and one can ask under what circumstances it is an isomorphism. The limit here is the categorical inverse limit. It is defined to be the subset of the product

$$\prod_{H\in\mathrm{Ob}(\mathcal{C})} R(H)$$

consisting of points in which the coordinates are compatible under all morphisms in $\mathcal{C}$. This means that for any morphism

$$H' \xrightarrow{h} H'',$$

Table 1.7: Generators of $\lim R(A)$ for Q_8

	A	B	C	D	1
A_1	$\lambda+\lambda^3-2$	λ^2-1	0	0	1
A_2	$\lambda+\lambda^3-2$	0	λ^2-1	0	1
A_3	$\lambda+\lambda^3-2$	0	0	λ^2-1	1
$\mathbf{Z}/(2)$	$2\alpha-2$	0	0	0	1
$\{e\}$	0	0	0	0	1

the coordinates $x_{H'} \in R(H')$ and $x_{H''} \in R(H'')$ must satisfy

$$x_{H'} = f^*(x_{H''}).$$

We are assuming that the objects in $\mathcal{C}$ are closed under subgroups, i.e., if a subgroup H is an object of $\mathcal{C}$, then so is any subgroup of H. We all assume that the morphisms in $\mathcal{C}$ are generated by inclusions and conjugations.

We will illustrate the categorical limit with the quaternion group Q_8 (whose character table is shown in 1.4) and the category of abelian subgroups. It has five abelian subgroups: the trivial subgroup, the subgroup of order 2, and three subgroups of order 4 (A_1, A_2 and A_3). An element in the limit

$$\lim_{\mathcal{C}} R(A)$$

is a 5–tuple of representations (one for each abelian subgroup) which are compatible under inclusions and conjugations. This means three things:

(i) All five must have the same degree, since they must have the same restriction to the trivial subgroup. It follows that the limit is spanned by the trivial representation of degree 1 and by elements in which each coordinate is a virtual representation of degree 0.

(ii) The representations of the three subgroups of order 4 must the same restriction to the subgroup of order 2. Let α denote the nontrivial representation of $\mathbf{Z}/(2)$ of degree 1.

(iii) The representations of the subgroups of order 4 must each be invariant under the nontrivial involution of that group, since it is induced by a conjugation in Q_8. If λ denotes a representation of $\mathbf{Z}/(4)$ of degree 1 with eigenvalue i, then the coordinate in the 5–tuple must be a linear combination of 1, λ^2 and $\lambda+\lambda^3$.

It follows that the limit is spanned by five generators whose restrictions to the five subgroups are as shown in Table 1.7. In it there is a row for each subgroup and a column for each generator.

Comparing this with Table 1.4, we see that the map of (1.6) sends α to $C+D+1$, β to $B+D+1$, γ to $B+C+1$ and σ to $A+2$. It follows that the map is one–to–one and has a cokernel of order 2.

Artin's theorem, which is proved in [Ser67, Chapter 9], says that in general this map is one–to–one, and the order of its cokernel is a product of primes dividing the order of G.

Theorem 1.8 (Artin) *Let $\mathcal{C}(G)$ denote the category whose objects are cyclic subgroups of G, with morphisms generated by inclusions and conjugations. Then the natural map*

$$R(G)\otimes \mathbf{Z}[|G|^{-1}] \longrightarrow \lim_{C\in\mathcal{C}(G)} R(C)\otimes \mathbf{Z}[|G|^{-1}]$$

is an isomorphism. The same is true if we replace $\mathcal{C}(G)$ by $\mathcal{A}(G)$, the category of abelian subgroups of G.

Corollary 1.9 *The natural maps*

$$K^*(BG)\otimes \mathbf{Z}[|G|^{-1}] \longrightarrow \lim_{C\in\mathcal{C}(G)} K^*(BC)\otimes \mathbf{Z}[|G|^{-1}]$$

and

$$K^*(BG)\otimes \mathbf{Z}[|G|^{-1}] \longrightarrow \lim_{A\in\mathcal{A}(G)} K^*(BA)\otimes \mathbf{Z}[|G|^{-1}]$$

are isomorphisms for every finite group G.

Note that this is trivially true if we replace K–theory by mod p cohomology. In that case the source and target of the map are both trivial.

2 Morava K–theories

We would like to generalize 1.1 and 1.2 to some other cohomology theories related to BP-theory. Recall that BP is a ring spectrum, which is a minimal wedge summand of the Thom spectrum MU localized at p, with

$$BP_* = \pi_*(BP) = \mathbf{Z}_{(p)}[v_1, v_2, \cdots],$$

where the dimension of v_n is $2p^n-2$. There are BP-module spectra $E(n)$ and $K(n)$ (the nth Morava K–theory at the prime p) for each $n\geq 0$ with

$$\begin{aligned} E(0)_* = K(0)_* &= \mathbf{Q} \\ K(n)_* &= \mathbf{Z}/(p)[v_n, v_n^{-1}] \\ E(n)_* &= \mathbf{Z}_{(p)}[v_1, v_2, \cdots v_n, v_n^{-1}] \\ &\quad \text{for } n>0. \end{aligned}$$

We will not elaborate here on the construction of these spectra. The interested reader can find more information in [Rav84] and [Wil82].

Some additional properties of these spectra should be noted. $E(0)=K(0)$ is the rational Eilenberg–Mac Lane spectrum. $E(1)$ is one of $p-1$ isomorphic summands of the complex K–theory spectrum localized at p, and $K(1)$ is its mod p analog. $E(2)$ is related in a similar way to elliptic cohomology; see [Bak89].

Hence 1.1 and 1.2 can easily be translated into descriptions of $E(1)^*(BG)$ and $K(1)^*(BG)$. We will give a partial generalization to $E(n)^*(BG)$ and $K(n)^*(BG)$.

The coefficient ring $K(n)_*$ is a graded field in the sense that every graded module over it is free. This makes it very convenient for computations. In particular there is a Künneth isomorphism

$$K(n)^*(X\times Y) = K(n)^*(X)\otimes K(n)^*(Y). \tag{2.1}$$

A consequence of the Nilpotence Theorem of [DHS88] is that the Morava K–theories, along with ordinary mod p cohomology, are essentially the *only* cohomology theories with Künneth isomorphisms.

For a finite complex X, we know that the rank of $K(n)^*(X)$ is finite, grows monotonically with n, and is bounded above by the rank of $H^*(X;\mathbf{Z}/(p))$. In [Rav82] it was shown that $K(n)^*(BG)$ for finite G also has finite rank. Our results indicate that this rank grows *exponentially* with n.

All of these theories are *complex oriented*. This means that they behave in the expected way on $\mathbf{C}P^\infty$, namely

$$K(n)^*(\mathbf{C}P^\infty) = K(n)^*(\text{pt.})[[x]]$$

with $x \in K(n)^2(\mathbf{C}P^\infty)$, and similarly for $E(n)^*(\mathbf{C}P^\infty)$.

We also have

$$K(n)^*(\mathbf{C}P^\infty \times \mathbf{C}P^\infty) = K(n)^*(\text{pt.})[[x \otimes 1, 1 \otimes x]].$$

The H–space multiplication

$$\mathbf{C}P^\infty \times \mathbf{C}P^\infty \xrightarrow{m} \mathbf{C}P^\infty$$

induces a homomorphism in $K(n)$-cohomology determined by its behavior on the class x, and we can write

$$m^*(x) = F(x \otimes 1, 1 \otimes x).$$

The power series F is a *formal group law*, that is it satisfies

(i) $F(u,0) = F(0,u) = u$,

(ii) $F(u,v) = F(v,u)$ and

(iii) $F(F(u,v),w) = F(u,F(v,w))$.

This particular formal group law is characterized by its p–*series*, i.e., by the image of x under the pth power map on $\mathbf{C}P^\infty$. This is given by

$$[p](x) = v_n x^{p^n}. \tag{2.2}$$

This formula enables us to compute $K(n)^*(B\mathbf{Z}/(p^i))$ in the following way. The space $B\mathbf{Z}/(p^i)$ is an S^1–bundle over $\mathbf{C}P^\infty$. As such, its $K(n)$-cohomology can be computed with a Gysin sequence similar to the one for ordinary cohomology. One needs to know the Euler class of the bundle, and this can be derived easily from (2.2). The result is

Proposition 2.3

$$K(n)^*(B\mathbf{Z}/(p^i)) = K(n)^*(\text{pt.})[x]/(x^{p^{ni}}).$$

Using, (2.1), this can easily be extended to finite abelian groups. Notice that for a finite abelian p-group A, the rank of $K(n)^*(BA)$ is the nth power of the order of A.

The reduction map $r : \mathbf{Z}/(p^{m+1}) \to \mathbf{Z}/(p^m)$ induces a homomorphism

$$K(n)^*(B\mathbf{Z}/(p^m)) \xrightarrow{r^*} K(n)^*(B\mathbf{Z}/(p^{m+1}))$$

which sends x to $v_n x^{p^n}$ and is therefore one–to–one. On the other hand, the inclusion map $i : \mathbf{Z}/(p^m) \to \mathbf{Z}/(p^{m+1})$ induces a surjection. From these we can easily deduce

Proposition 2.4 *Let $h : A \to A'$ be a homomorphism of finite abelian groups. Then $K(n)^*(h)$ is onto if h is one-to-one and $K(n)^*(h)$ is one-to-one if h is onto. If A and A' are also p-groups, the converse statements are also true.*

Notice that nothing like this holds for ordinary cohomology.

One can make similar computations with $E(n)$-theory. Again one has a formal group law F with a p-series satisfying

$$\begin{aligned} [p](x) &\equiv px && \bmod (x^2) \quad \text{and} \\ [p](x) &\equiv v_n x^{p^n} && \bmod (p, v_1, \cdots v_{n-1}). \end{aligned} \tag{2.5}$$

It follows that

$$\begin{aligned} [p^i](x) &\equiv p^i x && \bmod (x^2) \quad \text{and} \\ [p^i](x) &\equiv v_n^{(p^{ni}-1)/(p^n-1)} x^{p^{ni}} && \bmod (p; v_1, \cdots v_{n-1}). \end{aligned}$$

Furthermore, we have

$$E(n)^*(B\mathbf{Z}/(p^i)) = E(n)^*[[x]]/([p^i](x)). \tag{2.6}$$

After a certain completion, this is a free $E(n)^*$-module of rank p^{ni}, generated by $\{x^j : 0 \leq j < p^{ni}\}$. The analogue of 2.4 holds for $E(n)$-theory.

3 Main results and conjectures

Now we can state our main results, which are partial generalizations of 1.1, 1.2, and 1.9. We will give the second of these first.

1.2 identifies says $K(1)^*(BG)$ is concentrated in even dimensions and its rank is the number of conjugacy classes of elements in G of prime power order. We cannot prove that $K(n)^*(BG)$ is concentrated in even dimensions; this is Conjecture 3.5. We know it is true for finite abelian groups by 2.3. We will discuss this matter further at the end of this Section.

We know that $K(n)^*(BG)$ has finite rank, so we can speak of its Euler characteristic, i.e., the difference between the rank of its even dimensional part and that of its odd dimensional part. We denote this number by $\chi_{n,p}(G)$. From 2.3 we know that for a finite abelian group A,

$$\chi_{n,p}(A) = |A_{(p)}|^n, \tag{3.1}$$

the nth power of the order of the p-component of A.

Like $\chi_{1,p}(G)$ of 1.2, $\chi_{n,p}(G)$ can be described in terms of conjugacy classes. It is the number of *conjugacy classes of of commuting n-tuples* of elements of prime power order in G. More precisely, let

$$G_n = \{(g_1, g_2, \cdots g_n) : [g_i, g_j] = e \text{ for } 1 \leq i < j \leq n\},$$

i.e., the coordinates g_i lie in an abelian subgroup of G. We define $G_{n,p}$ similarly with the additional condition that the order of each g_i is some power of p. G acts on both G_n and $G_{n,p}$ by coordinate-wise conjugation.

Theorem 3.2 *Let $\chi_{n,p}(G)$ be the Euler characteristic of $K(n)^*(BG)$, i.e., the difference in ranks between the even and odd-dimensional components of $K(n)^*(BG)$. It is equal to the number of G–orbits in $G_{n,p}$.*

We have a formula for this number, which will be described and proven below in Section 4. We will outline the proof of Theorem 3.2 in Section 5.

In view of this result it is natural to consider characters as functions defined on orbits of $G_{n,p}$, generalizing the classical case of $n = 1$. A more sophisticated way of viewing $G_{n,p}$ is as follows. Let $\mathbf{Z}_p$ denote the p–adic integers. Observe that, as G–sets,

$$G_{n,p} = \mathrm{Hom}((\mathbf{Z}_p)^n, G);$$

Let $\mathrm{Cl}_{n,p}(G)$ be the ring of $\overline{\mathbf{Q}}_p$ valued conjugacy class functions on $G_{n,p}$. Then

$$\mathrm{Cl}_{n,p}(G) = \mathrm{Map}_G(\mathrm{Hom}(\mathbf{Z}_p^n, G), \overline{\mathbf{Q}}_p).$$

We can now state our main theorem, which is a partial generalization of Theorem 1.1.

Theorem 3.3 *Let A be the ring of integers in a finite extension K of the p–adic numbers, with maximal ideal (π). Given a ring homomorphism*

$$\varphi : E(n)^*(\text{ pt. }) \longrightarrow A \subset \overline{\mathbf{Q}}_p$$

such that $\varphi(I_n) \subset (\pi)$, there exist natural isomorphisms, for all finite groups G,

$$E(n)^*(BG) \otimes_{E(n)^*} \overline{\mathbf{Q}}_p \xrightarrow{\chi_G} \mathrm{Cl}_{n,p}(G),$$

which is an isomorphism after a suitable completion of the source.

This is our analogue of classical character theory:

$$\chi_G : R(G) \otimes \mathbf{C} \xrightarrow{\sim} \mathrm{Cl}(G),$$

where $\mathrm{Cl}(G)$ is the ring of complex valued class functions on G.

Our generalization of 1.9 is

Theorem 3.4 *For every finite group G and for each positive integer n and each prime p, the natural map*

$$E(n)^*(BG) \otimes \mathbf{Z}[|G|^{-1}] \longrightarrow \lim_{A \in \mathcal{A}(G)} E(n)^*(BA) \otimes \mathbf{Z}[|G|^{-1}]$$

is an isomorphism. The same is true if we replace $\mathcal{A}(G)$ by the category of abelian subgroups generated by at most n elements.

This theorem is actually true in much greater generality; $E(n)^*$ can be replaced by *any* complex oriented cohomology theory, provided we state it in terms of the category of $\mathcal{A}(G)$. This is proved in Section 2 of [HKR].

Our main conjecture about $K(n)^*(BG)$ is the following.

Conjecture 3.5 *$K(n)^*(BG)$ is concentrated in even dimensions.*

This is true for $n = 1$ by Atiyah's theorem.

Proposition 3.6 *3.5 is true for G if it is true for a p–Sylow subgroup $H \subset G$.*

Proof. The composite stable map

$$BG \xrightarrow{\mathrm{Tr}} BH \longrightarrow BG$$

induces multiplication by the index of H in G in ordinary homology. Since this index is prime to p, it follows that the map is an equivalence after localizing at p. This means that $BG_{(p)}$ is a retract of $BH_{(p)}$. The result follows. ∎

The conjecture holds for abelian p–groups by 2.3. Tezuka–Yagita [TY] have verified it for the nonabelian groups of order p^3. Theorem 5.4 of [Kuh89] implies that for any stable summand eBG of BG, $K(n)_*(BG)$ has positive Euler characteristic for sufficiently large n.

We can show that it holds for certain wreath products. In order to state this result, we need a preliminary definition.

Definition 3.7 *(a) For a finite group G, an element $x \in K(n)^*(BG)$ is* **good** *if it is a transferred Euler class of a complex subrepresentation of G, i.e., a class of the form $\mathrm{Tr}^*(e(\rho))$ where ρ is a complex representation of a subgroup $H < G$, $e(\rho) \in K(n)^*(BH)$ is its Euler class (i.e., its top Chern class, this being defined since $K(n)^*$ is a complex oriented theory), and $\mathrm{Tr} : BG \to BH$ is the transfer map.*

(b) G is **good** *if $K(n)^*(BG)$ is spanned by good elements as a $K(n)^*$–module.*

Recall that a representation ρ of a subgroup $H \subset G$ leads to an induced representation $\mathrm{Ind}_H^G(\rho)$ of G. Its degree is that of ρ times the index of H in G. One also has a stable transfer map

$$BG \xrightarrow{\mathrm{Tr}} BH.$$

The induced map K–theory sends the element corresponding (under Atiyah's isomorphism) to ρ to that corresponding to $\mathrm{Ind}_H^G(\rho)$.

However, $\mathrm{Tr}^*(e(\rho)) \neq e(\mathrm{Ind}_H^G(\rho))$ in general. For good G, $K(n)^*(BG)$ cannot be described only in terms of representations of G itself. If G is good then $K(n)^*(BG)$ is of course concentrated in even dimensions. We know of no groups which are not good in this sense, so we could strengthen the conjecture by saying that all finite groups are good.

With this definition we have

Theorem 3.8 *If a finite group G is good, then so is the wreath product $W = \mathbf{Z}/(p) \wr G$.*

A variant of this has been proved independently by Hunton [Hun90].

Corollary 3.9 *Let Σ_k denote the symmetric group on k letters. Then*

$$K(n)^*(B\Sigma_k)$$

is concentrated in even dimensions for all k.

Proof. The p–Sylow subgroup of Σ_k is a direct sum of iterated wreath products of $\mathbf{Z}/(p)$ with itself, and is therefore good by 3.8. The result follows by transfer arguments. ∎

Our conjecture is closely related to the one that says that $E(n)^*(BG)$ is torsion free and even–dimensional. Note that then the map of Theorem 1.8 would be monic *before* inverting $|G|$.

Part of Atiyah's proof goes as follows. We want to show that $K(1)_*(BG)$ is concentrated in even dimensions. It suffices to do this for p–groups. Each p–group G has a normal subgroup H of index p. We can assume inductively that $K(1)^*(BH)$ is concentrated in even dimensions. We can study the spectral sequence associated with the fibration

$$BH \longrightarrow BG \longrightarrow B\mathbf{Z}/(p)$$

converging to $K(1)^*(BG)$ with

$$E_2 = H^*(\mathbf{Z}/(p);\ K(1)^*(BH)),$$

where this cohomology has twisted coefficients based on the action of $\mathbf{Z}/(p)$ on H by conjugation. One wants to show that this spectral sequence collapses.

We know inductively that $K(1)^*(BH)$ has a basis corresponding to certain *irreducible* characters of H, i.e., characters associated with irreducible representations. These characters are clearly permuted by outer automorphisms of H, so it follows that $\mathbf{Z}/(p)$ acts on $K(1)^*(BH)$ by permutations. From this it follows easily that the E_2–term is concentrated in even dimensions and the spectral sequence collapses as desired.

The difficulty in generalizing this argument to $n > 1$ is that $\mathbf{Z}/(p)$ need not act on $K(n)^*(BH)$ via a permutation representation so we cannot say that the spectral sequence collapses.

4 Counting the orbits in $G_{n,p}$

In this section we will derive our formula for the number of G–orbits in $G_{n,p}$, i.e. the number of conjugacy classes of commuting n–tuples of elements whose order is a power of the prime p. This is an exercise in elementary group theory. Before proving it we will illustrate the formula with some examples. In this section no mention will be made of Morava K–theory. The notation $\chi_{n,p}(G)$ is used here only for convenience to denote the number of G–orbits in $G_{n,p}$. The connection of this number with Morava K–theory will be discussed in Section 5.

The formula for $\chi_{n,p}(G)$ is

Proposition 4.1 *The number of G–orbits in $G_{n,p}$ is*

$$\chi_{n,p}(G) = \sum_{A<G} \frac{|A|}{|G|}\mu_G(A)\chi_{n,p}(A)$$

where the sum is over all abelian subgroups $A < G$ and μ_G is a Möbius function defined recursively by

$$\sum_{A<A'} \mu_G(A') = 1$$

where the sum is over all abelian subgroups $A' < G$ which contain A. (In particular, $\mu_G(A) = 1$ when A is maximal.)

The Möbius function μ can be defined as above for any partially ordered set in which each element is dominated by only finitely many elements in the set. The classical arithmetic Möbius function is obtained (up to a factor of -1) in this way from the poset of proper subgroups of the integers.

If G is an abelian group, then $\mu_G(G) = 1$ and μ_G vanishes on all proper subgroups, since each is contained in precisely one maximal abelian subgroup. Thus the sum has but one terms and the formula is tautologous in this case.

Now we will consider the quaternion group Q_8 again. The lattice of abelian subgroups is

$$0 \longrightarrow \mathbf{Z}/(2) \longrightarrow 3(\mathbf{Z}/(4))$$

It follows that $\mu_{Q_8}(\mathbf{Z}/(2)) = -2$, while μ_{Q_8} vanishes on the trivial subgroup. (In general it can be shown that μ_G vanishes on any abelian subgroup not containing the center of G.)

Thus the formula gives

$$\begin{aligned} \chi_{n,2}(Q_8) &= \frac{3 \cdot 4 \cdot 4^n - 2 \cdot 2 \cdot 2^n}{8} \\ &= \frac{3 \cdot 4^n - 2^n}{2}. \end{aligned}$$

Note that when $n = 1$ this gives 5, the number of conjugacy classes in Q_8, and that for all n it gives an integer.

If we replace Q_8 by the dihedral group D_8 of order 8, the lattice of abelian subgroups has the same structure, but with two of the $\mathbf{Z}/(4)$s replaced by $(\mathbf{Z}/(2))^2$. This change will not alter any of the numbers in the formula, so we have

$$\chi_{n,2}(D_8) = \chi_{n,2}(Q_8).$$

Now we consider the symmetric group on three letters Σ_3. (It should be noted that the formula for $\chi_{n,p}(G)$ requires us to sum over *all* abelian subgroups, not just those which are p-groups.) Here there are four maximal abelian subgroups, three of order 2 and one of order 3. The intersection of any pair of them is the trivial subgroup. It follows that the value of μ_{Σ_3} on the trivial subgroup is -3. Hence the formula gives

$$\begin{aligned} \chi_{n,p}(\Sigma_3) &= \frac{3 \cdot 2\chi_{n,p}(\mathbf{Z}/(2)) + 3\chi_{n,p}(\mathbf{Z}/(3)) - 3}{6} \\ \chi_{n,2}(\Sigma_3) &= \frac{3 \cdot 2 \cdot 2^n + 3 - 3}{6} \\ &= 2^n \\ &= \chi_{n,2}(\mathbf{Z}/(2)) \\ \chi_{n,3}(\Sigma_3) &= \frac{3 \cdot 2 + 3 \cdot 3^n - 3}{6} \\ &= \frac{3^n + 1}{2} \end{aligned}$$

Recall that $B\Sigma_3$ localized at the prime 3 is one of two stable summands of $B\mathbf{Z}/(3)$, and we see that $\chi_{n,3}(\Sigma_3)$ is roughly half of $\chi_{n,3}(\mathbf{Z}/(3))$.

Next we consider the alternating group A_4. It has five maximal abelian subgroups, four of order three and one isomorphic to $(\mathbf{Z}/(2))^2$. The latter has three subgroups of order 2. Thus we have

$$\begin{array}{ccccc} 0 & \longrightarrow & 3(\mathbf{Z}/(2)) & \longrightarrow & (\mathbf{Z}/(2))^2 \\ \downarrow & & & & \\ 4(\mathbf{Z}/(3)) & & & & \end{array}$$

The Möbius function μ_{A_4} vanishes on each subgroup of order 2, and its value on the trivial subgroup is -4. Thus we have

$$\begin{aligned} \chi_{n,p}(A_4) &= \frac{4\cdot 3\chi_{n,p}(\mathbf{Z}/(3)) + 4\chi_{n,p}(\mathbf{Z}/(2))^2 - 4}{12} \\ \chi_{n,2}(A_4) &= \frac{4\cdot 3 + 4\cdot 4^n - 4}{12} \\ &= \frac{4^n + 2}{3} \\ \chi_{n,3}(A_4) &= \frac{4\cdot 3\cdot 3^n + 4 - 4}{12} \\ &= 3^n \\ &= \chi_{n,3}(\mathbf{Z}/(3)). \end{aligned}$$

Finally, we look at the symmetric group on 4 letters, Σ_4. It has three cyclic subgroups of order 4. In each case the subgroup of order 2 is contained in A_4. There are six additional subgroups of order 2, each generated by a transposition. There are three more noncyclic subgroups of order 4. In each of them one of the three subgroups of order 2 is contained in A_4 and the other two are not.

Thus the diagram of abelian subgroups is

$$\begin{array}{ccccc} 6(\mathbf{Z}/(2)) & & 3((\mathbf{Z}/(2))^2) & & \\ \uparrow & & \uparrow & & \\ 0 & \longrightarrow & 3(\mathbf{Z}/(2)) & \longrightarrow & (\mathbf{Z}/(2))^2 \\ \downarrow & & \downarrow & & \\ 4(\mathbf{Z}/(3)) & & 3(\mathbf{Z}/(4)) & & \end{array}$$

It follows that μ_{Σ_4} vanishes on the six subgroups of order 2 not contained in $_4$ and takes the value -2 on the three that are. Its value on the trivial subgroup is -4. Thus the formula gives

$$\begin{aligned} \chi_{n,p}(\Sigma_4) &= \frac{4\cdot 3\chi_{n,p}(\mathbf{Z}/(3)) + 4\cdot 4\chi_{n,p}(\mathbf{Z}/(2))^2 + 3\cdot 4\chi_{n,p}(\mathbf{Z}/(4))}{24} \\ &\quad - \frac{6\cdot 2\chi_{n,p}(\mathbf{Z}/(2)) + 4}{24} \\ \chi_{n,2}(\Sigma_4) &= \frac{4\cdot 3 + 4\cdot 4\cdot 4^n + 3\cdot 4\cdot 4^n - 6\cdot 2\cdot 2^n - 4}{24} \\ &= \frac{7\cdot 4^n - 3\cdot 2^n + 2}{6} \end{aligned}$$

$$\begin{aligned}\chi_{n,3}(\Sigma_4) &= \frac{4\cdot 3\cdot 3^n + 4\cdot 4 + 3\cdot 4 - 6\cdot 2 - 4}{24}\\ &= \frac{3^n+1}{2}\\ &= \chi_{n,3}(\Sigma_3)\end{aligned}$$

Proof of 4.1. We first treat the case $n = 0$. In this case we are counting conjugacy classes of 0–tuples, so the answer should be one. In particular $\chi_{0,p}(A) = 1$ for any abelian group A. If we multiply the right hand side by $|G|$ we get

$$\sum_{A<G} |A|\mu_G(A) = \sum_{A<G}\sum_{g\in A} \mu_G(A).$$

since $|A|$ is the number of elements in A. If we change the order of summation we get

$$\sum_{g\in G}\sum_{A\ni g} \mu_G(A). \tag{4.2}$$

where the second sum is over all abelian subgroups containing the element g. These groups also contain the cyclic subgroups generated by g, so the inner sum is 1 by the definition of μ_G. Hence the value of the expression in (4.2) is $|G|$ so $\chi_{0,p}(G) = 1$ for all G as expected.

For $n > 0$ we can write

$$\sum_{A<G} \frac{|A|}{|G|}\mu_G(A)\chi_{n,p}(A) = \sum_{A<G} \frac{1}{|G|} \sum_{g_0, g_1, \cdots, g_n \in A} \mu_G(A)$$

where the elements $g_1, \cdots, g_n \in A$ all have order a power of p, since $\chi_{n,p}(A) = |A_{(p)}|^n$. Changing the order of summation gives

$$\sum_{A<G} \frac{|A|}{|G|}\mu_G(A)\chi_{n,p}(A) = \sum_{g_0, g_1, \cdots, g_n \in G} \frac{1}{|G|} \sum_{A\ni g_0, g_1, \cdots, g_n} \mu_G(A)$$

where the outer sum is over all $(n+1)$–tuples (g_i) of commuting elements with all but g_0 having order a power of p. Again the inner sum is one because it is taken over all the abelian subgroups containing the one generated by the g_i, so we have

$$\sum_{A<G} \frac{|A|}{|G|}\mu_G(A)\chi_{n,p}(A) = \sum_{g_0, g_1, \cdots, g_n \in G} \frac{1}{|G|}.$$

Now G acts on the n–tuple $(g_1, \cdots, g_n)$ by coordinate–wise conjugation and the isotropy group is precisely the set of elements g_0 which commute with each of $g_1, \cdots, g_n$. We can write

$$\sum_{g_0, g_1, \cdots, g_n \in G} \frac{1}{|G|} = \sum_{(g_1, \cdots, g_n)\in G_{n,p}} \sum_{g_0} \frac{1}{|G|},$$

and the inner sum is one over the number n–tuples in $G_{n,p}$ conjugate to $(g_1, \cdots, g_n)$. It follows that our expression is the number of G–orbits in $G_{n,p}$ as claimed. ∎

5 The Euler characteristic of $K(n)^*(BG)$

In this section we will outline the proof of Theorem 3.2. Thus our aim is to prove the formula of 4.1 with the understanding that $\chi_{n,p}(G)$ denotes the Euler characteristic of $K(n)^*(BG)$. We already know $\chi_{n,p}(A)$ for finite abelian groups A by 2.3. The outline given here will be more pedestrian than the proof given in [HKR], more along the lines in which we first thought of it.

Let $\rho : G \to U(m)$ be a faithful unitary representation of G, so that G acts freely on $U(m)$. This gives a G–action on the flag manifold

$$F(m) = U(m)/T^n$$

(where T^n denotes the maximal torus in $U(m)$, i.e., the group of diagonal matrices) in which every isotropy group is abelian. The Borel construction

$$X = F(m) \times_G EG$$

(where EG is a free contractible G–space) is an $F(m)$–bundle over BG. It is easy to show that its $K(n)$–theoretic Euler characteristic, which we will denote by $\chi_{n,p}(X)$, is $m!$ times that of BG since $\chi_{n,p}(F(m)) = m!$.

For each abelian subgroup $A \subset G$, let $F(m)^{\langle A\rangle}$ denote the subspace of $F(m)$ consisting of points where the isotropy group is precisely A. Then $F(m)^{\langle A\rangle}$ is an open dense subset of $F(m)^A$, the subspace fixed by A. In fact we have

$$F(m)^A = \bigcup_{A' \supset A} F(m)^{\langle A'\rangle}.$$

We can describe the subspaces $F(m)^A$ explicitly as follows. Each $A \subset G$ determines an eigenspace decomposition of the vector space $V = \mathbf{C}^m$ on which G acts. Recall that a point in the flag manifold $F(m)$ is a decomposition of V into one dimensional subspaces. If each of these subspaces is contained in an eigenspace of A, then the flag is fixed by A. If the eigenspace decomposition has the form

$$V = V_1 \oplus V_2 \oplus \cdots V_k$$

where V_i has rank d_i, then $F(m)^A$ is a finite disjoint union of copies of submanifolds of the form

$$F(d_1) \times F(d_2) \times \cdots F(d_k),$$

the number of copies being

$$\frac{m!}{d_1!\, d_2! \cdots d_k!}.$$

(The fact that $F(m)^A$ has the same Euler characteristic as $F(m)$ is a special case of the Lefschetz Fixed Point Theorem.)

Each subspace $F(m)^A \subset F(m)$ is a complex submanifold equipped with a complex normal bundle. The compliment of the zero section in this bundle has trivial Euler characteristic, so we have

$$\chi_{n,p}(F(m)) = \chi_{n,p}(F(m)^A) + \chi_{n,p}(F(m) - F(m)^A),$$

from which we deduce that for each $A \subset G$

$$m! = \chi_{n,p}(F(m)^A) = \sum_{A \subset A'} \chi_{n,p}(F(m)^{\langle A' \rangle}),$$

the sum being over all abelian subgroups A' containing A.

We can solve these equations for $\chi_{n,p}(F(m)^{\langle A \rangle})$, and get

$$\chi_{n,p}(F(m)^{\langle A \rangle}) = m!\, \mu_G(A)$$

where μ_G is the Möbius function defined in Section 4.

The subspace $F(m)^{\langle A \rangle} \subset F(m)$ is not G–invariant because $g \in G$ sends $F(m)^{\langle A \rangle}$ to $F(m)^{\langle gAg^{-1} \rangle}$. Let

$$F(m)^{(A)} = \bigcup_{g \in G} F(m)^{\langle gAg^{-1} \rangle}.$$

It is invariant and each orbit in it is isomorphic to G/A.

It follows that

$$F(m)^{(A)} \times_G EG \simeq F(m)^{(A)}/G \times BA$$

and

$$\chi_{n,p}(F(m)^{(A)} \times_G EG) = \frac{|A|}{|G|} \chi_{n,p}(F(m)^{\langle A \rangle}) \chi_{n,p}(A).$$

Summing over all conjugacy classes (A) of abelian subgroups $A \subset G$, we get

$$\begin{aligned} m!\, \chi_{n,p}(G) &= \chi_{n,p}(F(m) \times_G EG) \\ &= \sum_{(A)} \chi_{n,p}(F(m)^{(A)} \times_G EG) \\ &= \sum_{A \subset G} \frac{|A|}{|G|} \chi_{n,p}(F(m)^{\langle A \rangle}) \chi_{n,p}(A) \\ &= \sum_{A \subset G} \frac{|A|}{|G|} m!\, \mu_G(A) \chi_{n,p}(A), \end{aligned}$$

which proves Theorem 3.2.

6 The Lubin–Tate construction

In this section we will describe the Lubin–Tate construction [LT65], which is a description of the maximal totally ramified abelian extension of a local field K of characteristic 0 (i.e., of a finite extension of the p–adic numbers $\mathbf{Q}_p$) using formal group laws. Accounts can also be found in [CGMM67] and [Haz78, Section 32]. We need these fields because they are the natural targets for our generalized characters, to be described in the next section.

Recall that the maximal abelian extension of the rationals $\mathbf{Q}$ can be obtained by adjoining all of the roots of unity in $\mathbf{C}$. These are the elements of finite order, or torsion points in the multiplicative group of the complex numbers.

Now suppose we are given a formal group law F defined over the ring of integers A of some finite extension K of the p–adic numbers $\mathbf{Q}_p$. This is power series $F(x, y)$ with certain properties described above. If x and y are elements in the maximal ideal of A, then this power series will converge, since A is complete. The same is true if x and y lie in the maximal ideal $\mathbf{m}$ of the completion of the algebraic closure $\overline{\mathbf{Q}}_p$ of K. In this way we obtain a group structure on $\mathbf{m}$. We will denote this group by $\mathbf{m}_F$.

When F is the additive formal group law $x+y$, then we get the usual additive group structure on $\mathbf{m}$. In this case there are no nontrivial elements of finite order.

When F is the multiplicative formal group law, we get the usual multiplicative groups structure on $1+\mathbf{m}$. The elements of finite order are roots of unity congruent to 1 modulo $\mathbf{m}$, i.e., the $(p^i)^{\text{th}}$ roots of unity for various i. The field obtained by adjoining all of these elements to $\mathbf{Q}_p$ is the maximal totally ramified abelian extension.

The main result of [LT65] is that with a suitable choice of the formal group law F, the field obtained by adjoining all of the elements of finite order in $\mathbf{m}_F$ is the maximal totally ramified abelian extension of the given field K. In order to specify this choice, we need the notion of a *formal A–module*, which is a formal group law over an A–algebra R with certain additional structure. (Recall that A is the ring of integers in K, a finite extension of $\mathbf{Q}_p$.).

Associated with any formal group law there are power series $[n](x)$ for integers n satisfying

(i) $[1](x) = x$,

(ii) $F([m](x),[n](x)) = [m+n](x)$ and

(iii) $[m]([n](x)) = [mn](x)$.

When F is defined over a $\mathbf{Z}_p$–algebra, one can use continuity to extend this definition to $[u](x)$ for any $u \in \mathbf{Z}_p$. *F is a formal A–module if we can define $[a](x)$ with similar properties for all $a \in A$.* An account of theory of formal A–modules can be found in [Haz78, Section 21].

Theorem 6.1 (Lubin–Tate) *Let A be the ring of integers in a finite extension K of the p–adic numbers $\mathbf{Q}_p$. Let $\pi \in A$ be a generator of the maximal ideal and let q be the cardinality of the residue field $A/(\pi)$. Let $f(x) \in A[[x]]$ be a power series with*

$$f(x) \equiv \pi x \bmod (x)^2 \quad \textit{and} \quad f(x) \equiv u x^q \bmod (\pi),$$

where u is a unit in A.

Then

(i) There is a unique formal A–module F over A for which $[\pi](x) = f(x)$.

(ii) The field obtained by adjoining the elements of finite order in $\mathbf{m}_F$ is the maximal totally ramified abelian extension L of K. These elements are the roots of the iterates of f. Let L_i denote the field obtained by adjoining the elements of order dividing $(q-1)q^{i-1}$ in $\mathbf{m}_F$, i.e., by the roots of $f^{\circ i}$, the ith iterate of f.

(iii) The Galois group $\mathrm{Gal}(L:K)$ is isomorphic to the group of units $A^\times$ in A. For x an element of finite order in $\mathbf{m}_F$, the image of x under the automorphism corresponding to $a \in A^\times$ is $[a](x)$. If x is a root of $f^{\circ i}$, then it it fixed by this automorphism if a is congruent to 1 modulo $(\pi)^i$.

When $A = \mathbf{Z}_p$, we can take

$$f(x) = (1+x)^p - 1 = \sum_{1 \leq k \leq p} \binom{p}{k} x^k$$

and the statements in the 6.1 can be readily verified. In this case F is the multiplicative formal group law $x + y + xy$ and the roots of the of $f^{\circ i}$ are elements of the form $\zeta - 1$, where ζ is a $(p^i)^{\text{th}}$ root of unity.

More generally, when f is a polynomial of degree q (the simplest example is $\pi x + x^q$), then $g_1(x) = f(x)/x$ is irreducible by Eisenstein's criterion. The same is true of the polynomials $g_i(x)$ defined inductively for $i > 1$ by $g_i(x) = g_{i-1}(f(x))$. L_i is the splitting field for $g_i(x)$.

7 Generalized characters and $E(n)^*(\mathbf{BG})$

In this section we will describe a homomorphism

$$E(n)^*(BG) \xrightarrow{\chi_G} \mathrm{Cl}_{n,p}(G), \tag{7.1}$$

the target being the ring of $\overline{\mathbf{Q}}_p$ valued conjugacy class functions on $G_{n,p}$, the set of commuting n-tuples of elements of prime power order in G.

As remarked in Section 3, an element $\gamma \in G_{n,p}$ is the same thing as a homomorphism

$$(\mathbf{Z}_p)^n \longrightarrow G.$$

Since G is finite, this factors through $(\mathbf{Z}/(p^i))^n$ for sufficiently large i. Thus we get a map

$$E(n)^*(BG) \xrightarrow{\gamma_i^*} E(n)^*(B(\mathbf{Z}/(p^i))^n).$$

Letting i go to ∞, we get a map

$$E(n)^*(BG) \xrightarrow{\gamma^*} \varinjlim_i E(n)^*(B(\mathbf{Z}/(p^i))^n).$$

Note that this direct limit is *not* the same as

$$E(n)^*(\varinjlim_i B(\mathbf{Z}/(p^i))^n),$$

which is far less interesting in this context.

We could define χ_G as in (7.1) if we had a suitable map

$$\varinjlim_i E(n)^*(B(\mathbf{Z}/(p^i))^n) \longrightarrow \overline{\mathbf{Q}}_p \tag{7.2}$$

This is where the Lubin–Tate construction comes into the picture.

We will illustrate with an easy example. Let the field K be the unramified extension of $\mathbf{Q}_p$ of degree n, and let π (the generator of its maximal ideal) be p. The cardinality q

of the residue field $A/(p)$ is p^n. The homomorphism φ as in Theorem 3.3 must send v_n to a unit in A since v_n is invertible in $E(n)^*$. Then the conditions on $f(x)$ in Theorem 6.1 are identical to those on $\varphi([p](x))$ in (2.5).

This means that (2.6) translates into

$$E(n)^*(B\mathbf{Z}/(p^i)) \otimes K = K[[x]]/(f^{\circ i}(x))$$

and we can extend φ to a homomorphism

$$E(n)^*(B\mathbf{Z}/(p^i)) \xrightarrow{\varphi} L_i$$

by sending x to a root of g_i. We can do this compatibly for all i and get a map

$$\varinjlim_i E(n)^*(B\mathbf{Z}/(p^i)) \xrightarrow{\varphi} L. \tag{7.3}$$

We will extend this further as in (7.2) in such a way that the characters given by (7.1) will be Galois equivariant in the sense of 1.5. The construction of φ in (7.3) depends on a choice of roots $r_i \in L$ of g_i satisfying

$$r_i = f(r_{i+1}).$$

Recall that the Galois group $\mathrm{Gal}(L : K)$ is isomorphic to the group of units $A^\times$ in A. We can regard $(\mathbf{Z}/(p^i))^n$ as $A/(p^i)$. Then we can define

$$\varinjlim_i E(n)^*(BA/(p^i)) \xrightarrow{\varphi} L$$

to be the equivariant extension of the φ of (7.3).

We can make this more explicit as follows. We have

$$E(n)^*(BA/(p^i)) = E(n)^*[[x_0, x_1, \cdots x_{n-1}]]/([p^i](x_j)).$$

Pick a $\mathbf{Z}_p$–basis of A of the form $\{a_0 = 1, a_1, \cdots a_{n-1}\}$. Then extend φ from $E(n)^*(B\mathbf{Z}/(p^i))$ to $E(n)^*(BA/(p^i))$ by defining

$$\varphi(x_j) = [a_j](r_i). \tag{7.4}$$

This map φ enables us to define

$$E(n)^*(BG) \otimes_{E(n)^*} K \xrightarrow{\chi_G} \mathrm{Cl}_{n,p}(G, L)^{A^\times}, \tag{7.5}$$

where $\mathrm{Cl}_{n,p}(G, L)^{A^\times}$ denotes the ring of Galois equivariant L–valued conjugacy class functions on $G_{n,p}$.

The Galois equivariance of these characters is not mentioned in [HKR]. If we replace K by L in the source of (7.5), we can drop the equivariant condition on the target. In the theorem stated in [HKR], L is replaced by $\overline{\mathbf{Q}}_p$.

We will now give an indication of why the map of (7.5) is an isomorphism after a suitable completion of the source. It is not hard to show that the target behaves well with respect to abelian subgroups, i.e. that it satisfies an analogue of Artin's Theorem

(1.8). (This is Lemma 4.11 of [HKR].) Thus Theorem 3.4 can be used to reduce to the case of abelian groups. Then a routine argument reduces it further to the case of finite cyclic p-groups.

Therefore we will examine the case $G = \mathbf{Z}/(p^i)$ in more detail. In this case the target of χ_G is a free K-module of rank p^{ni}. The same is true of the source after suitable completion, with the generators being the powers of the orientation class $x \in E(n)^2(BG)$. Thus χ_G can be represented by a $(p^{ni} \times p^{ni})$ matrix M_G over K, which must be shown to be nonsingular.

We will use the notation of (7.4). Recall that

$$G_{n,p} = \mathrm{Hom}(A, G).$$

Given $\theta \in \mathrm{Hom}(A, G)$, we need to compute

$$\chi_G(x)((\theta)) \;=\; \varphi(\theta^*(x)) \in L,$$

which must be a root of $f^{\circ i}$. Let θ_j for $0 \leq j \leq p^{ni} - 1$ denote the p^{ni} elements of $G_{n,p}$ and let

$$\lambda_j = \chi_G(x)(\theta_j) \in L.$$

These λ_j are the p^{ni} distinct roots of $f^{\circ i}$, and we have

$$\chi_G(x^k)(\theta_j) = \lambda_j^k \in L.$$

It follows that M_G is a Vandermonde matrix with entries

$$m_{j,k} = \lambda_j^k.$$

Therefore it is nonsingular, completing our outline of the proof of Theorem 3.3.

8 The wreath product theorem

In this section we will prove Theorem 3.8. Recall the definition of good groups given in 3.7. The following properties of such groups are immediate.

Proposition 8.1 *(a) G is good if its p-Sylow subgroup is good.*

(b) If G_1 and G_2 are good then so is their product $G_1 \times G_2$.

(c) Every finite abelian group is good. (In fact, if G is abelian then $K(n)^*(BG)$ is generated by Euler classes of representations of G itself.)

Before specializing to the wreath product situation we note

Lemma 8.2 *If $f : H \to G$ is a homomorphism and $x \in K(n)^*(BG)$ is good, then $f^*(x)$ is a linear combination of good elements in $K(n)^*(BH)$.*

Proof. Suppose $x = \mathrm{Tr}(e(\rho))$ where ρ is a representation of $K < G$. Then there is a pullback diagram of spaces of the form

$$\begin{array}{ccc} \coprod BH_\alpha & \xrightarrow{\coprod f_\alpha} & BK \\ \downarrow & & \downarrow \\ BH & \xrightarrow{f} & BG \end{array}$$

where each H_α is a subgroup of H. This happens because the pullback is a covering of BH whose degree is the index k of K in G. Therefore its higher homotopy groups vanish and it must be a disjoint union of the indicated form. The sum of the indices of the H_α must be k.

By naturality of the transfer,

$$f^*(x) = \sum_\alpha \mathrm{Tr}^* e(f_\alpha^*(\rho)). \quad \blacksquare$$

Corollary 8.3 *If x and y are good elements of $K(n)^*(BG)$ then their cup product xy is a sum of good elements.*

Proof. $x \times y \in K(n)^*(BG \times G)$ is good and $xy = \Delta^*(x \times y)$ where $\Delta : G \to G \times G$ is the diagonal map. $\blacksquare$

To prove the Theorem 3.8, we study the extension

$$G^p \longrightarrow W \longrightarrow \mathbf{Z}/(p)$$

and the associated spectral sequence

$$E_2 = H^*(\mathbf{Z}/(p); K(n)^*(BG^p)) \Rightarrow K(n)^*(BW). \tag{8.4}$$

$\mathbf{Z}/(p)$ acts on $K(n)^*(BG^p)$ by permuting the factors. Thus, as a module over $\mathbf{Z}/(p)$, we have

$$K(n)^*(BG^p) = F \oplus T$$

where F is a free $\mathbf{Z}/(p)$-module and T has trivial $\mathbf{Z}/(p)$-action. If $\{x_i\}$ is a basis of $K(n)^*(BG)$, then F has basis

$$\{x_{i_1} \otimes x_{i_2} \cdots \otimes x_{i_p}\}$$

where the subscripts $i_1, \ldots i_p$ are not all the same, and T has basis $\{Px_i\}$ where $Px_i = x_i \otimes x_i \cdots x_i$. Moreover

$$H^i(\mathbf{Z}/(p); F) = \begin{cases} F^{\mathbf{Z}/(p)} & \text{for } i = 0 \\ 0 & \text{for } i > 0 \end{cases}$$

and

$$H^*(\mathbf{Z}/(p); T) = T \otimes H^*(B\mathbf{Z}/(p))$$

where

$$H^*(B\mathbf{Z}/(p)) = E(u) \otimes P(x)$$

with $u \in H^1$ and $x \in H^2$.

We need to show that each element in

$$E_2^{0,*} = H^0(\mathbf{Z}/(p); T \oplus F)$$

is a permanent cycle. Once we have this, then we know that the spectral sequence has a differential of the form

$$d_{2p^n-1}(u) = v_n x^{p^n}$$

because this happens in the case when G is trivial, i.e., in the Atiyah–Hirzebruch spectral sequence for $K(n)^*(B\mathbf{Z}/(p))$. It would then follow that

$$E_{2p^n} = E_\infty = H^0(\mathbf{Z}/(p); F) \oplus (H^0(\mathbf{Z}/(p); T) \otimes K(n)^*(B\mathbf{Z}/(p)))$$

which is concentrated in even dimensions. We will also show that W is good. Thus we will need the following two lemmas.

Lemma 8.5 *Each element in $H^0(\mathbf{Z}/(p); F)$ is a permanent cycle in the spectral sequence (8.4).*

Lemma 8.6 *Each element in $H^0(\mathbf{Z}/(p); T)$ is a permanent cycle in the spectral sequence (8.4).*

Proof of Lemma 8.5. Let $x = x_{i_1} \otimes \cdots x_{i_p}$ be a basis element of F and let $\sigma(x)$ denote the sum of x and all of its conjugates under the action of $\mathbf{Z}/(p)$. Then $H^0(\mathbf{Z}/(p); F)$ is spanned by these $\sigma(x)$. Moreover $\sigma(x)$ is the image of x under the composite

$$K(n)^*(BG^p) \xrightarrow{\mathrm{Tr}^*} K(n)^*(BW) \xrightarrow{\mathrm{Res}^*} K(n)^*(BG^p).$$

(Res^* here denotes the restriction map induced by the inclusion $G^p \to W$.) An element in $E_2^{0,*} \subset K(n)^*(BG^p)$ is a permanent cycle iff it is the restriction of an element in $K(n)^*(BW)$. Hence this is true of each element of $H^0(\mathbf{Z}/(p); F) = F^{\mathbf{Z}/(p)}$. ■

Proof of 8.6. As in the proof above we need to show that for each basis element $x_i \in K(n)^*(BG)$, $Px_i \in K(n)^*(BG^p)$ is the restriction of an element $y \in K(n)^*(BW)$.

We can assume that x_i is a transferred Euler class $\mathrm{Tr}^*(e(\rho))$ for ρ a complex representation of some subgroup $H < G$. The representation $\rho \oplus \rho \oplus \cdots \rho$ of H^p extends to a representation $\hat{\rho}$ of $\mathbf{Z}/(p) \wr H$ and $e(\hat{\rho})$ restricts to $P(e(\rho))$.

The following diagram commutes.

$$\begin{array}{ccc} K(n)^*(BH^p) & \xleftarrow{\mathrm{Res}^*} & K(n)^*(B(\mathbf{Z}/(p) \wr H)) \\ \downarrow \mathrm{Tr}^* & & \downarrow \mathrm{Tr}^* \\ K(n)^*(BG^p) & \xleftarrow{\mathrm{Res}^*} & K(n)^*(BW) \end{array}$$

Hence we have

$$\begin{aligned} \mathrm{Res}^*\mathrm{Tr}^*(e(\hat{\rho})) &= \mathrm{Tr}^*\mathrm{Res}^* e(\hat{\rho}) \\ &= \mathrm{Tr}^*(Pe(\rho)) \\ &= P\mathrm{Tr}^*(e(\rho)) \\ &= Px_i \end{aligned}$$

so we can take $y = \mathrm{Tr}^*(e(\hat{\rho}))$. ■

Proof of Theorem 8.8. We have shown that each element in $E_2^{0,*}$ is a permanent cycle, so the spectral sequence has only one differential. It remains to show that W is good. We have

$$K(n)^*(BW) = H^0(\mathbf{Z}/(p); F) \oplus (H^0(\mathbf{Z}/(p); T) \otimes K(n)^*(B\mathbf{Z}/(p))).$$

$H^0(\mathbf{Z}/(p); F)$ is in the image of the transfer by Lemma 8.5, and Lemma 8.6 shows that $H^0(\mathbf{Z}/(p); T)$ is generated by elements of the form $\mathrm{Tr}^*(e(\hat{\rho}))$ where $\hat{\rho}$ is a representation of some subgroup of W. Recall that

$$K(n)^*(B\mathbf{Z}/(p)) = K(n)^*[x]/(x^{p^n})$$

where $x = e(\lambda)$, λ being a one-dimensional representation of of $\mathbf{Z}/(p)$. It follows from 8.3 that $\mathrm{Tr}^* e(\hat{\rho}) x^i$ is a sum of transferred Euler classes, completing the proof. ■

References

[Ati61] M. F. Atiyah. Characters and cohomology of finite groups. *Inst. Hautes Études Sci. Publ. Math.*, 9:23–64, 1961.

[Bak89] A. Baker. On the homotopy type of the spectrum representing elliptic cohomology. *Proceedings of the American Mathematical Society*, 107:537–548, 1989.

[CGMM67] E. B. Curtis, P. Goerss, M. E. Mahowald, and R. J. Milgram. J.-p. serre. In J. W. S. Cassells and A. Fröhlich, editors, *Algebraic Number Theory*, pages 128–161, Thompson Book Company, 1967.

[DHS88] E. Devinatz, M. J. Hopkins, and J. H. Smith. Nilpotence and stable homotopy theory. *Annals of Mathematics*, 128:207–242, 1988.

[Haz78] M. Hazewinkel. *Formal Groups and Applications.* Academic Press, New York, 1978.

[HKR] M. J. Hopkins, N. J. Kuhn, and D. C. Ravenel. Generalized group characters and complex oriented cohomology theories. To appear.

[Hun90] J. R. Hunton. The Morava K-theories of wreath products. *Math. Proc. Cambridge Phil. Soc.*, 107:309–318, 1990.

[Kuh87] N. J. Kuhn. The mod p K-theory of classifying spaces of finite groups. *Journal of Pure and Applied Algebra*, 44:269–271, 1987.

[Kuh89] N. J. Kuhn. Character rings in algebraic topology. In S. M. Salamon, B. Steer, and W. A. Sutherland, editors, *Advances in Homotopy*, pages 111–126, Cambridge University Press, Cambridge, 1989.

[LT65] J. Lubin and J. Tate. Formal complex multiplication in local fields. *Annals of Mathematics*, 81:380–387, 1965.

[Rav82] D. C. Ravenel. Morava K–theories and finite groups. In S. Gitler, editor, *Symposium on Algebraic Topology in Honor of José Adem*, pages 289–292, American Mathematical Society, Providence, Rhode Island, 1982.

[Rav84] D. C. Ravenel. Localization with respect to certain periodic homology theories. *American Journal of Mathematics*, 106:351–414, 1984.

[Ser67] J.-P. Serre. *Représentations Linéaires des Groupes Finis.* Hermann, Paris, 1967.

[TY] M. Tezuka and N. Yagita. Cohomology of finite groups and Brown–Peterson cohomology. To appear.

[Wil82] W. S. Wilson. *Brown–Peterson homology: an introduction and sampler. C. B. M. S. Regional Conference Series in Mathematics*, American Mathematical Society, Providence, Rhode Island, 1982.

M.J. Hopkins: Department of Mathematics
Massachusetts Institute of Technology
Cambridge, Massachusetts 02139
U.S.A.

N.J. Kuhn: Department of Mathematics
University of Virginia
Charlottesville, VA 22903
U.S.A.

D.C. Ravenel: Department of Mathematics
University of Rochester
Rochester, NY 14627
U.S.A.

1990 Barcelona Conference
on Algebraic Topology.

CLASSIFYING SPACES OF COMPACT SIMPLE LIE GROUPS AND P–TORI

KENSHI ISHIGURO

Zabrodsky shows [19] that if a map $f : BG \to BG'$ induces a trivial homomorphism in K-theory, then f is null homotopic. Here G and G' are compact Lie groups. We investigate a problem which is related to Zabrodsky's result. Namely, we study the question of whether a map $f : BG \to X$ is null homotopic if it is trivial in *mod p* cohomology. In addition, we point out that there are examples which show that K-theory does not eliminate the possibility of the existence of a non-trivial map from BG to BG'. For instance, the K-theory of both $BSO(3)$ and BS^3 are isomorphic to the invariant ring $K(BS^1)^{\mathbb{Z}/2}$ as λ-ring. Consequently the set of λ-maps from $K(BS^3)$ to $K(BSO(3))$ is infinite. However, the set of homotopy classes of maps from $BSO(3)$ to BS^3 consists of only a single element.

The answer to our question is false in general. There exist maps between classifying spaces which are trivial in *mod p* cohomology and are essential; a counter-example is given by an unstable Adams operation of type ψ^p on BG when $G = G'$ and the order of the Weyl group $W(G)$ is prime to p. However, if $|W(G)| \equiv 0$ *mod p*, the Lie group G is simple, and p is odd, then we will see that the answer is affirmative. The case $G = SO(n)$ at $p = 2$ is treated separately.

Theorem 1. *Let G be a compact connected simple Lie group and let p be an odd prime which divides the order of the Weyl group. Suppose X is a p-complete space such that $\pi_1(X)$ is finite, that $H^*(X;\mathbb{F}_p)$ is of finite type, and that the component of the space of based maps containing the constant map $map_*(B\mathbb{Z}/p, X)_0$ is weakly contractible. If $f : BG \to X$ is trivial in mod p cohomology, then the map f is null homotopic.*

The conditions on X are satisfied by the p-completion of the classifying spaces of compact Lie groups or connected *mod p* finite loop spaces, [9]. A result of Dwyer-Wilkerson [5] gives a sufficient condition, in terms of $H^*(X;\mathbb{F}_p)$, that $\mathrm{map}_*(B\mathbb{Z}/p, X)$ can be weakly contractible. Namely, X is a 1-connected p-complete space such that $H^*(X;\mathbb{F}_p)$ is of finite type and that the module of indecomposables $Q(H^*(X;\mathbb{F}_p))$ is locally finite as a module over the Steenrod algebra. We note that $\mathrm{map}_*(B\mathbb{Z}/p, X)_0 \underset{\omega}{\simeq} *$ implies $\mathrm{map}_*(B\pi, X)_0 \simeq *$ for any locally finite group π, [14 §9] and [5].

Our proof uses the classification of simple Lie groups. First we deal with the case of $SU(p)$, the special unitary group in p dimensions. The other cases are more or less reduced to this, using appropriate subgroups of G.

According to work of Lannes [13], the assumption $H^*(f;\mathbb{F}_p)=0$ implies that $f|BV \simeq 0$ for any elementary abelian p-subgroup V of G. It is natural to ask if the assumption $f|BV \simeq 0$ for any V implies $f\simeq 0$. We ask a general question: If $f|BV \simeq 0$ for some V with $Z(G)\cap V=\{1\}$, is f null homotopic? Here $Z(G)$ denotes the center of G. The answer to the general question is yes as long as the simple group G is not the exceptional Lie group G_2 at $p=3$.

Theorem 2. *(a) Let G, X and p be as in Theorem 1, except that G is not the exceptional Lie group G_2. If $f: BG\to X$ is a map with $f|BV\simeq 0$ for some elementary abelian p-subgroup V of G, the either V is included in the center of G or f is null homotopic.*
(b) There is a compact Lie group K with an essential map $f: BG \to (BK)^\wedge_3$ such that $f|B\mathbb{Z}/3\simeq 0$ for some subgroup $\mathbb{Z}/3$ of G_2.

Corollary 1. *Let G, X and p be as in Theorem 1.*
(a) If $f: BG\to X$ is a map with $f|BV\simeq 0$ for any elementary abelian p-subgroup V, then $f\simeq 0$.
(b)If the Krull dimension of $H^*(X;\mathbb{F}_p)$ is less than the rank of G, then any map $f: BG\to X$ is null homotopic.

The following is a generalization of a result of Adams-Mahmud [1, Proposition 2.12].

Corollary 2. *Let p be an odd prime and let G be a compact connected simple Lie group with Weyl group $W(G)$. Assume that X is a p-complete space such that $H^*(X;\mathbb{F}_p)\cong H^*(BT^n;\mathbb{F}_p)^W$, where W is a subgroup of $GL(n,\mathbb{F}_p)$. If $[BG,X]\neq 0$, there is a monomorphism $W(G)\to W$.*

In the case $p=2$, if $G=S^3$, Theorem 1 doesn't hold. The map $BS^3\to BSO(3)$ induced by the universal covering map is a counterexample. If $G=SO(n)$ with $n\geq 3$, but $n\neq 4$, we will show that Theorem 1, Theorem 2(a) and Corollary 1 hold at $p=2$. As an application, the necessary and sufficient condition that there be non-trivial maps between $BSO(2n+1)$ and $BSp(k)$. In particular, we obtain the following:

$$[BSO(2n+1),\ BSp(n)]=0 \text{ for any } n$$
$$[BSp(n),\ BSO(2n+1)]=0 \text{ for } n\geq 3.$$

These results can not be obtained using K-theory. If $n=1$ or 2, the homotopy set $[BSp(n),\ BSO(2n+1)]$ consists of the map induced by the covering projection together with unstable Adams operations on $BSO(2n+1)$, [12]. The groups $SO(2n+1)$ and $Sp(n)$ differ only at $p=2$, [7]. Our result illustrates how different they are.

This paper consists of three sections. In §1, we discuss maps which are trivial in *mod p* cohomology and prove Theorem 1. In §2 the finiteness of homomorphisms over the Steenrod algebra is studied. Theorem 2 and its corollaries are proved. In §3, the case $G=SO(n)$ at $p=2$ is treated. As a consequence, we show that most of the maps between $BSO(2n+1)$ and $BSp(n)$ are null homotopic.

The author would like to thank D. Notbohm and C. Wilkerson for their help.

§1. Maps which are trivial in *mod p* cohomology

We will prove Theorem 1 in this section. Let N_pT denote the inverse image in the normalizer of a maximal torus T of a p-Sylow subgroup W_p of the Weyl group of G. In order to prove that a map $f : BG \to X$ is null homotopic, it suffices to show $f|BN_pT \simeq 0$. If p is odd, the group extension $1 \to T \to N_pT \to W_p \to 1$ splits. The following lemmas allow us to consider T and W_p separately and to reduce the problem eventually to $SU(p)$. At $p = 2$, it is likely that $SO(3)$ plays the role of $SU(p)$.

Lemma 1. *Let $1 \to \kappa \to \pi \to \sigma \to 1$ be a group extension and let the component of the space of based maps containing the constant map $map_*(B\kappa, X)_0$ is weakly contractible. Then we have the following*

(i) If $f : B\pi \to X$ with $f|B\kappa \simeq 0$, then the map f factors through $B\sigma$ up to homotopy.

(ii) The map $[B\sigma, X] \to [B\pi, X]$ is injective.

Proof. (i) Recall that the mapping space $\mathrm{map}(B\pi, X)$ is homotopy equivalent to $\mathrm{map}_\sigma(E\sigma, \mathrm{map}(B\kappa, X))$. Hence we have the homotopy commutative diagram

$$\begin{array}{ccc} map(B\sigma, X) & \longrightarrow & map_\sigma(E\sigma, map(B\kappa, X)_0) \\ \downarrow & & \downarrow \\ map(B\pi, X) & \xrightarrow{\sim} & map_\sigma(E\sigma, map(B\kappa, X)) \end{array}$$

Since $f|B\kappa = 0$, the map f has a preimage in $\mathrm{map}_\sigma(E\sigma, \mathrm{map}(B\kappa, X)_0)$ up to homotopy. Our assumption that $\mathrm{map}_*(B\kappa, X)_0$ is weakly contractible implies that the canonical map $X \to \mathrm{map}(B\kappa, X)_0$ which sends each $x \in X$ to the constant map is a homotopy equivalence. Consequently the upper horizontal map is a homotopy equivalence. Thus there is an element in π_0 map $(B\sigma, X)$ which is sent to $[f] \in \pi_0\ \mathrm{map}(B\pi, X)$.

(ii) The homotopy fixed point set $\mathrm{map}_\sigma(E\sigma, \mathrm{map}(B\kappa, X))$ is decomposed as the disjoint union $\coprod_\alpha \mathrm{map}_\sigma(E\sigma, \mathrm{map}\ (B\kappa, X)_\alpha)$, where $\mathrm{map}(B\kappa, X)_\alpha$ is a connected component invariant under the σ-action, since $E\sigma$ is connected. This implies the desired result. //

For a map $f : B\pi \to X$ where π is a finite p-group and $\mathrm{map}_*(B\mathbf{Z}/p, X) \underset{\omega}{\simeq} *$, we define the kernel of f as $\ker f = \{x \in \pi | f|B\langle x\rangle \simeq 0\}$. Here $\langle x\rangle$ denotes the subgroup of π generated by x. In [6], it is shown that if π is an elementary p-abelian group, then $\ker f$ is a subgroup. The next result was shown by D. Notbohm.

Lemma 2. *For a map $f : B\pi \to X$, if π is a finite abelian p-group, then* $\ker f$ *is a subgroup of π.*

Proof: It is clear that the identity 0 is contained in $\ker f$ and that if $x \in \ker f$, then $-x \in \ker f$. It remains to show that if $x, y \in \ker f$, then $x + y \in \ker f$. We will show that the subgroup generated by x and y, denoted by $\langle x, y\rangle$, is included in $\ker f$. From the exact sequence of abelian groups

$$0 \to \langle x\rangle \to \langle x, y\rangle \to \langle \bar{y}\rangle \to 0$$

for suitable $\overline{y}$, we see $f|B\langle x,y\rangle$ factors through $B\langle\overline{y}\rangle$ by Lemma 1 (i), since $x \in \ker f$. Notice that $\langle y\rangle$ maps onto $\langle\overline{y}\rangle$. Since $y \in \ker f$, Lemma 1 (ii) implies $f|B\langle\overline{y}\rangle \simeq 0$. Consequently $f|B\langle x,y\rangle \simeq 0$. //

The following lemma is analogous to a special case of [11, Lemma 1.1].

Lemma 3. *Suppose H is a subgroup of a compact Lie group G. For a map $f : BG \to X$, if $f|BH \simeq 0$ and H is a conjugate to a subgroup H' of G, then $f|BH' \simeq 0$.*

Lemma 4. *Let p be a prime. Suppose G and X are as in Theorem 1. For a map $f : BG \to X$, if T^n is a maximal torus of G and $f|BT^j \simeq 0$ for some $1 \leq j \leq n$, then $f \simeq 0$.*

Proof: First we show $f|BT^n \simeq 0$. Let $M_k = \{x \in T^n | x^{p^k} = 1 \text{ and } f|B\langle x\rangle \simeq 0\}$. Lemma 2 says that the finite set M_k is a group and that M_{k-1} is a subgroup of M_k. Notice that $T^j \subset \overline{\bigcup_k M_k} \subset T^n$. If H_0 denotes the connected component of the topological group $\overline{\bigcup_k M_k}$ containing the identity, then H_0 is compact abelian connected Lie group. Hence H_0 is a torus. Each element of the Weyl group W is a continuous automorphism of T^n. Thus H_0 is invariant under the W-action. Since G is simple, there are no nontrivial invariant subtorus. Since $T^j \subset H_0$ for some $j \geq 1$, we see $H_0 = T^n$. Therefore $f|BT^n \simeq 0$.

By a result of Friedlander-Mislin [9], it suffices to show $f|B\gamma \simeq 0$ for any finite p-subgroup γ of G. Let N_pT^n denote the inverse image in the normalizer NT^n of a p-Sylow subgroup of $NT^n/T^n = W(G)$. According to [16, Theorem 1], any finite p-subgroup of the compact Lie group G is conjugate to a subgroup of N_pT^n. Define a finite p-subgroup γ_k of N_pT^n as the group extension $1 \to (\mathbf{Z}/p^k)^n \to \gamma_k \to W_p \to 1$, where W_p is the p-Sylow subgroup of $W(G)$. Since any p-subgroup γ is conjugate to a subgroup of γ_k for some k, it suffices to show $f|B\gamma_k \simeq 0$ for any k. The map $f|B\gamma_k$ factors through BW_p since $f|BT^n \simeq 0$. Notice that W_p is a group obtained from elementary abelian p-groups by means of a sequence of split extensions. By the induction on the order of p-group, Lemma 1 together with the fact that any element of G is conjugate to some element of the maximal torus T^n implies $f|BW_p \simeq 0$. Consequently $f|B\gamma_k \simeq 0$ for any k. //

Proof of Theorem 1. If this result holds when G is simply-connected, Lemma 1 (ii) shows the general case, using a universal covering. From Lemma 4, it suffices to find a subtorus T^j in G such that $f|BT^j \simeq 0$ for $j \geq 1$. We use the classification of compact 1-connected simple Lie group. The case of $G = SU(n)$ is first proved. The other cases are essentially reduced to this case, using appropriate subgroups, [3].

Case 1. $\quad G = SU(n)$ with $n \geq p$.

Consider the subgroup $SU(p) \subset SU(n)$. Let $\zeta = e^{2\pi i/p^k}$, a primitive p^k-th root of unity. Define the matrices in $SU(p)$ as follows:

$$a_k(1) = \begin{pmatrix} \zeta & & & & \\ & \zeta^{-1} & & & \\ & & 1 & & \\ & & & \ddots & \\ & & & & 1 \end{pmatrix}, \quad a_k(2) = \begin{pmatrix} 1 & & & & & \\ & \zeta & & & & \\ & & \zeta^{-1} & & & \\ & & & 1 & & \\ & & & & \ddots & \\ & & & & & 1 \end{pmatrix}$$

$$\ldots, a_k(p-1) = \begin{pmatrix} 1 & & & & \\ & \ddots & & & \\ & & 1 & & \\ & & & \zeta & \\ & & & & \zeta^{-1} \end{pmatrix}, \quad b = \begin{pmatrix} 0 & & & 1 \\ 1 & \ddots & & \\ & & \ddots & \\ & & 1 & 0 \end{pmatrix}.$$

If we take T^{p-1} to be the subgroup of diagonal matrices of $SU(p)$, then

$$\gamma_k = \left(\prod_{i=1}^{p-1} \mathbf{Z}/p^k \langle a_k(i) \rangle \right) \rtimes \mathbf{Z}/p\langle b \rangle$$

where

$$b \cdot a_k(i) \cdot b^{-1} = \begin{cases} a_k(i+1) & \text{if } 1 \leq i \leq p-2 \\ \prod_{j=1}^{p-1} a_k(j)^{-1} & \text{if } i = p-1. \end{cases}$$

If $k = 1$, Lemma 1 together with work of Lannes shows $f|B\gamma_1 \simeq 0$. Suppose

$$c_1 = \begin{pmatrix} \zeta^{p-1} & & & \\ & \zeta^{-1} & & \\ & & \ddots & \\ & & & \zeta^{-1} \end{pmatrix}.$$

If $d_1 = c_1 \cdot b$, then d_1 normalizes $\gamma_{k-1} = \left(\prod_{i=1}^{p-1} \mathbf{Z}/p^{k-1} \langle a_k(i)^p \rangle \right) \rtimes \mathbf{Z}/p\langle b \rangle$. Hence $\gamma_{k-1} \rtimes \mathbf{Z}/p\langle d_1 \rangle \subset \gamma_k$. Next suppose

$$c_2 = \begin{pmatrix} 1 & & & & & & \\ & \zeta & & & & & \\ & & \ddots & & & & \\ & & & \zeta^{\frac{p-1}{2}} & & & \\ & & & & \zeta^{-\frac{p-1}{2}} & & \\ & & & & & \ddots & \\ & & & & & & \zeta^{-1} \end{pmatrix}$$

If $d_2 = c_2 \cdot b$, then d_2 normalizes $\gamma_{k-1} \rtimes \mathbf{Z}/p\langle d_1 \rangle$. Consequently $(\gamma_{k-1} \rtimes \mathbf{Z}/p\langle d_1 \rangle) \rtimes \mathbf{Z}/p\langle d_2 \rangle \subset \gamma_k$. Let δ denote this subgroup. By a hypothesis of induction and Lemma 1, we see $f|B\delta \simeq 0$. Notice here that $c_2 \in \delta$. For

$$s = \begin{pmatrix} 0 & 1 & & & \\ -1 & 0 & & & \\ & & 1 & & \\ & & & \ddots & \\ & & & & 1 \end{pmatrix} \in SU(p),$$

let $c_3 = sc_2s^{-1}$. If ϵ denotes the group $\mathbf{Z}/p^k\langle c_2\rangle \oplus \mathbf{Z}/p^k\langle c_3\rangle$, then $a_k(1) = c_2^{-1}c_3 \in \epsilon$ and $f|B\epsilon \simeq 0$. Since $a_k(1)$ is conjugate to $a_k(j)$ for any j, Lemma 1 and Lemma 3 imply $f|B\left(\prod_{i=1}^{p-1}\mathbf{Z}/p^k\langle a_k(i)\rangle\right) \simeq 0$. Hence $f|B\gamma_k \simeq 0$. Therefore $f|BT^{p-1} \simeq 0$.

Case 2. $G = Spin(n)$ with $n \geq 2p$ or $G = Sp(n)$ with $n \geq p$.

For any odd prime p, we see $BSpin(n) \underset{p}{\simeq} BSO(n)$ and $BSO(2n+1) \underset{p}{\simeq} BSp(n)$. Thus we may assume $G = SO(n)$ with $n \geq 2p$. Note that there is an inclusion $SU(p) \subset SO(2p) \subset SO(n)$ since $n \geq 2p$. Since $H^*(f : \mathbf{F}_p) = 0$, it follows that $H^*(f|BSU(p); \mathbf{F}_p) = 0$. Case 1.1 shows $f|BSU(p) \simeq 0$. Thus $f|BT^{p-1} \simeq 0$.

Case 3. $G = G_2$ at $p = 3$.

Since $SU(3) \subset G_2$, we see $f|BT^2 \simeq 0$.

Case 4. $G = F_4$ at $p = 3$.

Since $Spin(9) \subset F_4$, we see $f|BT^4 \simeq 0$.

Case 5. $G = E_6$ at $p = 3, 5$.

Using the inclusion $SU(2) \times SU(6)/\mathbf{Z}_2 \subset E_6$, we define φ to be the homomorphism $SU(6) \to E_6$ that maps through the second factor. Case 1 shows $f \circ B\varphi \simeq 0$. If T^5 is a maximal torus of $SU(6)$, then $\varphi(T^5)$ is a subtorus of rank 5. Since $f \circ B\varphi|BT^5 \simeq 0$ and $BT^5 \underset{p}{\simeq} B\varphi(T^5)$, it follows that $f|B\varphi(T^5) \simeq 0$.

Case 6. $G = E_7$ at $p = 3, 5, 7$.

Since $SU(8)/\mathbf{Z}_2 \subset E_7$, we see $f|BT^7 = 0$.

Case 7. $G = E_8$ at $p = 3, 5, 7$.

Since $SU(9)/\mathbf{Z}_3 \subset E_8$, we see $f|BT^8 = 0$ when $p = 5, 7$. For $p = 3$, use $SU(5) \times SU(5)/\mathbf{Z}_5 \subset E_8$.

§2. Finiteness of homomorphisms over the Steenrod algebra $\mathcal{A}_p$

A ring homomorphism $R \to S$ is called finite if S is a finitely generated module over the ring R. It is known [15, Corollary 2.4] that for a group homomorphism $\varphi : G \to G'$, the induced homomorphism $(B\varphi)^* : H^*(BG'; \mathbf{F}_p) \to H^*(BG; \mathbf{F}_p)$ is finite if and only if $\ker\varphi$ is a finite group of order prime to p. Dwyer and Wilkerson [6, Proposition 4.4] observe a generalization when G is an elementary abelian p-group. Suppose V is an elementary abelian p-group. Their result implies that, for a map $f : BV \to X$, the induced homomorphism f^* is finite if and only if $\ker(f) = 0$.

Proposition 1. *Let G be a compact Lie group and let R be a connected $\mathcal{A}_p$-algebra. Suppose $\varphi : R \to H^*(BG; \mathbf{F}_p)$ is an $\mathcal{A}_p$-map. If, for any elementary abelian p-subgroup V of G, the composite $\varphi_V : R \to H^*(BG; \mathbf{F}_p) \to H^*(BV; \mathbf{F}_p)$ is finite, then φ is also finite.*

Proof: Suppose φ is not finite so that there is a non-nilpotent element x in $\mathbf{F}_p \otimes_R H^*(BG; \mathbf{F}_p)$. A result of Lannes and Schwartz [20] shows that there is a map of unstable algebras $\psi : \mathbf{F}_p \otimes_R H^*(BG; \mathbf{F}_p) \to H^*(BV; \mathbf{F}_p)$ for some elementary abelian p-group V with $\psi(x) \neq 0$. Consider the composite $H^*(BG; \mathbf{F}_p) \to \mathbf{F}_p \otimes_R H^*(BG; \mathbf{F}_p) \to H^*(BV; \mathbf{F}_p)$. This homomorphism over the Steenrod algebra is induced by a map

$BV \to BG$. Ajusting this map if necessary by using [6, Proposition 4.8], we may assume the composite homomorphism is finite. Hence the map $BV \to BG$ is assumed to be induced by a monomorphism $V \to G$. It is clear that $\varphi_V : R \to H^*(BV;\mathbf{F}_p)$ is not finite.//

Theorem 2'. *(a) Let G, X and p be as in Theorem 1, except that G is not the exceptional Lie group G_2. If $f : BG \to X$ is an essential map, then either $f^* : H^*(X;\mathbf{F}_p) \to H^*(BG;\mathbf{F}_p)$ is finite or $(f|BZ(G))^* : H^*(X;\mathbf{F}_p) \to H^*(BZ(G);\mathbf{F}_p)$ is not finite, where $Z(G)$ denotes the center of G. (b) There is a compact Lie group K with an essential map $f : BG \to (BK)^\wedge_3$ such that $f^* : H^*(BK;\mathbf{F}_3) \to H^*(BG_2;\mathbf{F}_3)$ is not finite.*

Remark: In the proof of Theorem 2'(a), it will be proved that if $f|BV \simeq 0$ and $V \not\subset Z(G)$ with $G \neq G_2$, then $f \simeq 0$. Hence Theorem 2 follows.

Proof of Theorem 2':

(a) We will show that if the induced homomorphism $f^* : H^*(X;\mathbf{F}_p) \to H^*(BG;\mathbf{F}_p)$ is not finite and $(f|BZ(G))^*$ is finite, then f is null homotopic. When f^* is not finite, Proposition 1 says that there is a nonzero elementary abelian p-subgroup V of G such that $(f|BV)^*$ is not finite. The result of Dwyer-Wilkerson [6] shows that there is a nonzero element $a \in V$ such that $f|B\mathbf{Z}/p\langle a\rangle \simeq 0$. Since $(f|BZ(G))^*$ is finite, we see $a \notin Z(G)$. First we assume G is 1-connected.

Case 1.1. $G = SU(n)$ with $n \geq p$.

Since $G = \bigcup_{g\in G} gTg^{-1}$, the element a is conjugate to a diagonal matrix:

$$b = \begin{pmatrix} \zeta_1 & & \\ & \ddots & \\ & & \zeta_n \end{pmatrix}$$

where $\zeta_i^p = 1$ for $1 \leq i \leq n$. Since $a \notin Z(G)$ and the Weyl group Σ_n acts on T^{n-1}, we may assume $\zeta_1 \neq \zeta_2$. If we let

$$c = \begin{pmatrix} \zeta_2 & & & & \\ & \zeta_1 & & & \\ & & \zeta_3 & & \\ & & & \ddots & \\ & & & & \zeta_n \end{pmatrix}$$

then

$$b^{-1}c = \begin{pmatrix} \zeta_1^{-1}\zeta_2 & & & & \\ & \zeta_2^{-1}\zeta_1 & & & \\ & & 1 & & \\ & & & \ddots & \\ & & & & 1 \end{pmatrix}$$

and $f|B\mathbf{Z}/p\langle b^{-1}c\rangle = 0$. Since $\zeta_1 \neq \zeta_2$, it follows that $\zeta_1^{-1}\zeta_2 \neq 1$. A subset of conjugacy classes of $b^{-1}\cdot c$ generates the elementary abelian p-subgroup $\overset{n}{\oplus}\mathbf{Z}/p$ in T^n. Consequently $H^*(f;\mathbf{F}_p) = 0$ and hence Theorem 1 shows $f = 0$.

Case 1.2.. $G = Spin(n)$ with $n \geq 2p$ or $G = Sp(n)$ with $n \geq p$.

Since p is odd, we may assume $G = SO(n)$ with $n \geq 2p$ again.

Suppose $n = 2m+1$. The element a is conjugate in $SO(n)$ to a matrix

$$b = \begin{pmatrix} A_1 & & & \\ & \ddots & & \\ & & A_m & \\ & & & 1 \end{pmatrix} \in SO(2) \times \cdots \times SO(2) \subset SO(2m+1)$$

where

$$A_i = \begin{pmatrix} \cos\theta_i & -\sin\theta_i \\ \sin\theta_i & \cos\theta_i \end{pmatrix}.$$

We may assume that the order of A_1 is p. Let $B = \begin{pmatrix} 0 & 1 \\ 1 & 0 \end{pmatrix}$ and let

$$c = \begin{pmatrix} I_2 & & & & \\ & B & & & \\ & & \ddots & & \\ & & & B & \\ & & & & (-1)^{m-1} \end{pmatrix} \in SO(2m+1).$$

Then

$$b \cdot cbc^{-1} = \begin{pmatrix} A_1^2 & & & & \\ & I_2 & & & \\ & & \ddots & & \\ & & & I_2 & \\ & & & & 1 \end{pmatrix}$$

Since p is odd, the order of A_1^2 is p. A subset of conjugacy classes of $b \cdot cbc^{-1}$ generates the elementary abelian p-subgroup $\overset{m}{\oplus}\mathbb{Z}/p$ in T^m. Consequently $H^*(f; \mathbb{F}_p) = 0$ and hence $f \simeq 0$ by Theorem 1.

Next suppose $n = 2m$. If m is odd, taking $c = \begin{pmatrix} I_2 & & & \\ & B & & \\ & & \ddots & \\ & & & B. \end{pmatrix} \in SO(2m)$, an analogous proof is applicable. If m is even (hence $m \geq 4$), let

$$c = \begin{pmatrix} I_2 & & & & \\ & I_2 & & & \\ & & B & & \\ & & & \ddots & \\ & & & & B \end{pmatrix}$$

An analogous argument shows

$$\begin{pmatrix} A_1^2 & & & & \\ & A_2^2 & & & \\ & & I_2 & & \\ & & & \ddots & \\ & & & & I_2 \end{pmatrix} \in \ker f | BV_0,$$

where $V_0 = \{x \in T^m | x^p = I_{2m}\}$. Since $m \geq 4$, we see the following are contained in $\ker f|BV_0$:

$$\begin{pmatrix} A_1^2 & & & & & \\ & I & & & & \\ & & A_3^2 & & & \\ & & & I & & \\ & & & & \ddots & \\ & & & & & I \end{pmatrix}, \begin{pmatrix} A_1^2 & & & & & & \\ & I & & & & & \\ & & I & & & & \\ & & & A_4^2 & & & \\ & & & & I & & \\ & & & & & \ddots & \\ & & & & & & I \end{pmatrix}$$

$$\begin{pmatrix} I & & & & & & & \\ & \ddots & & & & & & \\ & & I & & & & & \\ & & & A_i^2 & & & & \\ & & & & A_{i+1}^2 & & & \\ & & & & & I & & \\ & & & & & & \ddots & \\ & & & & & & & I \end{pmatrix} \quad \text{for } 5 \leq i \leq m-1.$$

Hence the product of these matrices

$$\begin{pmatrix} A_1^6 & & & \\ & A_2^2 & & \\ & & \ddots & \\ & & & A_m^2 \end{pmatrix}$$

is also contained in $\ker f|BV_0$. Consequently

$$\begin{pmatrix} A^6 & & & \\ & A_2^2 & & \\ & & \ddots & \\ & & & A_m^2 \end{pmatrix} \begin{pmatrix} A_1^{-2} & & \\ & \ddots & \\ & & A_m^{-2} \end{pmatrix} = \begin{pmatrix} A_1^4 & & & \\ & I & & \\ & & \ddots & \\ & & & I \end{pmatrix} \in \ker f|BV_0.$$

Since p is odd, the order of A_1^4 is p. We can show that $H^*(f;\mathbf{F}_p) = 0$ and hence $f \simeq 0$ by Theorem 1.

<u>Case 1.3.</u> $G = F_4$ at $p = 3$.

Since $Spin(9) \subset F_4$, Case 1.2 shows that $(f|BSpin(9))^*$ is trivial and hence $f|BT^4 \simeq 0$. Lemma 4 shows $f \simeq 0$.

<u>Case 1.4.</u> $G = E_6$ at $p = 3, 5$.

For $p = 3$, we use the subgroup $Spin(10) \times S^1/\mathbf{Z}_4 \subset E_6$. Let c be a generator of $Z(E_6) = \mathbf{Z}/3$. From [17, Theorem 2.27], we see $\mathbf{Z}/3\langle a\rangle \oplus \mathbf{Z}/3\langle c\rangle \subset T^6$ since E_6 is 1-connected. If $a = [\alpha_1, \beta_1], c = [\alpha_2, \beta_2] \in Spin(10) \times S^1/\mathbf{Z}_4$ where $\alpha_i \in Spin(10)$ and $\beta_i \in S^1$, then $\alpha_2 = 1$.

Since $H^2(B(\mathbf{Z}/3 \oplus \mathbf{Z}/3), \mathbf{F}_4) = 0$, the group extension of $\mathbf{Z}_4$ by $\mathbf{Z}/3 \oplus \mathbf{Z}/3$ splits. We may regard a and c as elements in $Spin(10) \times S^1$. We can show that the order of α_1 is divisible by 3. Thus the composite $H^*(X; \mathbf{F}_3) \to H^*(BE_6; \mathbf{F}_3) \to H^*(B(Spin(10) \times S^1/\mathbf{Z}_4); \mathbf{F}_3) \xrightarrow{\sim} H^*(BSpin(10) \times BS^1; \mathbf{F}_3) \to H^*(BSpin(10); \mathbf{F}_3)$ is trivial. Consequently $f|BT^5 \simeq 0$ and hence $f \simeq 0$.

For $p = 5$, we use $SU(3) \times SU(3) \times SU(3)/\mathbf{Z}_3 \subset E_6$. Since $H^2(B\mathbf{Z}/5, \mathbf{Z}_3) = 0$, there is $\tilde{a}$ in $SU(3) \times SU(3) \times SU(3)$ such that $\tilde{a}^5 = 1$ and the projection maps $\tilde{a}$ to a. Since $\tilde{a} \notin Z(SU(3) \times SU(3) \times SU(3))$, we can find $\tilde{b}$ which is conjugate to $\tilde{a}$ and $\mathbf{Z}/5\langle\tilde{a}\rangle \oplus \mathbf{Z}/5\langle\tilde{b}\rangle \subset SU(3) \times SU(3) \times SU(3)$. Consequently there is a subgroup $\mathbf{Z}/5 \oplus \mathbf{Z}/5$ of $SU(3) \times SU(3) \times SU(3)/\mathbf{Z}_3$ such that $f|B(\mathbf{Z}/5 \oplus \mathbf{Z}/5) \simeq 0$. Again, by [17, Theorem 2.27], we see $\mathbf{Z}/5 \oplus \mathbf{Z}/5 \subset T$. So $f \circ B\varphi|B\mathbf{Z}/5 \simeq 0$ for some $\mathbf{Z}/5 \subset SU(6)$, where $\varphi : SU(6) \to E_6$. Hence $H^*(X; \mathbf{F}_5) \to H^*(BSU(6); \mathbf{F}_5)$ is trivial, and $f|BT^5 \simeq 0$. Therefore $f \simeq 0$.

Case 1.5. $G = E_7$ at $p = 3, 5, 7$.

Since $SU(8)/\mathbf{Z}_2 \subset E_7$, we see $f|BT^7 \simeq 0$ and hence $f \simeq 0$.

Case 1.6. $G = E_8$ at $p = 3, 5, 7$.

If $p = 3$, use $SU(5) \times SU(5)/\mathbf{Z}_5 \subset E_8$ to show $f|BT^8 \simeq 0$ and hence $f \simeq 0$. If $p = 5$ or 7, use $SU(9)/\mathbf{Z}_3 \subset E_8$.

Next we consider the case that G is not 1-connected. Suppose $\widetilde{G}$ is the universal covering group of G. We need to consider the two cases.

Case 2.1. $\widetilde{G} = SU(n)$ with $n \geq 3$ and $n \equiv 0 \; mod \; p$.

Suppose p^i is the largest power of p that divides m, where $G = SU(n)/\mathbf{Z}_m$. Then $B\left(SU(n)/Z_{p^i}\right) \underset{p}{\simeq} B\left(SU(n)/\mathbf{Z}_m\right)$. Thus we may assume $G = SU(n)/\mathbf{Z}_{p^i}$. If $q : \widetilde{G} \to G$ is the covering map, then $q(Z(\widetilde{G})) \subset Z(G)$. Consequently, if $b \in q^{-1}(a)$, then $b \notin Z(\widetilde{G})$. The order of b is equal to p^j for some $j \geq 1$. Let ζ be a primitive p^j-th root of unity. Then b is conjugate to the matrix

$$c = \begin{pmatrix} \zeta^{\alpha_1} & & \\ & \ddots & \\ & & \zeta^{\alpha_n} \end{pmatrix} \in SU(n).$$

We may assume α_1 is a unit in the ring $\mathbf{Z}_{p^j}$ and $\alpha_1 \neq \alpha_2$. Notice that

$$d = \begin{pmatrix} \zeta^{\alpha_2} & & & & \\ & \zeta^{\alpha_1} & & & \\ & & \zeta^{\alpha_3} & & \\ & & & \ddots & \\ & & & & \zeta^{\alpha_n} \end{pmatrix} \in SU(n)$$

is conjugate to c in $SU(n)$ and that

$$c \cdot d^{-1} = \begin{pmatrix} \zeta^{\alpha_1 - \alpha_2} & & & & \\ & \zeta^{\alpha_2 - \alpha_1} & & & \\ & & 1 & & \\ & & & \ddots & \\ & & & & 1 \end{pmatrix} \in \ker(\widetilde{f}|B(\overset{n-1}{\oplus} \mathbf{Z}/p^j)),$$

where $\widetilde{f}$ is the composition $B\widetilde{G} \to BG \to X$. Since $\alpha_1 \neq \alpha_2$, the order of $\zeta^{\alpha_1-\alpha_2}$ is at least p. The subgroup of $\overset{n-1}{\oplus} \mathbb{Z}/p^j$ generated by conjugacy classes of $c \cdot d^{-1}$ includes $\overset{n-1}{\oplus} \mathbb{Z}/p$ in $SU(n)$. Thus $H^*(\widetilde{f}; \mathbb{F}_p) = 0$ and hence $\widetilde{f} \simeq 0$. This implies $f \simeq 0$.

<u>Case 2.2.</u> $\widetilde{G} = E_6$ at $p = 3$.

Suppose b is a preimage of the element a under the projection $E_6 \to E_6/\mathbb{Z}_3$. If the order of b is 3, we're done, since $b \notin Z(E_6)$. If the order of b is 9, there is $\widetilde{b} \in SU(2) \times SU(6)$ such that, under the projection $SU(2) \times SU(6) \xrightarrow{\varphi} SU(2) \times SU(6)/\mathbb{Z}_2$, this element is sent to b up to conjugation in E_6. One can see that if $\widetilde{b} = (\widetilde{b}_1, \widetilde{b}_2)$ where $\widetilde{b}_1 \in SU(2)$ and $\widetilde{b}_2 \in SU(6)$, then the order of $\widetilde{b}_2$ is equal to 9, since $Z(E_6) \subset SU(6)$. Hence $\widetilde{b}_2$ is conjugate to the matrix

$$\begin{pmatrix} \zeta_1 & & \\ & \ddots & \\ & & \zeta_6 \end{pmatrix} \in SU(6)$$

where $\zeta_i^9 = 1$. We may assume $\zeta_1 \neq \zeta_2$.
An argument similar to Case 2.1 shows that there is $\mathbb{Z}/3$ such that $\mathbb{Z}/3 \cap Z(SU(6)) = \{1\}$ and $f \circ B\varphi | B\mathbb{Z}/3 \simeq 0$. Thus $\widetilde{f} = 0$ and therefore $f \simeq 0$.

(b) First we will show that there is a proper subgroup H of the exceptional Lie group G_2 such that $(BH)^\wedge_3$ is homotopy equivalent to $(BG_2)^\wedge_3$. We regard G_2 as the automorphism group of the Cayley algebra $\mathbb{K}$ spanned by $e_0, e_1, \ldots, e_7$, [18, p. 690]. The unit vector e_0 is fixed by any element of G_2 and G_2 is regarded as a subgroup of $SO(7)$. The group of elements in G_2 which fixes e_1 is regarded as a subgroup of $SO(6)$, and is isomorphic to $SU(3)$. If τ is the element of G_2 such that $\tau(e_1) = -e_1, \tau(e_2) = e_2$ and $\tau(e_7) = e_7$, then τ normalizes $SU(3)$ and $SU(3) \rtimes \mathbb{Z}/2\langle\tau\rangle$ is a subgroup of G_2. Let $H = SU(3) \rtimes \mathbb{Z}/2\langle\tau\rangle$. We have the commutative diagram:

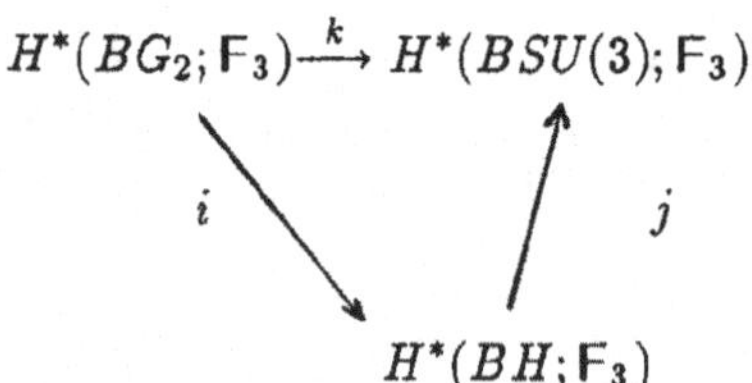

Here the homomorphisms i, j and k are induced by the inclusions. Since k is injective, so is i. Next we show j is injective. In fact, the Serre spectral sequence for the fibration $BSU(3) \to BH \to B\mathbb{Z}/2$ shows the desired result, since $E_2^{p,q}$ is the cohomology with local coefficients $H^p(B\mathbb{Z}/2, \mathcal{H}^q(BSU(3); \mathbb{F}_3))$ and $E_2^{p,q} = 0$ for $p > 0$. Notice that $\mathbb{Z}/2\langle\tau\rangle$ acts on $H^*(BSU(3); \mathbb{F}_3)$ and that the action is induced by an inner automorphism of H. Consequently $H^*(BH; \mathbb{F}_3) \subset H^*(BSU(3); \mathbb{F}_3)^{\mathbb{Z}/2} = H^*(BG_2; \mathbb{F}_3)$. Since i is injective, we see $\dim_{\mathbb{F}_3} H^n(BG_2; \mathbb{F}_3) = \dim_{\mathbb{F}_3} H^n(BH; \mathbb{F}_3)$ for any n. Hence i is bijective. Thus the map $(BH)^\wedge_3 \to (BG_2)^\wedge_3$ is a weak equivalence.

Next we notice that the center $Z(SU(3)) = \mathbb{Z}_3$ is a normal subgroup of H. Let $K = H/\mathbb{Z}_3$ and define a map f as follows

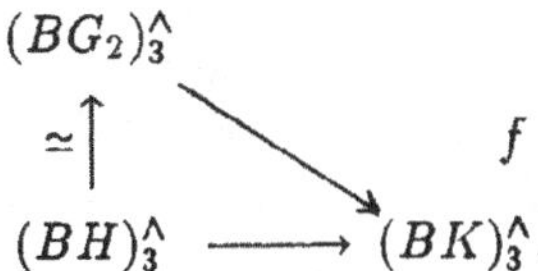

Then $f \not\simeq 0$ and $H^*(f;\mathbb{F}_3)$ is not finite since $f|B\mathbb{Z}/3 \simeq 0$. //

Proof of Corollary 1.

(a) If $G \neq G_2$, along the line of Theorem 2(a) we can show the desired result, since there is V such that $V \cap Z(G) = \{1\}$. If $G = G_2$ use $SU(3) \subset G_2$. We can show $f|BSU(3) \simeq 0$ and hence $f \simeq 0$.

(b) Since Krull $\dim(H^*(X;\mathbb{F}_p)) < rank(G)$, $H^*(BG;\mathbb{F}_p)$ is not an integral extension of the image of the homomorphism f^*. Hence f^* is not finite. If $G \neq G_2$, Theorem 2(a) shows $f \simeq 0$ or $f|BZ(G)^*$ is not finite. In the case $f|BZ(G)^*$ is not finite, we can find $\mathbb{Z}/p \subset Z(G)$ such that $f|B\mathbb{Z}/p \simeq 0$. Hence the map $f : BG \to X$ factor through $B(G/\mathbb{Z}_p)$. Let $\overline{f} : B(G/\mathbb{Z}_p) \to X$ be the map. Since $rank(G/\mathbb{Z}_p) = rank(G)$, we see $\overline{f} \simeq 0$ or $\overline{f}|BZ(G/\mathbb{Z}_p)^*$ is not finite. Continue this process. Since the center $Z(G)$ is a finite group, we see $f \simeq 0$. If $G = G_2$, use $SU(3) \subset G_2$, again. Since $rank(SU(3)) = rank(G_2) = 2$, it follows that $f|BSU(3) \simeq 0$ and hence $f \simeq 0$. //

The Krull dimension of the polynomial ring in n indeterminates is equal to n. In [15] Quillen showed that the Krull dimension of the ring $H^*(B\Gamma;\mathbb{F}_p)/\sqrt{0}$ is equal to the maximum rank of an elementary abelian p-subgroup for any compact Lie group Γ.

Lemma 5. *Let G be a compact Lie group and Let T be a maximal torus of the connected component of G containing the identity element. Assume that X is a p-complete space whose mod p cohomology ring is isomorphic to a ring of invariants $H^*(BT^n;\mathbb{F}_p)^W$. If $f : BG \to X$ is a map such that $f^* : H^*(X;\mathbb{F}_p) \to H^*(BG;\mathbb{F}_p)$ is finite, then there is a monomorphism $NT/T \to W$, where NT denotes the normalizer of T in G,*

Proof: Recall that the map $H^*(BG;\mathbb{F}_p) \to H^*(BT;\mathbb{F}_p)$ factors through then ring of invariants $H^*(BT;\mathbb{F}_p)^{NT/T}$. By a result of Adams-Wilkerson [2], we can find a homomorphism $\psi : T \to T^n$ such that the following diagram

$$\begin{array}{ccc} H^*(X;\mathbb{F}_p) \xrightarrow{f^*} H^*(BG;\mathbb{F}_p) & \longrightarrow & H^*(BT;\mathbb{F}_p)^{NT/T} \\ \downarrow & & \downarrow \\ H^*(BT^n;\mathbb{F}_p) & \xrightarrow[(B\psi)^*]{} & H^*(BT;\mathbb{F}_p) \end{array}$$

commutes. Here the vertical maps are the inclusions. Let $\phi = (B\psi)^*$. Since f^* is finite, so is ϕ. Thus ϕ is onto, [15]. Since ϕ is admissible, [1] and [2], we can define a map $\alpha : NT/T \to W$ by $s\phi = \phi\alpha(s)$. Since ϕ is onto, the map α is a monomorphism. //

Proof of Corollary 2: Suppose $f : BG \to X$ is a nonzero map. If f^* is finite, our result is immediate from Lemma 5. We now assume that f^* is not finite.

First we consider the case $|W(G)| \not\equiv 0 \ mod\ p$. Let T be a maximal torus of G and let $V = \{x \in T | x^p = 1\}$. Then $\ker f|BV$ is invariant under $W(G)$-action. Maschke's theorem implies $\ker f|BV = V$, since G is simple and f^* is not finite. Recall $BG \underset{p}{\simeq} BNT$ in this case. We see that the map f factors through $B(NT/V)$. Let $N_1 = NT/V$ and let $f_1 : BN_1 \to X$ be the induced map. Define N_k and $f_k : BN_k \to X$ inductively, if possible. There is a number k such that f_k^* is finite. Otherwise f would be a zero map. Let T_k be the maximal torus of N_k. Notice that NT_k/T_k includes a group isomorphic to $W(G)$. By Lemma 5, there is a homomorphism $W(G) \to W$.

Next suppose $W(G)| \equiv 0 \ mod\ p$. If $G \neq G_2$, Theorem 2 (a) implies that $f|BZ(G)^*$ is not finite, since f^* is not finite and p is odd. Thus there is $\mathbb{Z}_p$ in $Z(G)$ such that $f|B\mathbb{Z}_p \simeq 0$. The map f factors through $B(G/\mathbb{Z}_p)$. Since $W(G) = W(G/\mathbb{Z}_p)$, an inductive argument proves this case. It remains to show the case $G = G_2$ and $p = 3$. Since f^* is not finite, f factors through BK where $K = SU(3) \rtimes \mathbb{Z}_2/\mathbb{Z}_3$. (This group was discussed in the proof of Theorem 2 (b).) If $\overline{f} : BK \to X$ is the induced map, then $\overline{f}^*$ is finite. Let NT^2 be the normalizer of T^2 in K. Then NT^2/T^2 includes a group isomorphic to $W(G_2)$. Lemma 5 shows the desired result. //

§3. Maps between $BSO(2n+1)$ and $BSp(n)$.

Suppose G and G' are compact connected Lie groups. The result of Zabrodsksy [19] shows that a map $f : BG \to BG'$ is null homotopic if $K(f) = 0$. A result of Adams-Mahmud [1] implies $K(f) = 0$ if G is simple and $rank(G) > rank(G')$. This argument is applicable in many cases.

The groups $SO(2n+1)$ and $Sp(n)$ have, however, the same rank. In fact, the maximal torus T and the Weyl group W are the same, including the W-action on T. Since $K(BG) = K(BT)^W$, K-theory doesn't see the difference. For this reason, we use *mod* 2 cohomology. Another example that K-theory doesn't eliminate the possibility of existence of nonzero maps is given by quotient groups. Suppose that $\widetilde{G}$ is a compact simply-connected simple Lie group and that Z_1 and Z_2 are subgroups of the center of $\widetilde{G}$. One can show that

$$[B(\widetilde{G}/Z_1),\ B(\widetilde{G}/Z_2)] = 0$$

if there is a prime p such that the order of a p-Sylow subgroup of Z_1 is larger than that of a p-Sylow subgroup of Z_2. Extending such a map to a self-map of $B\widetilde{G}$, the proof would use the fact that any nonzero self-map of $(B\widetilde{G})^\wedge_p$ is a homotopy equivalence if $|W(\widetilde{G})| \equiv 0 \ mod\ p$.

Theorem 3.

Theorem 1 holds for $G = SO(n)$ with $n \geqq 3$, but $n \neq 4$, at $p = 2$.

Proof: The proof is analogoous to the proof of the case $G = SU(n)$ in Theorem 1.

Since $n \geq 3$ and $n \neq 4$, the Lie group $SO(n)$ is simple. By Lemma 4, it suffices to show $f|BT^j \simeq 0$ for some $j \geq 1$.

Suppose $G = SO(3)$. Then $\gamma_k = \mathbb{Z}/2^k\langle r\rangle \rtimes \mathbb{Z}/2\langle s\rangle$. This is a dihedral group. We can see that $\gamma_{k-1} = \mathbb{Z}/2^{k-1}\langle r^2\rangle \rtimes \mathbb{Z}/2\langle s\rangle$ and $\gamma_k = \gamma_{k-1} \rtimes \mathbb{Z}/2\langle rs\rangle$. By induction we obtain $f|B\gamma_k \simeq 0$ for any k. Consequently $f|BSO(3) \simeq 0$. In the general case, since $SO(3) \subset SO(n)$ and $SO(2) = T^1$, we see $f|BT^1 \simeq 0$. //

Theorem 4.
Theorem 2(a) holds for $G = SO(n)$ with $n \geq 3$, but $n \neq 4$, at $p = 2$.

Proof: An argument analogous to the proof of Theorem 2(a) shows that there is an element $a \in G$ such that $f|B\mathbb{Z}/2\langle a\rangle = 0$ and $a \notin Z(G)$.
Case 1. $G = SO(2m+1)$.
The element a is conjugate to

$$b = \begin{pmatrix} -I_{2k} & \\ & I_{2m+1-2k} \end{pmatrix}$$

Let V be the subgroup of diagonal matrices of G so that $V \cong \overset{2m}{\oplus}\mathbb{Z}/2$. Since $\begin{pmatrix} & 1 \\ -1 & \end{pmatrix}\begin{pmatrix} \alpha & \\ & \beta \end{pmatrix}\begin{pmatrix} & 1 \\ -1 & \end{pmatrix}^{-1} = \begin{pmatrix} \beta & \\ & \alpha \end{pmatrix}$, we see

$$c = \begin{pmatrix} -I_{2k-1} & & & \\ & 1 & & \\ & & -1 & \\ & & & I_{2m-2k} \end{pmatrix} \in \ker(f|BV).$$

Hence

$$bc = \begin{pmatrix} I_{2k-1} & & \\ & -I_2 & \\ & & I_{2m-2k} \end{pmatrix} \in \ker(f|BV).$$

Notice that the element bc together with its conjugacy classes generates V. Consequently $f|BV \simeq 0$ and hence $H^*(f;\mathbb{F}_2) = 0$. Theorem 3 shows $f \simeq 0$.
Case 2. $G = SO(2m)$.
Since $a \notin Z(G)$, this element is conjugate to

$$b = \begin{pmatrix} -I_{2k} & \\ & I_{2m-2k} \end{pmatrix}$$

for some $k < m$. An analogous proof is applicable. //

It is now easy to show that Corollary 1 holds for $G = SO(n)$ with $n \geq 3$, but $n \neq 4$, at $p = 2$. For $G = SO(4)$, Theorem 1 holds at $p = 2$. But Theorem 2(a) doesn't.

Theorem 5.
(a) $[BSO(2n+1)^\wedge_2, BSp(k)^\wedge_2] = 0$ if and only if $k \leq 2n$.
(b) Assume $n \geq 3$. Then $[BSp(n)^\wedge_2, BSO(k)^\wedge_2] = 0$ if and only if $k < 4n$.

Proof: (a) ($\Rightarrow$) If $k = 2n+1$, there is a nonzero map induced by the obvious inclusions $SO(2n+1) \subset O(2n+1) \subset U(2n+1) \subset Sp(2n+1)$.

($\Leftarrow$) For a map $f : BSO(2n+1)^\wedge_2 \to BSp(k)^\wedge_2$ the homomorphism f^* is finite only if $k \geqq 2n$. By Theorem 4 it remains to consider the case $k = 2n$. A result of Adams-Wilkerson [2] enables us to find an $\mathcal{A}_p$-map ϕ which makes the diagram commutative:

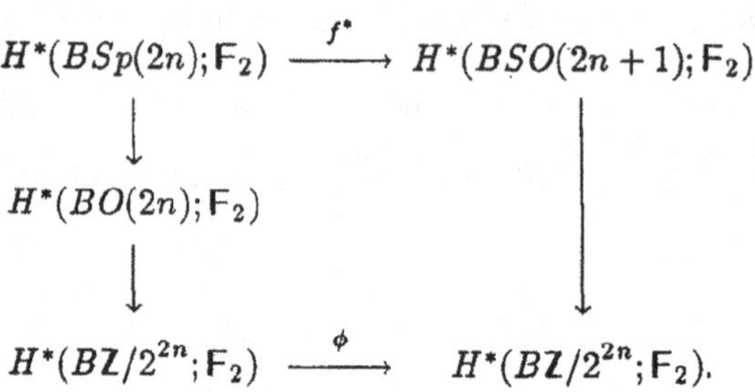

In this diagram, all vertical maps induced by the inclusions are injective. We note that $H^*(BSO(2n+1); \mathsf{F}_2) = H^*(B\mathsf{Z}/2^{2n}; \mathsf{F}_2)^{\Sigma_{2n+1}}$ and $H^*(BO(2n); \mathsf{F}_2) = H^*(B\mathsf{Z}/2^{2n}; \mathsf{F}_2)^{\Sigma_{2n}}$. For any $x \in H^*(BSp(2n); \mathsf{F}_2)$ there is $y \in H^*(BO(k); \mathsf{F}_2)$ such that $x = y^4$. Hence we can show that f^* extends to $H^*(BO(2n); \mathsf{F}_2)$. If ϕ were invertible, it would imply that $\phi^{-1}\Sigma_{2n+1}\phi \subset \Sigma_{2n}$. Hence, if $f|B\mathsf{Z}/2^{2n} = B\eta$ where η is a homomorphism $\mathsf{Z}/2^{2n} \to Sp(2n)$, then η is not injective. Consequently f^* is not finite. Theorem 4 shows $f \simeq 0$.

(b) ($\Rightarrow$) If $k = 4n$, there is a nonzero map induced by the inclusion $Sp(n) \subset U(2n) \subset SO(4n)$.

($\Leftarrow$) Suppose the map $\phi : H^*(B\mathsf{Z}/2^{k-1}, \mathsf{F}_2) \to H^*(B\mathsf{Z}/2^n; \mathsf{F}_2)$ covers the homomorphism $f : H^*(BSO(k); \mathsf{F}_2) \to H^*(BSp(n); \mathsf{F}_2)$. Over the vector space $H^1(B\mathsf{Z}/2^{k-1}; \mathsf{F}_2)$ of 1-dimensional elements, ϕ is expressed as a $(k-1) \times n$ matrix. Suppose $\phi = (\phi_1, \ldots, \phi_n)$ where ϕ_i is the i-th column. According to [1, Proposition 2.16], for each i and $\sigma \in \Sigma_k, \sigma\phi_i$ must appear among $\phi_1, \ldots, \phi_n$. As a subalgebra of $H^*(BO(n); \mathsf{F}_2) = \mathsf{F}_2[w_1, \ldots, w_n]$, we see $H^*(BSp(n); \mathsf{F}_2) = \mathsf{F}_2[w_1^4, \ldots, w_n^4]$. Thus, if $\phi_i \neq 0$, then ϕ_i must appear at least 4 times among $\phi_1, \ldots, \phi_n$. Consequently ϕ must be of the form

$$\begin{pmatrix} 1\ldots & 1 & 0\ldots & 0 \\ \cdot & \cdot & \cdot & \cdot \\ \cdot & \cdot & \cdot & \cdot \\ \cdot & \cdot & \cdot & \cdot \\ 1\ldots & 1 & 0\ldots & 0 \end{pmatrix}$$

up to permutation of columns, since $k < 4n$. This implies $f|BSO(n)^* = 0$. Theorem 4 shows $f|BSO(3) \simeq 0$ and hence $f|BT^1 \simeq 0$. Lemma 4 shows $f \simeq 0$.

Lemma 6. *Let G and H be compact connected Lie groups. If $f : BG \to BH$ is a map such that the p-completion $f_p^\wedge$ is null homotopic for some p, then $f \simeq 0$.*

<u>Proof:</u> Since H is connected, we have the following commutative diagram based on the

arithmetic square together with the projections:

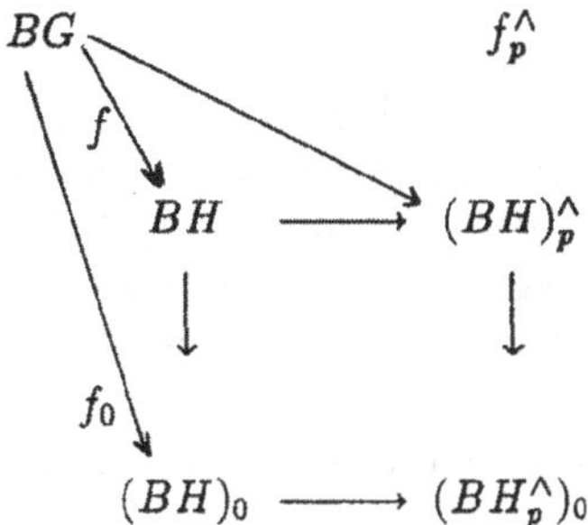

Notice that $(BH)_0$ and $(BH_p^\wedge)_0$ are products of Eilenberg-MacLane spaces. Since the map $H^*(BG;\mathbf{Q}) \to H^*(BG;\mathbf{Q}_p^\wedge)$ is injective, it follows that $f_0 \simeq 0$. Since G is connected, the map $K(BG) \to K(BG;\mathbf{Q})$ is injective. Thus we see $K(f) = 0$ and therefore $f \simeq 0$. //

We now see $[BSO(2n+1), BSp(n)] = 0$ for any n and $[BSp(n), BSO(2n+1)] = 0$ for $n \geq 3$. When $n = 1, mod$ 2 cohomology tells us that any map $f : BSp(1) \to BSO(3)_2^\wedge$ satisfies $f|BZ(Sp(1)) \simeq 0$. Using the homotopy equivalence $map(BSp(1), BSO(3)_2^\wedge) \simeq map_{SO(3)}(ESO(3), map(B\mathbf{Z}/2, BSO(3)_2^\wedge)$ one can show that there is a $1-1$ correspondence between $[BSp(1), BSO(3)_2^\wedge]$ and the self-maps of $BSO(3)_2^\wedge$ up to homotopy. This correspondence is induced by the universal coverinag map. For $n = 2$, an analogous argument is applicable.

References

1. J. F. ADAMS AND Z. MAHMUD, Maps between classifying spaces, *Invent. Math.* **35** (1976), 1-41.
2. J. F. ADAMS AND C. W. WILKERSON, Finite H-spaces and algebras over the Steenrod algebra, *Ann. of Math.* **111** (1980), 95-143.
3. A. BOREL AND J. DE SIEBENTHAL, Les sous-groupes fermés de rang maximum des groupes de Lie clos, *Comm. Math. Helv.* **23** (1949), 200-221.
4. M. CURTIS, A.WIEDERHOLD AND B. WILLIAMS, Normalizers of maximal tori, *Lecture Notes in Math., Springer-Verlag 418* (1974), 31-47.
5. W.G. DWYER AND C.W. WILKERSON, Spaces of null homotopic maps. preprint.
6. W.G. DWYER AND C.W. WILKERSON, A cohomology decomposition theorem. preprint.
7. E. M. FRIEDLANDER, Exceptional isogenies and the classifying spaces spaces of simple Lie groups, *Ann. of Math.* **101** (1975), 510-520.
8. E.M. FRIEDLANDER AND G. MISLIN, Locally finite approximation of Lie groups I, *Invent. Math.* **83** (1986), 425-436.
9. E.M. FRIEDLANDER AND G. MISLIN, Locally finite approximation of Lie groups II, *Math. Proc. Camb. Phil. Soc.* **100** (1986), 505-517.

10. K. ISHIGURO, Unstable Adams operations on classifying spaces, *Math. Proc. Camb. Phil. Soc.* **102** (1987), 71-75.

11. K. ISHIGURO, Rigidity of p-completed classifying spaces of alternating groups and classical groups over a finite field. preprint.

12. S. JACKOWSKI, J. MCCLURE AND B. OLIVER, Self-maps of classifying spaces of compact simple Lie groups, *Bull. A.M.S.* **22 no. 1** (1990), 65-71.

13. J. LANNES, Sur la cohomologie *modulo* p des p-groupes Abéliens élémentaires, in "Homotopy Theory, Proc. Durham Symp. 1985", *Cambridge Univ. Press* (1987), 97-116.

14. H.R. MILLER, The Sullivan conjecture on maps from classifying spaces, *Ann. of Math.* **120** (1984), 39-87.

15. D. QUILLEN, The spectrum of an equivariant cohomology ring I, *Ann. of Math* **94** (1971), 549-572.

16. J-P. SERRE, Sur les sous-groupes Abéliens des groupes de Lie compacts, *Séminaire "Sophus Lie" Expos/'e no. 24, E.N.S.* (1954/55).

17. R. STEINBERG, Torsion in reductive groups, *Advances in Math.* **15** (1975), 63-92.

18. G.W. WHITEHEAD, "*Elements of Homotopy Theory*," Springer, Berlin, 1978.

19. A. ZABRODSKY, Maps between classifying spaces, *Annals of Math. Studies* **113** (1987), 228-246.

20. J. LANNES AND L. SCHWARTZ, Sur la structure des A-modules instables injectifs, *Topology* **28** (1989), 153-169.

Department of Mathematics
Hofstra University
Hempstead, NY 11550
U.S.A.

1990 Barcelona Conference
on Algebraic Topology.

ON PARAMETRIZED BORSUK–ULAM THEOREM FOR FREE Z_p–ACTION

MAREK IZYDOREK AND SŁAWOMIR RYBICKI

0. Introduction

Let $\pi : E \to B$, $\pi' : E' \to B$ be vector bundles over the same space B. Let $SE \subset E$ be the sphere bundle of E. Consider a fibre–preserving free actions of a cyclic group $G = Z_p$ on E and E', where p is prime. Here and subsequently "free action" means a free action of Z_p on $E - \{0\}$ resp. $E' - \{0\}$.

From now on, the sphere bundle SE is assumed to be a G–subspace of E.

Let us choose a fibre–preserving, G–equivariant map $f : SE \to E'$, $(\pi' f = \pi)$. Parametrized Borsuk–Ulam theorems (for the standard antipodal action of the group Z_2) are concerned with describing the size of the set $X_f = \{x \in SE : f(x) = 0\}$. Such studies were initiated by Jaworowski [**J1,J2**] and later developed and extended by Nakaoka [**N1,N2**], Fadell and Husseini [**FH**], Dold [**D**] and recently by Izydorek and Jaworowski [**IJ**].

A. Dold suggested in [**D**] to consider G–bundles for cyclic groups of order which is greather than two and then ask for the totality X_f of solutions of the equation $f(x) = 0$.

In this paper Dold's ideas will be developed. In particular some lower estimations for cohomological dimension of the set X_f will be given. The problem of parametrized Borsuk–Ulam theorem for a large class of multivalued maps will also be studied.

Throughout the paper we assume that the spaces considered here are paracompact. We use the Čech cohomology theory H^* with coefficients mod p (the coefficient group Z_p will be suppressed from the notation).

1. Preliminaries

Let $\pi : E \to B$ be a vector budle with a fibre–preserving free action of a cyclic group Z_p, p is an odd prime number and let SE be its sphere bundle which is a G–subspace of E. It follows from the Lefschetz fixed point theorem that the fiber dimension of E has to be even otherwise there is no free action of Z_p on SE which preserves fibres.

By identyfying all points of the same orbit in SE we obtain the lens–bundle $\overline{\pi} : SE/G = \overline{S}E \to B$ and p–sheeted covering $SE \to \overline{S}E$.

Now, denote by $EG \to BG$ the classiyfying principal G–bundle and consider a classyfying map $\overline{h} : \overline{S}E \to BG$, which is covered by an equivariant map $h : SE \to EG$.

It is well known, that the cohomology ring of $BG = BZ_p$ with coefficients in Z_p is the polynomial ring over two generators with one relation, namely

$$H^*(BZ_p; Z_p) = Z_p[a,b]/(a^2 = 0);\ a \in H^1(BZ_p);\ b \in H^2(BZ_p).$$

This allows us to define two elements:

$$\mathcal{A} = \overline{h}^*(a) \in H^1(\overline{S}E) \text{ and } \mathcal{B} = \overline{h}^*(b) \in H^2(\overline{S}E) \tag{1}$$

Now we can apply the Leray–Hirsch theorem to the fibre bundle $\overline{\pi} : \overline{S}E \to B$ and find that $H^*(\overline{S}E)$ is freely generated as an $H^*(B)$-module, by

$$1,\ \mathcal{A},\ \mathcal{B},\ \mathcal{AB},\ \ldots, \mathcal{B}^{n-1},\ \mathcal{AB}^{n-1}. \tag{2}$$

That is why all the coefficients $\gamma_i \in H^i(B)$ in the equality

$$\mathcal{B}^n + \gamma_1 \mathcal{B}^{n-1}\mathcal{A} + \gamma_2 \mathcal{B}^{n-1} + \cdots + \gamma_{2n-1}\mathcal{A} + \gamma_{2n} = 0$$

are uniquely determined, where $2n$ is fibre dimension of $E \to B$.

Putting additionally $\gamma_0 = 1$ and $\gamma_i = 0$ for $i > 2n$ we get a family of elements which we call *the Grothendieck-Chern classes* of the vector bundle $E \to B$.

We are now in a position to define a polynomial of two variables

$$\Gamma(E; x, y) = y^n + \sum_{i=1}^{n} [\gamma_{2i-1} \cdot x + \gamma_{2i}] \cdot y^{n-i} \in H^*(B)[x, y]$$

which (using the same convention as in [D]) will be called *the Grothendieck–Chern polynomial* associated with the vector bundle $E \to B$.

2. Main Results

Let $\pi : E \to B$, $\pi' : E' \to B$ be vector bundles with fibre–preserving free actions of a cyclic group $G = Z_p$ of fibre dimensions $2n$ and $2m$ respectively. Assume as above that $SE \subset E$ is the sphere bundle which is a G–subspace of E and consider a G–equivariant map $f : SE \to E'$ which is fibre–preserving ($\pi' f = \pi$).

We can define a set $X_f = \{x \in SE : f(x) = 0\}$ which is a G–equivariant subset of SE. Put $\mathcal{A}_f = \overline{\iota}^*(\mathcal{A}) \in H^1(\overline{X}_f)$ and $\mathcal{B}_f = \overline{\iota}^*(\mathcal{B}) \in H^2(\overline{X}_f)$, where $\overline{\iota} : \overline{X}_f \to \overline{S}E$ is the map induced on the orbit spaces by the natural inclusion $\iota : X_f \to SE$.

We can substitute these elements for the indeterminates x and y respectively and obtain homomorphisms of $H^*(B)$–algebras

$$\sigma : H^*(B)[x, y] \to H^*(\overline{S}E) \to H^*(\overline{X}_f),$$

$$x \to \mathcal{A},\ y \to \mathcal{B},\ \text{resp. } x \to \mathcal{A}_f,\ y \to \mathcal{B}_f.$$

It is easily seen that $\ker \sigma$ is an ideal generated by x^2 and $\Gamma(E; x, y)$.

Theorem 2.1. *If $q(x,y) \in H^*(B)[x,y]$ is such that $q(\mathcal{A}_f, \mathcal{B}_f) = 0$ then*

$$q(x,y) \cdot \Gamma(E'; x,y) - \Gamma(E; x,y) \cdot q'(x,y) = x^2 \cdot q''(x,y)$$

for some polynomials $q'(x,y), q''(x,y) \in H^*(B)[x,y]$. In other words

$$q(x,y) \cdot \Gamma(E'; x,y) = \Gamma(E; x,y) \cdot q'(x,y)$$

in the ring $H^*(B)[x,y]/(x^2 = 0)$.

Corollary 2.1. *If $2n$ and $2m$ are fibre dimensions of E, E', respectively, then $q(\mathcal{A}_f, \mathcal{B}_f) \neq 0$ for all polynomials $q(x,y)$ whose degree with respect to y is smaller than $n - m$.*
One can also say, that the $H^*(B)$-homomorphism

$$H^*(B)[x,y]/(x^2 = 0 = y^{n-m}) \to H^* \overline{X}_f,$$

$$x \to \mathcal{A}_f, \quad y^i \to \mathcal{B}_f^i$$

is monomorphic.
In particular, if $n > m$ then

$$\text{coh.}\dim(\overline{X}_f) \geq \text{coh.}\dim(B) + 2 \cdot (n-m) - 1,$$

where coh. dim *denotes the cohomological dimension.*

Proof of Theorem 2.1: Practically Dold's proof of Th. 1.14. in [D] can be repeated. If $q(\mathcal{A}_f, \mathcal{B}_f) = 0$ then (by continuity of Čech–cohomology) $q(x,y)$ vanishes in an open neighbourhood $V \subset \overline{S}E$ of X_f.
By exactness of

$$H^*(\overline{S}E, V) \to H^*(\overline{S}E) \to H^*(V),$$

there is $v \in H^*(\overline{S}E, V)$ such that $j^*(v) = q(\mathcal{A}, \mathcal{B})$.
On the other hand, the map $f : SE - X_f \to E' - \{0\}$ induces a homomorphism

$$\overline{f}^* : H^*(\overline{S}E') \to H^*(\overline{S}E - \overline{X}_f)$$

and $\Gamma(E'; \mathcal{A}_f^c, \mathcal{B}_f^c) = \Gamma(E'; \overline{f}^*(\mathcal{A}'), \overline{f}^*(\mathcal{B}')) = \overline{f}^*(\Gamma(E'; \mathcal{A}', \mathcal{B}')) = \overline{f}^*(0) = 0$, where $\mathcal{A}' = \overline{k}^*(a) \in H^1(\overline{S}E')$, $\mathcal{B}' = \overline{k}^*(b) \in H^2(\overline{S}E')$ for classyfying map $\overline{k} : \overline{S}E' \to BG$ and $\mathcal{A}_f^c = \overline{\imath}^*(\mathcal{A}) \in H^1(\overline{S}E - \overline{X}_f)$, $\mathcal{B}_f^c = \overline{\imath}^*(\mathcal{B}) \in H^1(\overline{S}E - \overline{X}_f)$ for the canonical inclusion $\overline{\imath} : \overline{S}E - \overline{X}_f \to \overline{S}E$. By exactness as above, there is $z \in H^*(\overline{S}E, \overline{S}E - \overline{X}_f)$ such that $j^*z = \Gamma(E'; \mathcal{A}, \mathcal{B})$.
Now, $v \cup z = 0$, hence

$$q(\mathcal{A}, \mathcal{B}) \cdot \Gamma(E'; \mathcal{A}, \mathcal{B}) = j^*v \cup j^*z = j^*(v \cup z) = 0.$$

But $H^*(\overline{S}E) = H^*(B)[x,y]/(x^2 = 0 = \Gamma(E; x,y))$, hence

$$q(x,y) \cdot \Gamma(E'; x,y) = q'(x,y) \cdot \Gamma(E; x,y) + x^2 \cdot q''(x,y). \quad \blacksquare$$

Now, consider a vector bundle $M \to B$ which is the Whitney sum of p copies of a vector bundle $E'' \to B$ of fibre dimension m.

We can equip M with a fibre-preserving action of $G = Z_p$ putting

$$g(m_1, ..., m_p) = (m_p, m_1, ..., m_{p-1}) \tag{3}$$

for a fixed generator $g \in G$ and $(m_1, ..., m_p) \in M$.

Denote by Δ a subspace of M consisted of all points $(m_1, ..., m_p) \in M$ such that $m_1 = \cdots = m_p$.

Hence, we obtain a subbundle $\Delta \to B$ of $M \to B$ which we will call *diagonal bundle*.

Each fibre M_b of M can be represented as a direct sum $\Delta_b \oplus IG_b$, where the second summand is the orthogonal complement of Δ_b and is called *augumentation G-module* (see **[M1]**).

In fact, the bundle $M \to B$ is just the Whitney sum of the diagonal bundle $\Delta \to B$ and the *augumentation bundle* $IG \to B$. Note, that IG is a G-subspace of M and Z_p acts freely on the sphere bundle $SIG \subset IG$.

Since the fibre dimension of $\Delta \to B$ equals m we get the fibre dimension of $IG \to B$ which is equal to $(p-1) \cdot m$.

Let $E \xrightarrow{\pi} B$ be a vector bundle as in Col. 2.1. For an arbitrary vector bundle $E'' \xrightarrow{\pi''} B$ of the fibre dimension m consider a fibre-preserving map $f : SE \to E''$ $(\pi'' f = \pi)$ and define $Y_f = \{x \in SE : f(x) = f(gx) \text{ for all } g \in G\}$.

Now we come to the following.

Corollary 2.2. *Let $E \xrightarrow{\pi} B \xleftarrow{\pi''} E''$ be as above. For an arbitrary fibre-preserving map $f : SE \to E''$ one has*

$$\operatorname{coh.dim} Y_f \geq \operatorname{coh.dim} B + 2n - (p-1) \cdot m - 1.$$

Proof: Consider a fibre-preserving G-equivariant map $F : SE \to M$ defined by the formula

$$F(x) = (f(x), f(gx), \ldots, f(g^{p-1}x)).$$

Notice that $Y_f = F^{-1}(\Delta)$, hence our conclusion immediately follows from Corollary 2.1 applied to $E \to B$ and $IG \to B$. ■

If B is a single point then Corollary 2.2 is the classical version of Z_p-Borsuk-Ulam theorem proved in **[M2]** by Munkholm.

Now, we turn to multi-valued fibre-preserving maps.

We recall the following two definitions.

Definition 2.1. *Let X, Y be spaces and let φ be a multi-valued map from X to Y, i.e., a function which assings to each $x \in X$ a nonempty subset $\varphi(x)$ of Y. We say that φ is* upper semicontinuous *(u.s.c.), if each $\varphi(x)$ is compact and if the following condition holds: For every open subset V of Y containing $\varphi(x)$ there exists an open subset U of X containing x such that for each $x' \in U$, $\varphi(x') \subset V$.*

For instance, if X and Y are compact then f is upper semicontinuous iff its graph is closed in $X \times Y$.

Definition 2.2. *An u.s.c. map φ from X to Y is said to be Z_p-*admissible *(briefly admissible), if there exists a space Γ and two singlevalued continuous maps $\alpha : \Gamma \to X$ and $\beta : \Gamma \to Y$ such that*

(i) α is a Vietoris map, i.e., it is surjective, proper and each set $\alpha^{-1}(x)$ is Z_p-acyclic,

(ii) for each $x \in X$ the set $\beta(\alpha^{-1}(x))$ is contained in $\varphi(x)$.

We will say that the pair (α, β) is a "selected pair" for φ.

For instance, if each $\varphi(x)$ is acyclic (and if φ is u.s.c.) then φ is admissible.

Our aim is to show that Corollaty 2.2 is valid also for multivalued maps in the following sense.

For vector bundles $E \xrightarrow{\pi} B$ and $E'' \xrightarrow{\pi''} B$ of the fibre dimension $2n$ and m respectively consider a multi-valued, admissible and fibre preserving map $\varphi : SE \to E''$ ($\pi'' \cdot \varphi = \pi$ i.e., for each $x \in SE$ $\varphi(x)$ is contained in the fibre $\pi''^{-1}(\pi(x))$.

Recall, that SE is a G-space with an action of G which is free and let us define a set

$$Y_\varphi = \{x \in SE : \varphi(x) \cap \varphi(gx) \cap \cdots \cap \varphi(g^{p-1}x) \neq \emptyset\},$$

where $g \in G$ is a fixed generator.

Theorem 2.2. *The following inequality holds*

$$\operatorname{coh.dim} Y_\varphi \geq \operatorname{coh.dim} B + 2n - (p-1) \cdot m - 1. \quad \blacksquare$$

If B is a single point then we get theorem which has been proved by the first author in [**I**, Th. 2.1].

First we show, that Theorem 2.1 is valid not just for maps $f : SE \to E'$, but also in the following, more general setting. Suppose that Z is any space with a free action $g : Z \to Z$ of a group $G = Z_p$, $p > 2$ prime, g^p =id and $v : Z \to SE$ is an equivariant Vietoris map. Let $s : Z \to E'$ be a single-valued, equivariant map which makes the diagram

$$\begin{array}{ccc} Z & \xrightarrow{s} & E' \\ {\scriptstyle v}\downarrow & & \downarrow{\scriptstyle \pi'} \\ SE & \xrightarrow{\pi} & B \end{array}$$

commutative, $\pi' s = \pi v$.

Set $Z_s = \{z \in Z : s(z) = 0\}$.

Note, that $H^*(\overline{Z}_s)$ are $H^*(B)$-algebras via the homomorphism $\overline{v}^* \cdot \pi^* : H^*(B) \to H^*(\overline{Z})$, where $\overline{v} : \overline{Z} \to \overline{SE}$ is the homomorphism induced by v on orbit spaces.

Put $\mathcal{J} = \overline{v}^*(\mathcal{A})$, $\mathcal{R} = \overline{v}^*(\mathcal{B})$, $\mathcal{J}_s = \overline{\imath}^*(\mathcal{J})$ and $\mathcal{R}_s = \overline{\imath}^*(\mathcal{J})$, where $\mathcal{A}$, $\mathcal{B}$ are defined by equalities (1) and $\overline{\imath}^*$ es the homomorphism induced by the canonical inclusion $\overline{\imath} : \overline{X}_s \to \overline{X}$.

Lemma 2.1. *If $q(x,y) \in H^*(B)[x,y]$ vanishes on $(\mathcal{J}_s, \mathcal{R}_s)$, $q(\mathcal{J}_s, \mathcal{R}_s) = 0$, then there are polynomials $q'(x,y)$, $q''(x,y) \in H^*(B)[x,y]$ such that*

$$q(x,y) \cdot \Gamma(E'; x,y) - \Gamma(E; x,y) \cdot q'(x,y) = x^2 \cdot q''(x,y).$$

Proof: It is well known (and easily seen) that $\overline{v} : \overline{X} \to \overline{S}E$ is a Vietoris map since such is v. Now, note that proof of Theorem 2.1 can be adopted to present setting due to the fact that the homomorphism $\overline{v}^*$ induced by the Vietoris map $\overline{v}$ is an isomorphism.

Thus, as far as the cohomology is concerned, the arrows $SE \leftarrow Z \to E'$ work just as well as a single arrow $SE \to E'$. ∎

Corollary 2.3. *Assume that $E'' \xrightarrow{\pi''} B$ is a vector bundle of fibre dimension m and consider a map $s : Z \to E''$ such that $\pi'' s = \pi v$.*
Then the cohomological dimension of the set

$$Z_s = \{z \in Z : s(z) = s(gz) = \cdots = s(g^{p-1}z)\}$$

is greather than or equal to $\mathrm{coh.dim}\, B + 2n - (p-1) \cdot m - 1$.

Now, let as note that Y_φ is a G–invariant subspace of SE so, as previous we can consider elements $\mathcal{A}_\varphi = \overline{\iota}^*(\mathcal{A}) \in H^1(\overline{Y}_\varphi)$ and $\mathcal{B}_\varphi = \overline{\iota}^*(\mathcal{B}) \in H^1(\overline{Y}_\varphi)$, where this time $\overline{\iota} : \overline{Y}_\varphi \to \overline{S}E$ is covered by the inclusion $\iota : Y_\varphi \to SE$.

Proof of Theorem 2.2: Given an admissible multi–valued map from SE to E'', choose a space Γ and (single–valued) maps α and β such that (α, β) is a "selected pair" for φ (see Def. 2.2).
Let

$$Z = \{(\gamma_1, \ldots, \gamma_p) \in \Gamma \times \cdots \times \Gamma : \alpha(\gamma_1) = g\alpha(\gamma_2) = \cdots = g^{p-1}\alpha(\gamma_p)\}.$$

Consider the following commutative diagram.

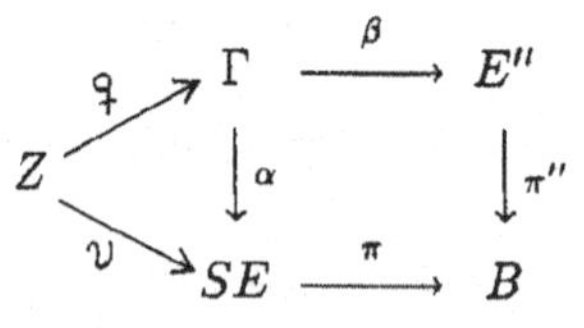

$$\begin{array}{ccccc} & \overset{q}{\nearrow} & \Gamma & \xrightarrow{\beta} & E'' \\ Z & & \downarrow{\scriptstyle\alpha} & & \downarrow{\scriptstyle\pi''} \\ & \underset{v}{\searrow} & SE & \xrightarrow{\pi} & B \end{array}$$

Here q is the first projection $(\gamma_1, \ldots, \gamma_p) \to \gamma_1$, and $v = \alpha \circ q$. Then v is a Vietoris map since for each $x \in SE$

$$v^{-1}(x) = \alpha^{-1}(x) \times \alpha^{-1}(gx) \times \cdots \times \alpha^{-1}(g^{p-1}x)$$

is acyclic as the Cartesian product of acyclic sets.

The space Z admits a free action of Z_p,

$$(\gamma_1, \ldots, \gamma_p) \to (\gamma_p, \gamma_1, \ldots, \gamma_{p-1}),$$

and $v : Z \to SE$ becomes then an equivariant map.

Let $s : \beta \circ q : Z \to E''$. Notice that if

$$s(\gamma_1, \ldots, \gamma_p) = s(\gamma_p, \gamma_1, \ldots, \gamma_{p-1}) = \cdots = s(\gamma_2, \ldots, \gamma_p, \gamma_1)$$

for some $(\gamma_1, \ldots \gamma_p) \in Z$ then $\varphi(\alpha(\gamma_1)) \cap \cdots \cap \varphi(\alpha(\gamma_p)) \neq \emptyset$. Thus

$$v(Z_s) \subset Y_\varphi \quad \text{and} \quad (\overline{v}|\overline{Z}_s)^*(\mathcal{A}_\varphi) = \mathcal{J}_s \quad \text{and} \quad (\overline{v}|\overline{Z}_s)^*(\mathcal{B}_\varphi) = \mathcal{R}_s.$$

Therefore

$$q(\mathcal{J}_s, \mathcal{R}_s) = q((\overline{v}|\overline{Z}_s)^*\mathcal{A}_\varphi, \quad (\overline{v}|\overline{Z}_s)^*\mathcal{B}_\varphi) = (\overline{v}|\overline{Z}_s)^*(q(\mathcal{A}_\varphi, \mathcal{B}_\varphi))$$

so $q(\mathcal{J}_s, \mathcal{R}_s) \neq 0$ implies $q(\mathcal{A}_\varphi, \mathcal{B}_\varphi)) \neq 0$ and by Lemma 2.1 and Corollary 2.3 we have got the desired result. ■

Since the covering dimension of Y_φ is not less than its cohomological dimension we come to

Corollary 2.4. *If B is a closed d-dimensional manifold then*

$$\operatorname{cov.dim} Y_\varphi \geq \operatorname{coh.dim} Y_\varphi \geq d + 2n - (p-1) \cdot m - 1. \quad ■$$

References

[D] Dold, A., *"Parametrized Borsuk-Ulam theorems"*, Comment. Math. Helv. **63**, (1988), 275–285.

[FH] Fadell, E.R. and Husseini, S.Y., *"Cohomological index theory with applications to critical point theory and Borsuk-Ulam theorems"*, Seminaire de Math. Sup. 108, Universite de Montreal, Montreal, 1989, 10–54.

[G] Gorniewicz, L., *"Homological methods in fixed point theory for multivalued maps"*, Dissertationes Math. **129**, (1976), 1–71.

[I1] Izydorek, M., *"Nonsymetric version of Bourgin–Yang theorem for multi-valued maps and free Z_p-actions"*, J. Math. Anal. and Appl., **vol. 137, n. 2**, (1989), 349–353.

[I2] Izydorek, M., *"Remarks on Borsuk-Ulam theorem for multi-valued maps"*, Bull. Acad. Sci. Pol., Math., **35**, (1987), 501–504.

[IJ] Izydorek, M. and Jaworowski, J., *"Parametrized Borsuk-Ulam theorems for multi-valued maps"*, (to appear).

[J1] Jaworowski, J., *"A continuous version of the Borsuk-Ulam theorem"* Proc. Amer. Math. Soc. **82**, (1981), 112–114.

[J2] Jaworowski, J., *"Fibre preserving maps of sphere bundles into vector space bundles"*, Proc. of the Fixed Point Theory Conference, Sherbrooke, Quebec 1980, LN in Math. **886**, Springer–Verlag 1981, 154–162.

[M1] Munkholm, H.J., *"On the Borsuk-Ulam theorem for Z_{p^α} actions on S^{2n-1} and maps $S^{2n-1} \to R^m$"*, Osaka J. Math. **7**, (1970), 451–456.

[M2] Munkholm, H.J., *"Borsuk-Ulam type theorems for proper Z_p-actions on (*mod p *homology) n-spheres"*, Math. Scand. **24**, (1969), 167–185.

[N1] Nakaoka, M., *"Equivariant point theorems for fibre-preserving maps"*, Osaka J. Math. **21**, (1984), 809–815.

[N2] Nakaoka, M., *"Proceedings of the Tianjin Fixed Point Conference 1988"*, LN in Math., **1411**, Springer–Verlag 1989.

Department of Mathematics
Technical University of Gdańsk
ul. Majakowskiego 11/12
80–952 Gdańsk,
Poland.

1990 Barcelona Conference
on Algebraic Topology.

RÉALISATION TOPOLOGIQUE DE CERTAINES ALGÈBRES ASSOCIÉES AUX ALGÈBRES DE DICKSON

ALAIN JEANNERET AND ULRICH SUTER

Abstract

Topological realisation of certain algebras associated to the Dickson algebras.

We discuss the topological realisation of certain $\mathbb{Z}/2$-algebras $A(n)$ over the mod 2 Steenrod algebra $\mathcal{A}(2)$. If an associative H-space $X(n)$ satisfies $H^*(X(n);\mathbb{Z}/2) \cong A(n)$, the mod 2 cohomology of its classifying space is isomorphic to the algebra of invariants of the canonical $\mathrm{Gl}_n(\mathbb{Z}/2)$-action on a graded polynomial algebra in n variables of degree 1.

1. Introduction. Dans cette note, on envisage de déterminer les entiers $n \geq 2$ pour lesquels l'algèbre $A(n)$ définie sous (1) peut être réalisée comme algèbre de cohomologie modulo 2 d'un espace topologique. La $\mathbb{Z}/2$-algèbre graduée

$$(1) \qquad A(n) = \mathbb{Z}/2\,[v_1]\,/\,(v_1^4) \otimes E(v_2, v_3, ..., v_{n-1}) \quad \text{où degré} \quad (v_k) = 2^n - 2^{n-k} - 1$$

est munie d'une unique structure de module sur l'algèbre de Steenrod $\mathcal{A}(2)$ imposée par l'égalité $Sq^{2^{n-1}-1}v_1 = v_1^2$ et les relations d'Adem $Sq^{2^i-1} = Sq^1 Sq^2 ... Sq^{2^{i-1}}$. Par exemple, $A(2) \cong H^*(SO(3);\mathbb{Z}/2)$ et $A(3) \cong H^*(G_2;\mathbb{Z}/2)$, où G_2 est le groupe de Lie exceptionnel.

Les algèbres $A(n)$ sont liées à certaines algèbres de Dickson $D(n)$ définies comme suit: le groupe linéaire $GL_n(\mathbb{Z}/2)$ agit canoniquement sur l'algèbre de polynômes graduée $\mathbb{Z}/2\,[t_1, ..., t_n]$, où degré $(t_k) = 1$ $(k = 1, ..., n)$, et par définition, $D(n)$ est l'algèbre des invariants. Dickson a démontré que $D(n) = \mathbb{Z}/2\,[w_1, ..., w_n]$, où degré $(w_k) = 2^n - 2^{n-k}$. L'algèbre $D(n)$ est un module sur $\mathcal{A}(2)$ et la question suivante est naturellement liée au problème de Steenrod:

Existe-t-il un CW-complexe $Y(n)$ tel que $H^*(Y(n);\mathbb{Z}/2) \cong D(n)$?

Les exemples connus sont: $D(1) \cong H^*(B\mathbb{Z}/2;\mathbb{Z}/2), D(2) \cong H^*(BSO(3);\mathbb{Z}/2)$ et $D(3) \cong H^*(BG_2;\mathbb{Z}/2)$. Dans [6] L. Smith et M. Switzer montrent qu'un tel espace n'existe pas si $n \geq 6$. Les cas où $n = 4$ et $n = 5$ sont plus délicats à traiter.

Si $Y(n)$ existe, on obtient facilement à l'aide de la suite spectrale d'Eilenberg–Moore que l'espace des lacets $\Omega Y(n) = X(n)$ vérifie:

$$H^*(X(n); \mathbf{Z}/2) \cong A(n) \tag{2}$$

La question précédente peut donc se reformuler en ces termes:

Existe-t-il un H-espace associatif $X(n)$ vérifiant la condition (2)?

J. Lin et F. Williams ont répondu par la négative au cas où $n = 5$ [4]. Leur démonstration utilise la machinerie des opérations cohomologiques secondaires et tertiaires. A l'aide de la K-théorie complexe, il est possible de démontrer, sans faire appel aux opérations cohomologiques d'ordres supérieurs, le résultat plus général suivant:

Théorème. *Pour $n \geq 5$, l'algèbre $A(n)$, définie sous (1), ne peut pas être réalisée comme algèbre de cohomologie modulo 2 d'un espace topologique.*

Notons que ce théorème exclut l'existence de $X(n)$, pour $n \geq 5$, comme espace topologique et non seulement comme H-espace mais qu'il n'apporte aucune information au cas où $n = 4$ pour lequel il existe deux résultats contradictoires [2], [5].

Le reste de cette note est consacré à l'esquisse de la démonstration du théorème. Elle peut être résumée comme suit: si $X(n)$ satisfait (2), on constate d'abord que la cohomologie entière de $\Omega X(n)$ est sans 2-torsion, et on montre ensuite, grâce aux résultats d'Atiyah [1], que les opérations de Steenrod Sq^k et l'opération d'Adams Ψ^2 ne sont pas compatibles pour $\Omega X(n)$, où $n \geq 5$.

Nous n'expliciterons que le cas où $n = 5$, les autres se traitant de manière identique. Dans la suite de cette note, nous supposerons que X est un CW-complexe fini simplement connexe vérifiant:

$$H^*(X; \mathbf{Z}/2) \cong \mathbf{Z}/2\,[x_{15}]/(x_{15}^4) \otimes E(x_{23}, x_{27}, x_{29}) \tag{3}$$

Un tel complexe a le même type d'homotopie rationnelle que le produit de sphères $S^{15} \times S^{23} \times S^{27} \times S^{59}$. Faisant appel à la technique du "mixing homotopy types ", nous supposerons également que, pour les premiers impairs p, la cohomologie modulo p de X est isomorphe à l'algèbre extérieure $E_{\mathbf{Z}/p}(z_{15}, z_{23}, z_{27}, z_{59})$.

Nous tenons à remercier J. Lin avec qui nous avons eu de nombreuses et fructueuses discussions.

2. Calculs de cohomologie ordinaire.

Soit $\rho_* : H^*(X; \mathbf{Z}) \to H^*(X; \mathbf{Z}/2)$ l'homomorphisme de réduction modulo 2. A l'aide de la formule des coefficients universels on montre l'existence d'éléments $y_{15}, y_{23}, y_{27}, z_{59} \in H^*(X; \mathbf{Z})$ pour lesquels $\rho_*(y_k) = x_k$ $(k = 15, 23, 27)$ et $\rho_*(z_{59}) = x_{29} \cdot x_{15}^2$; de plus,

$$H^*(X; \mathbf{Z}) \cong P(y_{15}) \otimes E_{\mathbf{Z}/2}(y_{23}, y_{27}, z_{59})/(y_{15}^2 \otimes z_{59}),$$

où $P(y_{15}) = \mathbf{Z}\,[y_{15}]/(y_{15}^4, 2y_{15}^2) \cong \mathbf{Z} \oplus \mathbf{Z} \oplus \mathbf{Z}/2 \oplus \mathbf{Z}/2$ est engendré additivement par $1, y_{15}, y_{15}^2, y_{15}^3$.

Nous allons décrire la cohomologie modulo 2 de ΩX au moyen de la suite spectrale d'Eilenberg–Moore. Notre référence, dans ce contexte, est le livre de R. Kane [**3**, ch. VII] dont nous adoptons aussi les notations. Le terme E_2 de la suite s'obtient à l'aide de la résolution bar $BH^*(X;\mathbf{Z}/2)$; dans notre situation on obtient

$$E_2 \cong \mathrm{Tor}_{H^*(X;\mathbf{Z}/2)}(\mathbf{Z}/2,\mathbf{Z}/2) \cong E(u_{14}) \otimes \Gamma(u_{22}, u_{26}, u_{28}, u_{58}),$$

où $u_k = sx_k$ ($k = 14, 22, 26, 28$) sont des éléments "suspension" représentés par $[x_k]$ et u_{58} est un élément "transpotcnt" représenté par $[x_{15}^2|x_{15}^2] = [x_{15}|x_{15}^3]$. Le degré total de tous les générateurs étant pair, on conclut que la suite spectale est triviale, i.e. $E_2 \cong E_\infty$, et on obtient ainsi les isomorphismes additifs:

$$H^*(\Omega Z;\mathbf{Z}/2) \cong E^0(H^*(\Omega X;\mathbf{Z}/2)) \cong E(u_{14}) \otimes \Gamma(u_{22}, u_{26}, u_{28}, u_{58})$$

Lemme 1. *Soit u_{56} l'unique élément non nul dans $H^{56}(\Omega X;\mathbf{Z}/2)$. Alors:*

$$Sq^2 u_{56} \neq 0.$$

Démonstration: La suite spectrale d'Eilenberg–Moore est une suite de modules sur $\mathcal{A}(2)$ et dans $\mathrm{Tor}_{H^*(X;\mathbf{Z}/2)}(\mathbf{Z}/2,\mathbf{Z}/2)$ on a l'élément $u_{56} = \gamma_2(u_{28}) = [x_{29}|x_{29}]$ ainsi que les relations $Sq^2[x_{29}|x_{29}] = \sum_{i+j=2} [Sq^i x_{29}|Sq^j x_{29}] = [x_{15}^2|x_{15}^2] = u_{58} \neq 0$.

Notons encore que la cohomologie entière de ΩX est sans torsion (fait résultant des égalités $H^{2n+1}(\Omega X;\mathbf{Z}/2) = 0$ pour $n \geq 0$), et que l'on peut choisir des générateurs $v_k \in H^k(\Omega X;\mathbf{Z}) \cong \mathbf{Z}$ ($k = 14, 22, 26, 28, 56$) tels que:

$$(4) \qquad \rho_*(v_k) = u_k \quad \text{et} \quad v_{14}^2 = v_{28},\ v_{14}^4 = v_{28}^2 = 2v_{56}$$

3. Calculs de K–théorie.

Une étude soigneuse de la suite spectrale d'Atiyah–Hirzebruch en K–théorie mod 2 et entière pour l'espace X montre que, dans les deux cas, la seule différentielle non-triviale est d_3. On obtient ainsi $E_4 \cong E_\infty \cong E^0(K^*(X))$, et le calcul explicite fournit que $K^*(X) \cong E(\xi_{15}, \xi_{23}, \xi_{27}, \xi_{59})$, les générateurs ξ_κ satisfaisants aux conditions suivantes: dans l'ordre indiqué, les éléments $\xi_{15}, \xi_{23}, \xi_{27}, \xi_{59}, 2\xi_{59}$ correspondent, dans le sous-quotient E_∞ de $H^*(X;\mathbf{Z})$, aux éléments représentés par $y_{15}, y_{23}, 2y_{27}, y_{15}^2 \cdot y_{27}, z_{59}$; la filtration rationnelle de ξ_κ est égale à k. On définit $\eta_\kappa \in \widetilde{K}^0(\Sigma X)$ par

$$\eta_\kappa = \sigma(\xi_{\kappa-1}) \quad (k = 16, 24, 28, 60),$$

où $\sigma : \widetilde{K}^1(X) \cong \widetilde{K}^0(\Sigma X)$ et on envisage de déterminer l'opération d'Adams Ψ^2 sur ces éléments.

Pour tout CW–complexe Y et tout entier $q \geq 0$ soit $\mathcal{F}^q\widetilde{K}(Y) \subseteq \widetilde{K}(Y) = \widetilde{K}^0(Y)$ le sous-groupe des éléments de filtration $\geq q$, c'est-à-dire $\mathcal{F}^q\widetilde{K}(Y) = \mathrm{Ker}\{\widetilde{K}(Y) \to \widetilde{K}(Y^{q-1})\}$. Les éléments $\eta_\kappa (k = 16, 24, 28, 60)$ forment une base du groupe abelien libre $\widetilde{K}(\Sigma X)/\mathcal{F}^{62}\widetilde{K}(\Sigma X)$.

Lemme 2. *Pour un choix convenable de* $\eta_{28} \in \widetilde{K}(\Sigma X)$ *on a dans* $\widetilde{K}(\Sigma X)/\mathcal{F}^{62}\widetilde{K}(\Sigma X)$:

$$\text{(i)} \qquad \Psi^2(\eta_{16}) \equiv 2\eta_{28} \qquad (mod\, 4)$$
$$\text{(ii)} \qquad \Psi^2(\eta_{28}) \equiv 0 \qquad (mod\, 4)$$

Démonstration: Remarquons préalablement que $\Psi^2(\eta) \equiv \eta^2 = 0$ (mod 2) pour tout $\eta \in \widetilde{K}(\Sigma X)$. A homotopie près, il existe un sous-complexe $Z = S^{16} \cup e^{24} \cup e^{28}$ de ΣX qui porte la cohomologie entière jusqu'à la dimension 28. Soit $i : Z \to \Sigma X$ l'inclusion. La K-théorie de Z est libre à trois générateurs, $\alpha_{16} = i^*(\eta_{16}), \alpha_{24} = i^*(\eta_{24})$ et α_{28}, où $2\alpha_{28} = i^*(\eta_{28})$. Par le résultat d'Atiyah [1], on obtient $\Psi^2(\alpha_{16}) = 2^8\alpha_{18} + 2^4 a\ \alpha_{24} + 2^2 b\ \alpha_{28}$, où $a \equiv b \equiv 1$ (mod 2), il suit que

$$\text{(5)} \qquad \Psi^2(\eta_{16}) = 2^8\eta_{16} + 2^4 a\ \eta_{24} + 2\ b\ \eta_{28} + 2\ c\ \eta_{60}$$

Cette dernière égalité implique directement (i) dans le cas où c est pair; si c est impair on remplace d'abord le générateur η_{28} par $\widetilde{\eta}_{28} = \eta_{28} + \eta_{60}$.
Pour la démonstration de (ii), on note que $\Psi^m(\eta_{28}) = m^{14}\eta_{28} + d_m\eta_{60}, \Psi^m(\eta_{60}) = m^{30}\eta_{60}$ et on considère l'égalité $\Psi^3\Psi^2(\eta_{28}) = \Psi^2\Psi^3(\eta_{28})$.

Remarque: dans le cas où $n = 4$ la relation $\Psi^3\Psi^2 = \Psi^2\Psi^3$ implique que $d_2 \equiv 0$ (mod 2) mais ne permet pas d'affirmer que ce coefficient est divisible par 4.

Considérons maintenant l'application canonique $\varepsilon : \Sigma^2\Omega X \to \Sigma X$. On définit $\mu_\kappa \in \widetilde{K}(\Omega X)$ $(k = 14, 22, 26)$ par $\sigma^2(\mu_\kappa) = \varepsilon^*(\eta_{\kappa+2})$ et on rappelle que $2\sigma^2\Psi^2(\mu_\kappa) = \varepsilon^*\Psi^2(\eta_{\kappa+2})$. La cohomologie entière $H^*(\Omega X; \mathbf{Z})$ étant sans torsion, il en est de même pour $\widetilde{K}(\Omega X)/\mathcal{F}^{60}\widetilde{K}(\Omega X)$. En calculant dans ce dernier groupe on obtient du lemme 2 d'abord que $\mu_{14}^2 \equiv \Psi^2(\mu_{14}) \equiv \mu_{26}$ (mod 2) et ensuite que $\mu_{14}^4 \equiv \Psi^2(\mu_{14}^2) \equiv \Psi^2(\mu_{26}) \equiv 0$ (mod 2). Il existe donc un élément $\omega \in \widetilde{K}(\Omega X)$ tel que

$$\mu_{14}^4 \equiv 2\,\omega \qquad (\text{mod } \mathcal{F}^{60}\widetilde{K}(\Omega X)).$$

La relation (5) implique que $\Psi^2(\mu_{14}^4) \equiv 2^{28}\mu_{14}^4$ (mod $\mathcal{F}^{60}\widetilde{K}(\Omega X)$) et on obtient ainsi:

$$\Psi^2(\omega) \equiv 2^{28}\omega \qquad (\text{mod } \mathcal{F}^{60}\widetilde{K}(\Omega X)).$$

La suite spectrale d'Atiyah–Hirzebruch de ΩX en K-théorie est triviale, i.e. $H^*(\Omega X; \mathbf{Z}) \cong E^0(K^*(\Omega X))$. L'élément μ_{14} représente v_{14}, et il suit de (4) que ω représente $v_{56} \in H^{56}(\Omega X; \mathbf{Z})$. A l'aide du théorème d'Atiyah [1] on déduit que $Sq^2 u_{56} = 0$; ce qui contredit le lemme 1 et achève la démonstration du théorème.

Réferences bibliographiques.

[1] M.F. ATIYAH. Power operations in K-theory, *Quart J. Math. Oxford* (2) 17 (1966), 165–193.

[2] W.G. DWYER et C.W. WILKERSON. A new finite loop space at the prime 2, *Preprint.*

[3] R.M. KANE. The homology of Hopf spaces, *North Holland Math. Study* # 40 (1988).

[4] J.P. LIN et F. WILLIAMS. On 14–connected finite H–spaces, *Israel J. Math.* 66 (1989), 274–288.

[5] J.P. LIN et F. WILLIAMS. On 6–connected finite H–spaces with two torsion, *Topology* 28 (1989), 7–34.

[6] L. SMITH et R.M. SWITZER. Realizability and nonrealizability of Dickson Algebras as cohomology rings, *Proc. Amer. Math. Soc.* 89 (1983), 303-313.

Institut de Mathématiques et d'Informatique,
Université de Neuchâtel,
Chantemerle 20,
CH-2000 Neuchâtel,
Suisse.

1990 Barcelona Conference
on Algebraic Topology.

NORMALIZED OPERATIONS IN COHOMOLOGY

LUCIANO LOMONACO

1. Introduction.

In this paper we exibit a geometric construction of the normalized iterated total squaring operation in mod 2 cohomology S_m, and point out its relation with invariant theory. The paper is set out as follows. In §2. we recall some notation from invariant theory. In §3. we summarize the material we need related to extended powers of spectra. In §4. we actually construct S_m and the ordinary iterated total squaring operation T_m and show, using a purely algebraic computation, that T_m is obtained from S_m by inverting the Euler class of the principal bundle associated to the reduced regular representation of the elementary abelian 2-group $\Sigma_2 \times \cdots \times \Sigma_2$.

2. Invariant theory.

We will employ the standard notation used by W. Singer in [11]. In particular, we set

2.1 $$P_m = F_2[t_1, \dots, t_m] \quad , \quad m \geq 1$$

2.2 $$e_m = \prod_{\substack{\lambda_1,\dots,\lambda_m=0,1 \\ \sum \lambda_i > 0}} (\lambda_1 t_1 + \cdots + \lambda_m t_m)$$

2.3 $$\Phi_m = P_m[e_m^{-1}]$$

P_m is isomorphic to

$$H^*\Big(B(\underbrace{\Sigma_2 \times \cdots \times \Sigma_2}_{m \text{ times}})\Big)$$

and this isomorphism gives P_m the structure of an $\mathcal{A}$-algebra, where $\mathcal{A}$ indicates the mod 2 Steenrod algebra. Such an action of $\mathcal{A}$ on P_m extends to Φ_m (e.g., see [11],[12]). Under the above mentioned isomorphism, e_m corresponds to the product of all the non-trivial line bundles over

$$B(\Sigma_2 \times \cdots \times \Sigma_2) \approx RP^\infty \times \cdots \times RP^\infty .$$

Hence we call e_m the Euler class of P_m.

Let us write GL_m for $GL_m(F_2)$ and let $T_m \leq GL_m$ be its upper triangular subgroup. GL_macts on P_m, e_m is fixed in the GL_m action, and such action commutes with the $\mathcal{A}$-action. So GL_m acts on Φ_m too, and we write

$$\Delta_m = \Phi_m^{T_m} \qquad ; \qquad \Gamma_m = \Phi_m^{GL_m}$$

for the corresponding rings of invariants.

Proposition 2.4.

(i) $$\Delta_m = F_2[v_1^{\pm 1}, \dots, v_m^{\pm 1}] \quad ;$$

(ii) $$\Gamma_m = F_2[Q_{m,0}^{\pm 1}, Q_{m,1}, \dots, Q_{m,m-1}] \; .$$

Here

2.5 $$V_{k+1} = \prod_{\lambda_1,\dots,\lambda_k=0,1} (\lambda_1 t_1 + \dots + \lambda_k t_k + t_{k+1})$$

2.6 $$v_{k+1} = \frac{V_{k+1}}{e_k} \; .$$

Moreover $Q_{m,0} = e_m$ and $Q_{m,j}$ is defined inductively by the formula

2.7 $$Q_{m,j} = Q_{m-1,0}Q_{m-1,j}v_m + Q^2_{m-1,j-1}$$

with the convention that

2.8 $$Q_{m,j} = \begin{cases} 0 & \text{if } j < 0 \text{ or } m < j \; ; \\ 1 & \text{if } m = j \geq 1 \; . \end{cases}$$

3. Extended powers of spectra.

In [1], [7] a detailed construction of the functor D_G (G a finite group) is given on the category of CW-spectra, and the cohomology of $D_G E$ is described, for any spectrum E. All the nice properties enjoyed by this functor on CW-complexes still hold when working on spectra.

We recall that, for $H \leq G \leq \Sigma_t$, we have natural transformations

$$D_H \longrightarrow D_G \; .$$

We are interested in the cases

$$G \quad = \quad \Sigma_2 \times \cdots \times \Sigma_2 \quad , \quad \Sigma_2 \int \ldots \int \Sigma_2 \quad , \quad \Sigma_{2^m} \quad , \qquad m \geq 1 \, .$$

We will write D_t for D_{Σ_t}. We have

$$D_{\Sigma_2 \int \ldots \int \Sigma_2} = D_2 \ldots D_2 \, .$$

Let X be a CW-complex, and let us use the same symbol to indicate its suspension spectrum. In [3], and later in [1], [5], [6], the system

3.1 $$\cdots \longrightarrow S^k D_G S^{-k} X \xrightarrow{\psi_k} S^{k-1} D_G S^{1-k} X \longrightarrow \cdots$$

has been studied. Here ψ_k is the following composite

$$S^k D_G S^{-k} X \xrightarrow{\approx} S^{k-1} \wedge S^1 \wedge D_G S^{-k} X \xrightarrow{\beta} S^{k-1} D_G(S^1 \wedge S^{-k} X) \xrightarrow{\approx} S^{k-1} D_G S^{1-k} X$$

where

$$\beta : Y \wedge D_G S^{-k} X \longrightarrow D_G(Y \wedge S^{-k} X)$$

is defined in [3] for each CW-complex Y and corresponds, essentially, to a diagonal embedding.

We recall that for $X = S^0$ we get

3.2 $$S^k D_G S^{-k} \approx BG^{-k\rho}$$

where ρ is the reduced regular representation of G and $BG^{-k\rho}$ is the Thom spectrum associated with the virtual representation

$$-k\rho = -(\underbrace{\rho \oplus \cdots \oplus \rho}_{k \text{ times}}) \qquad k \geq 0$$

(e.g., see [1]).

4. The iterated total squaring operation.

From now on, H^* will always indicate the mod 2 reduced cohomology functor. Let X be a CW-complex and let $x \in H^n(X)$. Then a representative of x is a map

$$\xi : X \longrightarrow K(Z/2, n)$$

where $K(Z/2,n)$ is an Eilenberg-MacLane space of type (Z/2,n). We define the composite

4.1 $$\bar{\xi} : B\Sigma_2^+ \wedge X \xrightarrow{\Delta} D_2 X \xrightarrow{D_2 \xi} D_2 K(Z/2,n) \xrightarrow{\mu} K(Z/2,2n)$$

where Δ, μ are the usual classical maps (e.g., see [2]). We set

$$T_1(x) = [\bar{\xi}] \in H^{2n}(B\Sigma_2^+ \wedge X) .$$

As it is well known

4.2 $$T_1(x) = \sum t_1^{n-j} \otimes Sq^j(x) \quad , \quad t_1 \in H^1(B\Sigma_2) .$$

The map

4.3 $$T_1 : H^*(X) \longrightarrow H^*(B\Sigma_2^+ \wedge X) \cong F_2[t_1] \otimes H^*(X)$$

is called the *ordinary total squaring operation* and it is a ring homomorphism which doubles the degrees. The ordinary total squaring operation T_1 can be iterated. We set

4.4 $$\begin{aligned}\tilde{\xi} : B(\underbrace{\Sigma_2 \times \cdots \times \Sigma_2}_{m \text{ times}})^+ \wedge X \xrightarrow{\pi\wedge id} B(\underbrace{\Sigma_2 \int \ldots \int \Sigma_2}_{m \text{ times}})^+ \wedge X \xrightarrow{\Delta} \\ \underbrace{D_2 \ldots D_2}_{m \text{ times}} X \xrightarrow{D_2 \ldots D_2 \xi} \underbrace{D_2 \ldots D_2}_{m \text{ times}} K(Z/2, n) \xrightarrow{D_2 \ldots D_2 \mu} \\ \underbrace{D_2 \ldots D_2}_{(m-1) \text{ times}} K(Z/2, 2n) \longrightarrow \cdots \xrightarrow{\mu} K(Z/2, n \cdot 2^m) \quad .\end{aligned}$$

Here π is the covering map which makes the following diagram commute, where E is a contractible space with a free Σ_{2^m}-action and ρ_1, ρ_2 are the universal bundles.

$$\begin{array}{ccc} E & \xrightarrow{id} & E \\ {\scriptstyle \rho_1} \swarrow & & \searrow {\scriptstyle \rho_2} \\ B(\Sigma_2 \times \cdots \times \Sigma_2) & \xrightarrow{\pi} & B(\Sigma_2 \int \ldots \int \Sigma_2) \end{array}$$

Then we set

$$T_m(x) = [\tilde{\xi}] \in H^{n \cdot 2^m}(B(\Sigma_2 \times \cdots \times \Sigma_2)^+ \wedge X) \quad .$$

Clearly

4.5 $$T_m(x) = \sum_{i_1, \ldots, i_m \geq 0} \left(t_1^{n \cdot 2^{m-1} - i_1} Sq^{i_1}\right)\left(t_2^{n \cdot 2^{m-2} - i_2} Sq^{i_2}\right) \ldots \left(t_m^{n - i_m} Sq^{i_m}\right)(x) \quad .$$

$T_m(x)$ is therefore in $P_m \otimes H^*(X)$, after identifying P_m with $H^*(B(\Sigma_2 \times \cdots \times \Sigma_2))$.

We would like to outline the geometric construction of the normalized version of T_m. Consider

$$S^{-n}\xi : S^{-n}X \longrightarrow H \quad .$$

Here H is the mod 2 Eilenberg-MacLane spectrum, whose r-th space is $K(Z/2,r)$. We recall that H is an H_∞-ring spectrum, i.e., there are maps

$$\xi_r : D_r H \longrightarrow H$$

satisfying a certain set of conditions (see [1]). We form the composite

$$\Phi(S^{-n}\xi) : D_2 S^{-n} X \xrightarrow{D_2 S^{-n}\xi} D_2 H \xrightarrow{\xi_2} H$$

and consider its suspension

$$S^n \Phi S^{-n}\xi : S^n D_2 S^{-n} X \longrightarrow S^n H \quad .$$

For each $k \geq n$ we define

4.6 $\quad \psi_k(\xi) : S^k D_2 S^{-k} X \xrightarrow{\psi_k} S^{k-1} D_2 S^{1-k} \longrightarrow \cdots \longrightarrow S^n D_2 S^{-n} X \xrightarrow{S^n \Phi S^{-n}\xi} S^n H \quad .$

The sequence $\psi(\xi) = \{\psi_k(\xi)\}_{k\geq n}$ defines an element

$$[\psi(\xi)] \in \operatorname*{colim}_{k\to\infty} H^n(S^k D_2 S^{-k} X) \quad .$$

From the construction, we see that $[\psi(\xi)]$ is the element of $\operatorname{colim} H^n(S^k D_2 S^{-k})$ determined by

$$\sigma^n[1 \times S^{-n}\xi \times S^{-n}\xi] \in H^n(S^n D_2 S^{-n} X)$$

where σ indicates the suspension isomorphism. We consider now the homomorphism

$$\begin{aligned} \alpha^* : \operatorname{colim} H^*(S^k D_2 S^{-k} X) &\longrightarrow \operatorname{colim} H^*((S^k D_2 S^{-k}) \wedge X) \\ &\approx \operatorname{colim} H^*(S^k D_2 S^{-k}) \otimes H^*(X) \end{aligned}$$

induced by the sequence of maps

$$\alpha_k : (S^k D_2 S^{-k}) \wedge X \longrightarrow S^k D_2 S^{-k} X$$

described, for example, in [3]. It is known that

$$\operatorname{colim} H^*(S^k D_2 S^{-k}) \cong F_2[t_1^{\pm 1}]$$

and

$$\alpha^* \sigma^n [1 \otimes S^{-n}\xi \otimes S^{-n}\xi] = \sum_{j\geq 0} t_1^{-j} \otimes Sq^j(x) \quad .$$

The resulting ring homomorphism

4.7
$$\begin{aligned} S_1 : H^* &\longrightarrow F_2[t_1^{\pm 1}] \otimes H^*(X) \\ \xi &\longmapsto \alpha^*(\psi(\xi)) = \sum_{j\geq 0} t_1^{-j} \otimes Sq^j(x) \end{aligned}$$

is the *normalized* total squaring operation. We point out that, with the notation introduced in §2, we have

$$F_2[t_1^{\pm 1}] = \Phi_1 \quad .$$

A geometric construction of the normalized iterated total squaring operation

4.8 $$S_m : H^*(X) \longrightarrow \Phi_m \otimes H^*(X)$$

is obtained by substituting the quadratic construction functor D_2 with the 2^m-adic construction functor D_{2^m}. More precisely, we define, for each $k \geq n$,

$$\psi_k(\xi) : S^k D_{2^m} S^{-k} X \longrightarrow \cdots \longrightarrow S^n D_{2^m} S^{-n} X \xrightarrow{S^n D_{2^m}\xi} S^n D_{2^m} H \xrightarrow{S^n \xi_{2^m}} S^n H \quad .$$

The sequence $\{\psi_k(\xi)\}_{k\geq n}$ defines an element

$$[\psi(\xi)] = [\psi_k(\xi)]_{k\geq n} = \sigma^n[1 \otimes (S^{-n}\xi)^{2^M}] \in \operatorname{colim} H^n(S^k D_{2^m} S^{-k} X) \quad .$$

As $\Sigma_2 \int \cdots \int \Sigma_2$ embeds (as a 2-Sylow subgroup) in Σ_{2^m}, there is a natural transformation

$$\pi : D_2 \ldots D_2 \longrightarrow D_{2^m}$$

and

$$\pi^* : \operatorname{colim} H^* S^k D_{2^m} S^{-k} \longrightarrow \operatorname{colim} H^*(S^k D_2 \ldots D_2 S^{-k} X)$$

takes

$$\sigma^n[1 \otimes (S^{-n}\xi)^{2^m}] \longmapsto \sigma^n[1 \otimes (S^{-n}\xi)^{2^m}] \quad .$$

Finally we set

$$S_m(x) = \widetilde{\alpha}^* \sigma^n[1 \otimes (S^{-n}\xi)^{2^m}]$$

where $\widetilde{\alpha}^*$ is induced by the sequence of maps

$$\widetilde{\alpha}_k : (S^k D_{2^m} S^{-k}) \wedge X \longrightarrow S^k D_{2^m} S^{-k} X$$

or

$$\widetilde{\alpha}_k : (S^k D_2 \ldots D_2 S^{-k}) \wedge X \longrightarrow S^k D_2 \ldots D_2 S^{-k} X \quad .$$

$S_m(x)$ can be regarded as an element of

$$\operatorname{colim} H^*\big((S^k D_{2^m} S^{-k}) \wedge X\big) \cong \operatorname{colim} H^*(S^k D_{2^m} S^{-k}) \otimes H^*(X)$$

or

$$\operatorname{colim} H^*\big((S^k D_2 \ldots D_2 S^{-k}) \wedge X\big) \cong \operatorname{colim} H^*(S^k D_2 \ldots D_2 S^{-k}) \otimes H^*(X) \quad .$$

Notice that

4.9 $$S^k D_{2^m} S^{-k} \approx B\Sigma_{2^m}^{-k\rho}$$

as explained in §3, and the map

$$\psi_k : S^k D_{2^m} S^{-k} \longrightarrow S^{k-1} D_{2^m} S^{1-k}$$

corresponds,under the above equivqlence, to the transfer map

$$t : B\Sigma_{2^m}^{-k\rho} \longrightarrow B\Sigma_{2^m}^{(1-k)\rho}$$

induced by the stable map

$$-k\rho \hookrightarrow (1-k)\rho \quad .$$

Moreover, the following diagram commutes

4.10
$$\begin{array}{ccc} H^*(B\Sigma_{2^m}^{-k\rho}) & \stackrel{t^*}{\longleftarrow} & H^*(B\Sigma_{2^m}^{(1-k)\rho}) \\ \cong \downarrow & & \downarrow \cong \\ H^*(B\Sigma_{2^m}) & \stackrel{e(\rho)}{\longleftarrow} & H^*(B\Sigma_{2^m}) \end{array}$$

where the vertical maps are Thom isomorphisms and the bottom map is the multiplication by the Euler class of ρ. Therefore

4.11
$$\begin{aligned} &\operatorname{colim} H^*(S^k D_{2^m} S^{-k}) \\ &\cong H^*(B\Sigma_{2^m})[e(\rho)^{-1}] \\ &\cong H^*(B\Sigma_2 \times \cdots \times \Sigma_2)^{GL_m}[e(\rho)^{-1}] \quad \text{(see [4], thm 5.1, p.248)} \\ &\cong P_m^{GL_m}[e_m^{-1}] \\ &\cong \Gamma_m \subseteq \Phi_m \quad . \end{aligned}$$

$\widetilde{\alpha}^*$ can be iewed as an iteration of α^* and we easily see that

4.12
$$S_m(x) = \sum_{i_1,\dots,i_m \geq 0} (t_1^{i_1} Sq^{i_1}) \dots (t_m^{i_m} Sq^{i_m})(x) \in \Phi_m \otimes H^*(X) \quad .$$

A characterization of $S_m(x)$ as an element of $\Delta_m \otimes H^*(X)$ is given in the following proposition.

Proposition 4.13.

$$S_m(x) = \sum_{i_1,\dots,i_m \geq 0} v_1^{-i_1} \dots, v_m^{-i_m} \otimes Sq^{i_1} \dots Sq^{i_m} \quad .$$

The verification of this formula is not difficult, and can be found in [8], §2, prop.1, p.11, or [9], prop.2.1. This expression and the fact that $H^*(X)$ is an unstable $\mathcal{A}$-module assure us that the summation in 4.12 is in fact finite. The following expression of $S_m(x)$ in its GL_m-invariant form, which can be found in [9], theorem 3.13, is given using the Milnor basis $\mathcal{B}$ of $\mathcal{A}$.

Proposition 4.14. *If we write R for the multi-index $(r_1, \dots, r_k)$ and*

$$\xi_R = \xi_1^{r_1} \dots \xi_k^{r_k} \in \mathcal{A}_*$$

we have

4.15 $$S_m(x) = \sum_{r_1,\dots,r_k \geq 0} Q_{m,0}^{-(r_1+\dots+r_k)} Q_{m,1}^{r_1} \dots Q_{m,k}^{r_k} \otimes \xi_R^*(x)$$

where ξ_R^* denotes the element of $\mathcal{A}$ dual to $\xi_R \in \mathcal{B}$ with respect to the basis of admissible monomials in $\mathcal{A}$.

This is a normalized version of a result of Mui (see [10], Theorem 1, p.346). The construction of T_m and S_m suggests that $S_m(x)$ is obtained from $T_m(x)$ by inverting the Euler class e_m. We will verify this fact algebraically.

Proposition 4.16.

$$S_m(x) = e_m^{-n} \cdot T_m(x) \quad \forall\, x \in H^n(X) \quad .$$

Proof. We think of $P_m \otimes H^*(X)$ as embedded in $\Phi_m \otimes H^*(X)$. We want to prove that

4.17 $$T_m(x) = e_m^n \cdot \sum_{i_1,\dots,i_m \geq 0} (t_1^{-i_1} Sq^{i_1}) \dots (t_m^{-i_m} Sq^{i_m})(x) \quad .$$

If we set $m = 1$, 4.17 is clearly true. We now proceed by induction. We assume that $k \geq 2$ and that 4.17 holds for $m = k-1$. Here we need to introduce some notation. Let us consider $k-1$ symbols $\alpha_2, \dots, \alpha_k$. Then there is an obvious isomorphism

$$\omega : P_{k-1} \longrightarrow F_2[\alpha_2, \dots, \alpha_k]$$

$$t_i \longmapsto \alpha_{i+1} \quad .$$

We write $e_{k-1}(\alpha_2, \dots, \alpha_k)$ for the element $\omega(e_{k-1})$. Using the above notation, for our inductive hypothesis we have

4.18 $$\sum_{i_2,\dots,i_k \geq 0} (t_2^{n \cdot 2^{k-2} - i_2} Sq^{i_2}) \dots (t_k^{n - i_k} Sq^{i_k})(x)$$
$$= (e_{k-1}(t_2, \dots, t_k))^n \cdot \sum_{i_2,\dots,i_k \geq 0} (t_2^{-i_2} Sq^{i_2}) \dots (t_k^{-i_k} Sq^{i_k})(x) \quad .$$

Therefore

$$T_k(x) = \sum_{i_1 \geq 0} (t_1^{n \cdot 2^{k-1} - i_1} Sq^{i_1}) \Bigg(\prod_{\substack{\lambda_2, \ldots, \lambda_k = 0,1 \\ \sum \lambda_i > 0}} (\lambda_2 t_2 + \cdots + \lambda_k t_k)^n$$

$$\cdot \sum_{i_2, \ldots, i_k \geq 0} (t_2^{-i_2} Sq^{i_2}) \ldots (t_k^{-i_k} Sq^{i_k})(x) \Bigg) .$$

Hence

4.19
$$\begin{aligned} T_k(x) &= t_1^{n \cdot 2^{k-1}} \cdot \sum t_1^{-i_1} Sq^{i_1} (\prod (\lambda_2 t_2 + \cdots + \lambda_k t_k)^n \\ &\quad \cdot \sum (t_2^{-i_2} Sq^{i_2}) \ldots (t_k^{-i_k} Sq^{i_k})(x)) \\ &= t_1^{n \cdot 2^{k-1}} \cdot \sum t_1^{-i_1} Sq^{i_1} (\prod (\lambda_2 t_2 + \cdots + \lambda_k t_k)^n) \\ &\quad \cdot \sum (t_1^{-i_1} Sq^{i_1}) \ldots (t_k^{-i_k} Sq^{i_k})(x) \quad . \end{aligned}$$

Here we have used the fact that $S_1 = \sum t_1^{-i_1} Sq^{i_1}$ is a ring homomorphism. Thus we want to prove that

$$e_k = t_1^{2^{k-1}} \cdot \sum (t_1^{-i_1} Sq^{i_1}) (\prod (\lambda_2 t_2 + \cdots + \lambda_k t_k)) \quad .$$

Since $S_1(t_i) = t_i + t_1^{-1} t_2^2$, we have

$$\begin{aligned} & t_1^{2^{k-1}} \cdot \sum (t_1^{-i_1} Sq^{i_1}) (\prod (\lambda_2 t_2 + \cdots + \lambda_k t_k)) \\ &= t_1^{2^{k-1}} \cdot \prod \big(\lambda_2 (t_2 + t_1^{-1} t_2^2) + \cdots + \lambda_k (t_k + t_1^{-1} t_k^2)\big) \\ & t_1 \cdot \prod \big(\lambda_2 (t_1 t_2 + t_2^2) + \cdots + \lambda_k (t_1 t_k + t_k^2)\big) \\ &= t_1 \cdot \prod \big(t_1 (\lambda_2 t_2 + \cdots + \lambda_k t_k) + (\lambda_2 t_2 + \cdots + \lambda_k t_k)^2\big) \quad (\text{as } \lambda^2 = \lambda \text{ in } F_2) \\ &= t_1 \cdot \prod (\lambda_2 t_2 + \cdots + \lambda_k t_k)(t_1 + \lambda_2 t_2 + \cdots + \lambda_k t_k)^2 = e_k \end{aligned}$$

where, in all the above products, $\lambda_2, \ldots, \lambda_k \geq 0$ and $\sum \lambda_i > 1$.

References.

[1] R. R. Bruner, J. P. May, J. E.McClure, M. Steinberger, *H_∞-ring spectra and their applications.* Lecture Notes in Maths., vol. 1176, Springer, 1986.

[2] B. Gray, *Homotopy theory - An introduction to Algebraic Topology.* Academic Press, 1975.

[3] J. D. S. Jones, S. A. Wegmann, *Limits of stable homotopy and cohomotopy groups.* Math. Proc. Cambridge Phil. Soc., **94** (1983), 473-482.

[4] N. J. Kuhn, *Chevalley group theory and the transfer in the homotopy of symmetric groups.* Topology **24** (1985), 247-264.

[5] J. Kulich, *Homotopy models for desuspensions.* Ph. D. Thesis, Northwestern Univ., Illinois, U.S.A., 1985.

[6] J. Kulich, *A quotient of the iterated Singer construction.* Algebraic Topology, Contemporary Math. **96** 1989.

[7] L. G. Lewis, J. P. May, M. Steinberger, *Equivariant stable homotopy theory.* Lecture Notes in Maths., vol. 1213, Springer, 1986.

[8] L. Lomonaco, *Invariant theory and the total squaring operation.* Ph. D. Thesis, Univ. of Warwick, U.K. 1986.

[9] L. Lomonaco, *The iterated total squaring operation.* Preprint.

[10] H. Mui, *Dickson invariants and the Milnor basis of the Steenrod algebra.* Eger International Colloquium in Topology, 1983.

[11] W. Singer, *Invariant theory and the Lambda algebra.* Trans. Amer. Math. Soc., **280** (1981), 673-693.

[12] C. Wilkerson, *Classifying spaces, Steenrod operations and algebraic closure.* Topology, **16** (1977), 227-237.

Dipartimento di Matematica e Applicazioni
Università di Napoli
Italy

1990 Barcelona Conference
on Algebraic Topology.

CONCISE TABLES OF JAMES NUMBERS AND SOME HOMOTOPY OF CLASSICAL LIE GROUPS AND ASSOCIATED HOMOGENEOUS SPACES

ALBERT T. LUNDELL

The following tables express the James numbers and homotopy groups of the homogeneous spaces involved in the Bott maps, with the exception of the Grassmann manifolds, in as concise a manner as we know, consistant with displaying the various periodicities involved. Throughout the tables of groups ∞ has been used to represent the infinite cyclic group $\mathbf{Z}$, and an integer n has been used to represent the cyclic group $\mathbf{Z}/n\mathbf{Z}$. We also use 0 to denote the trivial group. This leads to the rather unaesthetic $1 \oplus 1 \oplus 1 = 0$ as a direct summand of $\pi_{2n+8}(U(n))$ for $n > 4$ and odd. We think that the advantage of clearly displaying the various periodicities overrides these aesthetic objections. Of course the generators of the groups and the effects of various maps on them are of utmost importance. What information is available on such matters can be found in the partially annotated references following each table.

We believe the tables to be error-free. We have incorporated the corrections we knew about for the existing literature, and checked our conversion to the format of these tables. A few "first order" computations have been done on results that appear in the literature, i.e., the numbers or groups were not explicitely given, but the propositions of the paper seemed clear enough that an easy calculation gave the result without a thorough reading of the paper. Of course we are responsible for all errors, and would appreciate knowing about any that a reader discovers. We believe our citations to be correct, but if not we would like to know about this as well.

Throughout the tables, (a,b) is the greatest common divisor of integers a and b. The Bernouilli number B_m is indexed as

$$t/(e^t - 1) = \sum_{m \geq 0} B_m t^m/m!.$$

We use $\text{denom}(B_m)$ to denote the integer b where $B_m = a/b, (a,b) = 1$, and $b > 0$. In the case of a group having several identical summands, we use the notation n^r, thus $2^4 = 2 \oplus 2 \oplus 2 \oplus 2$.

We sent a preliminary version of these tables to most of the authors cited, and received in return much valuable information. We want to thank them for this help.

Stable homotopy groups

	k (mod 8)							
	0	1	2	3	4	5	6	7
$\pi_k(U(n))$ $k \leq 2n-1$	0	∞	0	∞	0	∞	0	∞
$\pi_k(SO(n))$ $0 < k \leq n-2$	2	2	0	∞	0	0	0	∞
$\pi_k(SO(2n)/U(n))$ $0 < k \leq 2n-2$	2	0	∞	0	0	0	∞	2
$\pi_k(U(2n)/Sp(n))$ $k \leq 4n-1$	0	∞	0	0	0	∞	2	2
$\pi_k(Sp(2n)/Sp(n) \times Sp(n))$ $0 < k \leq 4n+2$	∞	0	0	0	∞	2	2	0
$\pi_k(Sp(n))$ $k \leq 4n+1$	0	0	0	∞	2	2	0	∞
$\pi_k(Sp(n)/U(n))$ $k \leq 2n$	0	0	∞	2	2	0	∞	0
$\pi_k(U(n)/SO(n))$ $k \leq n-1$	0	∞	2	2	0	∞	0	0
$\pi_k(SO(2n)/SO(n) \times SO(n))$ $0 < k \leq n-1$	∞	2	2	0	∞	0	0	0

Reference: R. Bott, Ann. of Math. 70 (1959), 313-337.

Complex James Numbers $W\{n,k\}$

k	$W\{n,k\}$		
2	2/(n,2)		
3	16/(n,8)(n,2)(n+5,8) 3/(n,3)		
4	16/(n,8)(n,2)(n+5,8) 3/(n,3)	(1)	
5	128/(n,64)(n+2,16)(n+1,16)(n+3,8) 9/(n,9)(n+1,3) 5/(n,5)		
6	256/(n,64)(n+2,16)(n,2)(n+1,32)(n+11,16) 9/(n,9)(n+1,3) 5/(n,5)		
7	2048/(n,128)(n+50,64)(n+4,8)(n,2)(n+19,32)(n+1,16)(n+1,8) 81/(n,81)(n+1,27)(n+2,9) 5/(n,5) 7/(n,7)		
8	2048/(n,128)(n+50,64)(n+4,8)(n,2)(n+19,32)(n+1,16)(n+1,8) 81/(n,81)(n+1,27)(n+2,9) 5/(n,5) 7/(n,7)	(2)	
n-4	(n-1)!2/(n-1,4)(n-3, denom(B_{n-3}))denom(B_{n-3})		if $n \equiv 1 \pmod 2$.
n-3	(n-1)!/denom(B_{n-2}) (n-1)!/(n-3, denom(B_{n-3}))denom(B_{n-3})		if $n \equiv 0 \pmod 2$ if $n \equiv 1 \pmod 2$.
n-2	(n-1)!4/(n,4)denom(B_{n-2}) (n-1)!2/(n+1,4)		if $n \equiv 0 \pmod 2$ if $n \equiv 1 \pmod 2$.

(1) Multiplied by 2 if n = 6.
(2) Multiplied by 2 if n = 10, 14; by 1 or 2 if n = 13.

References:

For k=2,3,4: F. Sigrist, Comment. Math. Helv. 43 (1968),121-131.
For k=5,6,7,8: K. Knapp, Habilitationsschrift, Bonn 1979.
H. Oshima, Proc. Camb. Phil. Soc. 92 (1982), 139-161.
H. Oshima, Osaka J. Math 21 (1984), 765-772.
For k=n-2,n-3: G. Walker, Quart. J. Math. 32 (1981), 467-489.
G. Walker, The James numbers $b_{n,n-3}$ for n odd, (preprint) 1988.

Quaternionic James Numbers X{n, k}

k	X(n,k)
2	8/(n,8) 3/(n,3)
3	32/(n,32)(n+1,4) 9/(n,9)(n+2,3) 5/(n,5)
4	256/(n,128)(n+2,8)(n+25,32) 81/(n,81)(n+14,27)(n+1,9) 5/(n,5) 7/(n,7)
5	2048/(n,1024)(n+34,64)(n+1,32)(n+3,8) 243/(n,81)(n,3)(n+1,27)(n+14,27) 25/(n,25)(n+3,5) 7/(n,7)
n-1	(2n-1)!(n-1,2)/denom(B_{2n-2})

References:
H. Oshima, Proc. Camb. Phil. Soc. 92 (1982), 139-161.
G. Walker, Quart. J. Math. 32 (1982), 467-489.
For $k = 4$, $n \equiv 30 \pmod{64}$, $n \equiv 31 \pmod{32}$: K. Morisugi, Osaka J. Math. 23 (1986), 867-880.
For $k = n - 1$, 2-component: K. Morisugi, *ibid.*
For $k = n - 1$, odd component: G. Walker, *ibid.*

	$\pi_{2n-1+k}(SU(n))$	
k=1	$n!$	
k=2	$(n,2)$	
k=3	$(n+1)!/W\{n+2,2\} \oplus (n,2)$	if $n > 2$;
	12	if $n=2$.
k=4	$(n,8)(n,2)(n+3,8)/2$	
	$\oplus(n,3)$	
k=5	$(n+2)!/W\{n+3,3\}$	if $n > 2$;
	2	if $n=2$.
k=6	$(n+4,8)(n+1,8)(n+1,2)/2$	
	$\oplus\ (n+1,3)$	
k=7	$(n+3)!/W\{n+4,4\} \oplus (n,4)(n+3,4)/2$	if $n > 2$;
	15	if $n=2$.
k=8	$(n+6,16)(n,8)(n,4)(n,2)^3(n+5,64)(n+7,16)(n+3,4)^2/32 \oplus (n+1,4)(n,2)^2/2$	
	$\oplus(n,3)(n+5,9)$	
	$\oplus(n,5)$	if $n > 3$;
	6	if $n=3$;
	2	if $n=2$.
k=9	$(n+4)!/W\{n+5,5\} \oplus (n,2) \oplus (n,2) \oplus (n,4)/2 \oplus (n+5,8)/2$	if $n > 4$;
	$504 \oplus 2 \oplus 2 \oplus 2 \oplus 2$	if $n=4$;
	$84 \oplus 2$	if $n=3$;
	$2 \oplus 2$	if $n=2$.

k=10	$(n+6,64)(n+8,16)(n+4,8)(n,4)(n+7,32)(n+5,8)(n+1,8)(n+1,2)^2/32$ $\oplus (n,8)(n+2,4)/8 \oplus (n+2,4) \oplus (n,2)$ $\oplus (n+6,9)$ $\oplus (n+1,5)$ $\oplus (n+6,7)$	if $n > 4$;
	$40 \oplus 2 \oplus 2 \oplus 2$	if n=4;
	36	if n=3;
	$12 \oplus 2$	if n=2.

References:

For k=1: A. Borel & F. Hirzebruch, Amer. J. Math. 81 (1959), 315-382.

For k=2,3: H. Toda, Mem. Fac. Sci. Kyoto 32 (1959), 103-119.
M. Kervaire, Ill. J. Math. 4 (1960), 161-169.

For k=4,5,6: H. Matsunaga, Mem. Fac. Sci. Kyushu Univ. 15 (1961), 72-81, (with correction sheet).
H. Matsunaga, Osaka J. Math. 1 (1964), 15-24.

For k=7,8: H. Matsunaga, Mem. Fac. Sci. Kyushu Univ. 17 (1963), 29-62, (with corrections due to Mosher).
H. Matsunaga, Mem. Fac. Sci. Kyushu Univ. 16 (1962).
R. Mosher, Topology 7 (1968),179-193.

For k=9: H. Oshima, Proc. Camb. Phil. Soc. 92 (1982), 139-161.

For k=10: H. Oshima, Osaka J. Math. 17 (1980), 495-511.
R. Mosher, Ill. J. Math. 13 (1969), 192-197.

Also see: H. Imanishi, J. Math. Kyoto Univ. 7 (1968), 221-243.
J. Mukai, Math. J. Okayama Univ. 24 (1982),179-200.

$\pi_{4n+1+k}(Sp(n))$

k		
k=1	$(2n+1)!(n+1,2)$	
k=2	2	
k=3	$2 \oplus (n,2)$	
k=4	$(n+2,8) \oplus (n,2)$ $\oplus (n+2,3)$	
k=5	$(2n+3)!(n+2,2)/X\{n+2,2\}$	
k=6	2	
k=7	$2 \oplus 2$	
k=8	$(n,4)(n,2)^8(n+3,32)(n+7,8)(n+1,4)^3/64 \oplus (n+3,8)/2 \oplus (n+1,2)$ $\oplus (n+3,9)(n+2,3)$ $\oplus (n+3,5)$ $12 \oplus 2$	 if $n > 1$; if $n = 1$.
k=9	$(2n+5)!(n+3,2)/X\{n+3,3\} \oplus (n+3,4)(n,2)/2 \oplus 2$	
k=10	$(n+3,4)(n,2)/2 \oplus 2 \oplus 2$ $2 \oplus 2$	$n > 1$ $n = 1$
k=11	$2 \oplus 2$ $\oplus (n,3)(n+2,3)$ $2 \oplus 2 \oplus 2$ 6	 $n > 2$ $n = 2$ $n = 1$
k=12	$(n+4,128)(n,8)^2(n,4)(n+6,8)^3(n+2,4)^2(n-3,32)(n+1,2)^{14}/4096$ $\oplus (n+4,8)^2(n+2,8)^3(n,4)(n+1,2)^9/512 \oplus (n,4)(n+1,2)/2 \oplus (n,2)$ $\oplus (n+4,81)(n+18,27)(n+5,9)(n,3)(n+2,3)/3$ 30	 $n > 1$ $n = 1$

k=13	$(2n+7)!(n+4,2)/X\{n+4,4\}\oplus (n,4)(n+1,2)/2\oplus 2$	$n>2$
	$(2n+7)!(n+4,2)/X\{n+4,4\}\oplus 2\oplus 2$	$n=2$
	30	$n=1$
k=14	$(n,4)(n+1,2)/2\oplus 2\oplus 2$	$n>2$ [(1)]
	$2\oplus 2\oplus 2$	$n=2$
	$6\oplus 2$	$n=1$
k=15	$(n,4)(n+3,4)/2\oplus (n,2)^2(n+3,4)/2\oplus 2\oplus 2$	$n\geq 3$ [(1)]
	$2\oplus 2$	$n=2$
	$12\oplus 2\oplus 2$	$n=1$

(1) The 3-component is outside the range of the computations of Imanishi.

References:

For k = 1: R. Bott, Comment. Math. Helv. 34 (1960), 249-256.

For k = 2,3: B. Harris, Trans. Amer. Math. Soc. 106 (1963), 174-184.
M. Mimura & H. Toda, J. Math. Kyoto Univ. 3 (1964), 251-273.

For k = 4,5,6,7: M. Mimura, C.R. Acad. Sci. Paris 262 (1966), 20-21.

For k = 8: H. Oshima, Proc. Cambridge Phil. Soc. 92 (1982), 139-161.

For $9\leq k\leq 14$: K. Morisugi, J. Math. Kyoto Univ. 27 (1987), 367-380. (With corrections by the author,Preprint 1989.)

For $8\leq k\leq 14$, odd components: H. Imanishi, J. Math. Kyoto Univ. 7 (1968), 221-243.

For k = 15, n = 3, 2 component: K. Morisugi, to appear in J. Math. Kyoto Univ.

For k = 15, n > 3, 2 component: K. Morisugi, Bull. Fac. Ed. Wakayama Univ. (1988), 19–23.

For n=2: M. Mimura and H. Toda, J. Math. Kyoto Univ. 3 (1964), 217-250.
N. Oda, Fukuoka Univ. Sci. Reports 8 (1978), 77-89.
H. Toda, J. Math. Kyoto Univ. 5 (1965), 87-142.

For n=1: H. Toda, Composition methods in homotopy groups of spheres, Ann. of Math. Studies 49, Princeton, 1962.

$\pi_{n-2+k}(SO(n)), n > 9$

	n (mod 8)			
	0	1	2	3
k=1	$\infty \oplus \infty$	$2 \oplus 2$	$\infty \oplus 2$	2
k=2	$2 \oplus 2 \oplus 2$	$2 \oplus 2$	4	∞
k=3	$2 \oplus 2 \oplus 2$	8	∞	2
k=4	$8 \oplus 8$ $\oplus 3$	$\infty \oplus 2$	4 $\oplus 3$	$2 \oplus 2$
k=5	$\infty \oplus 2$	0	2	16 8, n=11
k=6	0	2	16 8, n=10	$\infty \oplus 2$
k=7	$2 \oplus 2$	$16 \oplus 2$ $8 \oplus 2$, n=9	$\infty \oplus 2 \oplus 2$	$2 \oplus 2$
k=8	$2 \oplus (n,16) \oplus 2(n+8,16)$ $\oplus 15$	$\infty \oplus 2^3$	$16 \oplus 2 \oplus 2$ $\oplus 15$	$2 \oplus 2$
k=9	$\infty \oplus 2^5$	2^5	$4 \oplus 2$	8

k=10	2^8	$8 \oplus 2 \oplus 2$	8 $n > 10$	$\infty \oplus 2 \oplus 2$
k=11	$8 \oplus 2^3$ $\oplus 3$	8	$\infty \oplus 3$ $n > 10$	$2 \oplus 2$ $n > 11$
k=12	$8 \oplus 8$ $\oplus 63$	∞	8 $\oplus 63$ $n > 10$	$2 \oplus 2 \oplus (n+13,128)/8$ $n > 11$
k=13	∞	0	$2 \oplus (n+14,128)/8$ $n > 10$	$2(n+13,128)$ $n > 11$
k=14	3	$2 \oplus (n+15,128)/8$	$2(n+14,128)$ $\oplus 3$ $n > 10$	$\infty \oplus 2$ $n > 11$

References:

For $1 \leq k \leq 6$: M. Kervaire, Ill. J. Math. 4 (1960),161-169.

For the ambiguity in Kervaire's calculation:

M.Barratt and M.Mahowald, Bull. Amer. Math. Soc. 70 (1964),758-760.

M.Mahowald, Proc. Amer. Math. Soc. 19 (1968),639-641.

For $1 \leq k \leq 14$, Mahowald *ibid.* shows that for m large large, $n \geq 13$ and $j < 2n - 1$,

$$\pi_j(SO(n)) \cong \pi_j(SO(n+m)) \oplus \pi_{j+1}(V_{n+m,m}).$$

$\pi_{n-2+k}(SO(n)), n > 9$

	n (mod 8)			
	4	5	6	7
k=1	$\infty \oplus \infty$	2	∞	2 0, n=7
k=2	$2 \oplus 2$	2	4	∞
k=3	$2 \oplus 2$	8	∞	$2 \oplus 2$
k=4	$4 \oplus 16$ $\oplus 3$ $4 \oplus 24$, n=12	$\infty \oplus 2$	$4 \oplus 2$ $\oplus 3$	$2 \oplus 2$
k=5	$\infty \oplus 2$	2	2	8
k=6	2	2	8	$\infty \oplus 2$
k=7	2	8	$\infty \oplus 4$	0
k=8	$16 \oplus 4$ $\oplus 15$	$\infty \oplus 2 \oplus 2$	16 $\oplus 15$	$2 \oplus (n+9,16)/8$ 2, n=7
k=9	$\infty \oplus 2^4$	$2 \oplus 2 \oplus 2$	$4 \oplus (n+10,32)/8$	$2 \oplus 2(n+9,16)$

k=10	2^5	$4 \oplus 2 \oplus$ (n+11,64)/8	$2 \oplus 2$(n+10,32)	$\infty \oplus 2^3$
k=11	$2^3 \oplus (n+12,64)/8$ $\oplus 3$	2(n+11,64)	$\infty \oplus 2$ $\oplus 3$	2^5
k=12	$8 \oplus 2$(n+12,128) $\oplus 63$ $n > 12$	$\infty \oplus 2$	$8 \oplus 2^3$ $\oplus 63$	$8 \oplus 2 \oplus 2$
k=13	$\infty \oplus 2$ $n > 12$	$2 \oplus 2 \oplus 2$ $n > 13$	$8 \oplus 2$	8
k=14	2^3 $\oplus 3$ $n > 12$	$8 \oplus 2$ $n > 13$	8 $\oplus 3$ $n > 14$	∞

For computation of $\pi_{j+1}(V_{n+m,m})$ with m large:

$1 \le k \le 14$: C. Hoo and M. Mahowald, Bull. Amer. Math. Soc. 71 (1965), 661-667.

$15 \le k \le 30$, 2-torsion: M. Mahowald, Mem. Amer. Math. Soc. 72 (1967). (Not tabulated.)

For odd torsion, recall that

$$tors_{odd}(\pi_j(SO(2n+1))) = tors_{odd}(\pi_j(Sp(n)))$$

$$tors_{odd}(\pi_j(SO(2n))) = tors_{odd}(\pi_j(SO(2n-1))) \oplus tors_{odd}(\pi_j(S^{2n-1}))$$

See: B.Harris, Ann. of Math. 74 (1961),407-413.

$\pi_k(SO(n)), n \leq 9$ and $\pi_k(SU(3))$ $(= *)$

	3	4	5	6	7	8	9	10	11	12
n=3	∞	2	2	12	2	2	3	15	2	2^2
n=4	$\infty \oplus \infty$	2^2	2^2	12^2	2^2	2^2	3^2	15^2	2^2	2^4
n=5	∞	2	2	0	∞	0	0	120	2	2^2
n=6	∞	0	∞	0	∞	24	2	$120 \oplus 2$	4	60
n=7	∞	0	0	0	∞	2^2	2^2	8	$\infty \oplus 2$	0
n=8	∞	0	0	0	$\infty \oplus \infty$	2^3	2^3	$24 \oplus 8$	$\infty \oplus 2$	0
n=9	∞	0	0	0	∞	2^2	2^2	8	$\infty \oplus 2$	0
n=*	∞	0	∞	6	0	12	3	30	4	60

	13	14	15	16	17	18	19	20
n=3	$4 \oplus 2 \oplus 3$	$4 \oplus 2^2 \oplus 21$	2^2	$2 \oplus 3$	$2 \oplus 15$	$2 \oplus 15$	$2^2 \oplus 3$	$4 \oplus 2^2 \oplus 3$
n=4	$4^2 \oplus 2^2 \oplus 3^2$	$4^2 \oplus 2^4 \oplus 21^2$	2^4	$2^2 \oplus 3^2$	$2^2 \oplus 15^2$	$2^2 \oplus 15^2$	$2^4 \oplus 3^2$	$4^2 \oplus 2^4 \oplus 3^2$
n=5	$4 \oplus 2$	$16 \oplus 105$	2	2^2	$8 \oplus 5$	$8 \oplus 2 \oplus 315$	2^2	2^3
n=6	4	$16 \oplus 2 \oplus 105$	$8 \oplus 2 \oplus 9$	$8 \oplus 2^4 \oplus 63$	$8 \oplus 2^3 \oplus 5$	$8 \oplus 4 \oplus 2 \oplus 315 \oplus 3$	$4 \oplus 2 \oplus 3$	$4 \oplus 2 \oplus 15$
n=7	2	$8^2 \oplus 2 \oplus 315$	2^4	2^7	$8^2 \oplus 2^2$	$16 \oplus 8 \oplus 2 \oplus 945$	2	2^2
n=8	2^2	$8^3 \oplus 2 \oplus 315 \oplus 15$	2^7	2^{11}	$8^3 \oplus 2^3 \oplus 3$	$16 \oplus 8^2 \oplus 2^2 \oplus 945 \oplus 63$	2	$2^3 \oplus 3$
n=9	2	$8 \oplus 2$	$\infty \oplus 2^3$	2^6	$8 \oplus 2^2$	$16 \oplus 8 \oplus 2 \oplus 2835$	2	2
n=*	$2 \oplus 3$	$4 \oplus 2 \oplus 21$	$4 \oplus 9$	$4 \oplus 2 \oplus 63 \oplus 3$	$2^2 \oplus 15$	$2^2 \oplus 15 \oplus 3$	$4 \oplus 2 \oplus 3^2$	$4 \oplus 2 \oplus 15 \oplus 3$

For $n \leq 6$ one has local isomorphisms of Lie groups which yield:

$$\pi_j(SO(3)) \cong \pi_j(SU(2)) \cong \pi_j(Sp(1)) \qquad \pi_j(SO(4)) \cong \pi_j(SO(3)) \oplus \pi_j(S^3)$$

$$\pi_j(SO(5)) \cong \pi_j(Sp(2)) \qquad \pi_j(SO(6)) \cong \pi_j(SU(4)),$$

for $1 < j$. Thus these tables contain $\pi_k(SU(n))$, $n \leq 4$ and $\pi_k(Sp(n))$, $n \leq 2$.

$\pi_k(SO(n)), n \leq 9$ and $\pi_k(SU(3))$ $(= *)$

	21	22	23	24	25	26	27	28
n=3	$4 \oplus 2^2 \oplus 3$	$4 \oplus 2^2 \oplus 33$	2^2	2	$2 \oplus 105$	$4 \oplus 1365 \oplus 3$	$2^3 \oplus 3$	$4 \oplus 2^3$
n=4	$4^2 \oplus 2^4 \oplus 3^2$	$4^2 \oplus 2^4 \oplus 33^2$	2^4	2^2	$2^2 \oplus 105^2$	$4^2 \oplus 1365^2 \oplus 3^2$	$2^6 \oplus 3^2$	$4^2 \oplus 2^6$
n=5	$32 \oplus 2$	$32 \oplus 2^2 \oplus 165$	2^3	2^2	$8 \oplus 2 \oplus 105$	$8 \oplus 2 \oplus 15015$	2^3	$4 \oplus 2^3$
n=6	$16 \oplus 2$	$16 \oplus 4 \oplus 2^2 \oplus 165$	$8 \oplus 2^4 \oplus 3$	$8 \oplus 2^3 \oplus 33$	$8 \oplus 2^3 \oplus 105 \oplus 3$	$8 \oplus 2^2 \oplus 15015 \oplus 3$	$4^2 \oplus 2 \oplus 315$	$8 \oplus 4 \oplus 2^2 \oplus 4095 \oplus 3$
n=7	$8 \oplus 4 \oplus 3$	$8^2 \oplus 2^4 \oplus 10395$	$G \oplus 2^5$ (1)					
n=8	$8^2 \oplus 4^2 \oplus 3^2$	$8^3 \oplus 2^7 \oplus 10395 \oplus 15$	$G \oplus 2^9$ (1)					
n=9	$4 \oplus 3$	$8^2 \oplus 2^2 \oplus 155925$	$G \oplus 2^2$ (1)					
n=*	$2 \oplus 3$	$2^2 \oplus 33$	$4 \oplus 2 \oplus 3$	$4 \oplus 33$	$2^2 \oplus 105 \oplus 3$	$2^2 \oplus 1365 \oplus 3^2$	$4 \oplus 2 \oplus 315 \oplus 3$	$4^2 \oplus 2 \oplus 4095 \oplus 3$

	29	30	31	32	33	34	35	36
n=3	$4 \oplus 2^3 \oplus 3^2$	$4 \oplus 2^3 \oplus 3^2$	2^4	2^3	$4 \oplus 2 \oplus 15$	$2^4 \oplus 255$	$4 \oplus 2^2 \oplus 3$	$2^3 \oplus 3^2$
n=4	$4^2 \oplus 2^6 \oplus 3^4$	$4^2 \oplus 2^6 \oplus 3^4$	2^8	2^6	$4^2 \oplus 2^2 \oplus 15^2$	$2^8 \oplus 255^2$	$4^2 \oplus 2^4 \oplus 3^2$	$2^6 \oplus H \oplus 3^4$
n=5	$16 \oplus 2^3 \oplus 315$	$16 \oplus 8 \oplus 2^2 \oplus 4095$	2^5	$8 \oplus 2^3$	$8 \oplus 4 \oplus 2^3 \oplus 15$	$8 \oplus 4 \oplus 2 \oplus 255$	2^3	$4 \oplus 2^5 \oplus 3$
n=6	$16 \oplus 2^4 \oplus 315 \oplus 3$	$16 \oplus 8^2 \oplus 2^4 \oplus 4095$	$8 \oplus 4 \oplus 2^7 \oplus 3$	$8^2 \oplus 2^6 \oplus 3$	$8 \oplus 4^2 \oplus 2^5 \oplus 15$	$8^2 \oplus 2^5 \oplus 255 \oplus 3$	$8 \oplus 4 \oplus 2^3 \oplus 15 \oplus 3$	$8 \oplus 4 \oplus 2^6 \oplus 255 \oplus 3^2$
n=*	$4 \oplus 2 \oplus 3^3$	$4 \oplus 2^3 \oplus 3^2$	$4^2 \oplus 2^2 \oplus 3$	$4^2 \oplus 2^2 \oplus 3$	$4 \oplus 2^2 \oplus 15$	$4 \oplus 2^4 \oplus 255 \oplus 3$	$4^2 \oplus 2^2 \oplus 15 \oplus 3^2$	$4 \oplus 2^2 \oplus 255 \oplus 3^3$

(1) G is of order 4.

References:

For n=3,4, k ≤ 23: H. Toda, Composition methods in the homotopy groups of spheres.

2-component, k=23: M. Mimura & H. Toda, J. Math Kyoto Univ. 3 (1963), 37-58.

k=24,25: M. Mimura, J. Math. Kyoto Univ. 4 (1965), 301-326.

k=26,27: M. Mimura, M. Mori & N. Oda, Proc. Japan Acad. 50 (1974), 277-280.

28 ≤ k ≤ 31: N. Oda, Proc. Japan Acad. 53 (1977), 202-205.

32 ≤ k ≤ 36: N. Oda, Proc. Japan Acad. 53 (1977), 215-218.

3 ≤ k ≤ 36: E. Curtis & M. Mahowald, preprint.

odd component: H. Toda, J. Math. Kyoto Univ. 5 (1965), 87-142.

For n=5,6,*, 3 ≤ k ≤ 23: M. Mimura & H. Toda, J. Math. Kyoto Univ. 3 (1964), 217-250.

M. Mimura & H. Toda, J. Math Kyoto Univ. 3 (1964), 251-273.

24 ≤ k ≤ 36: N. Oda, Fukuoka Univ. Sci. Reports 8 (1978), 77-89.

odd component: H. Toda, J. Math. Kyoto Univ. 5 (1965), 87-142.

H. Toda, J. Math, Kyoto Univ. 8 (1968), 101-130.

S. Oka, J. Sci. Hiroshima Univ. 33 (1969), 161-195.

For n=7,8,9, 3 ≤ k ≤ 23: M. Mimura, J. Math. Kyoto Univ. 6 (1967), 131-176.

$\pi_{4n-1+k}(U(2n)/Sp(n))$

k	
k=1	(2n)!/(n+1,2)
k=2	2
k=3	$2 \oplus (n+1,2)$
k=4	$(n,24) \oplus (n+1,2)$
k=5	(2n+2)!(n,24)(n+1,2)/48
k=6	2

References:
For k = 1,2: B. Harris, Ann. of Math. 76 (1962), 295-305.
B. Harris, Trans. Amer. Math Soc. 106 (1963), 174-184.
For k = 3,4,5,6: M. Mimura, C.R. Acad. Sci. Paris 262 (1966), 20-21.

$\pi_{2n+k}(Sp(n)/U(n))$

k		
k=1	n!(n+2,4)/2	n ≡ 0 (mod 2)
	∞⊕ (n+3,4)/2	n ≡ 1 (mod 2)
k=2	$(n,2)^2(n+3,4)/2$	
k=3	∞⊕ (n,4)/2 ⊕ 2	n ≡ 0 (mod 2)
	(n+1)!(n+3,4)/4	n ≡ 1 (mod 2)
k=4	(n,8)(n,2)(n+3,8)/2 ⊕ (n,4)(n+1,2)/2 ⊕ (n,3)	
k=5	(n+2)!(n,24)(n,4)/96	n ≡ 0 (mod 2)
	∞⊕ (n+1,4)/2	n ≡ 1 (mod 2)
k=6	(n+4,8)(n+1,8)(n+1,2)/2 ⊕ (n,2)(n+1,4)/2 ⊕ (n+1,3)	
k=7	∞⊕ 2	n ≡ 0 (mod 2)
	(n+3)!(n+1,24)(n+1,4)/96	n ≡ 1 (mod 2)

Reference:
H. Kachi, J. Fac. Sci. Shinshu Univ. 13 (1978), 35-41.

$\pi_{2n-2+k}(SO(2n)/U(n))$

k=1	$\infty \oplus (n,4)/2$	$n \equiv 0 \pmod 2$
	$(n-1)!(n+3,4)/4$	$n \equiv 1 \pmod 2$
k=2	$(n,2) \oplus (n,4)/2$	
k=3	$n!(n,4)(n+1,2)^3/4 \oplus 2$	$n \equiv 0 \pmod 2$
	$\infty \oplus (n+1,4)/2$	$n \equiv 1 \pmod 2$, $n > 3$
	∞	$n = 3$
k=4	$(n,4)^2(n,2)(n-1,8)(n-3,4)/4$	
	$\oplus (n-1,3)$	$n > 2$
	12	$n = 2$
n=5	$\infty \oplus (n+2,4)/2$	$n \equiv 0 \pmod 2$, $n > 2$
	2	$n = 2$
	$(n+1)!(n-1,24)(n+1,4)/96$	$n \equiv 1 \pmod 2$
k=6	$(n,8)(n+2,4)(n+1,4)^2(n+1,2)/4$	
	$\oplus (n,3)$	

References:
For k = 1,2,3: B. Harris, Trans. Amer. Math. Soc. 106 (1963), 174-184.
For k = 3, n ≡ 2 (mod 4): H. Oshima, Osaka J. Math. 21 (1984), 473-475.
For k = 4,5,6: H. Kachi, J. Fac. Sci. Shinshu Univ. 13 (1978),] 103-120.
(With corrections by the author.)

$\pi_{n-1+k}(SU(n)/SO(n))$

	n (mod 8)							
	0	1	2	3	4	5	6	7
k=1	∞	$\infty \oplus 2$	$\infty \oplus 2$	$2 \oplus 2$	∞	$\infty \oplus 2$	∞	2
k=2	$\infty \oplus 2 \oplus 2$	$2 \oplus 2$	$2 \oplus 4$	0	$\infty \oplus 2 \oplus 2$	2	4	0
k=3	$8 \oplus 8 \oplus 2$	2	$\infty \oplus 2$	$2 \oplus 2$	$16 \oplus 4$	2	$\infty \oplus 4$	$2 \oplus 2$
k=4	$8 \oplus 8 \oplus 2$ $\oplus 3$	2	$\infty \oplus 2$	$2 \oplus 2$	$16 \oplus 4$ $\oplus 3$ $8 \oplus 12$, n=12	2	$\infty \oplus 4$ $\oplus 3$	$2 \oplus 2$
k=5	2	∞	2	16 8, n=11	2	∞	2	$2 \oplus 8$
k=6	∞	2	16 8, n=10	2	∞	2	$8 \oplus 2$	2

Reference: H. Kachi, J. Fac. Sci. Shinshu Univ. 13 (1978), 27-34.
For correction of the ambiguity in these calculations:
M. Mahowald, Proc. Amer. Math. Soc. 19 (1968), 639-641.

Remarks. The tables for the homotopy of SO(n) and SU(n)/SO(n) have a different format than the remaining tables. One reason is that we are in the metastable range for the homotopy of SO(n) where the odd torsion is of period 2 (see the remarks following the tables for SO(n)). In the other cases, the p-torsion, when periodic, has period depending on p. Moreover, there are often non-periodic aspects to the homotopy groups of the other spaces. In our preliminary version of these tables, we wrote the 2-torsion of the homotopy of SO(n) in the same format as for the other groups, but in a more restricted range. In extending the range, the formulae became unwieldy, so we resorted to the present format. On the other hand, for the homotopy of U(n), for example, there is odd periodicity and non-periodicity (the summands involving n!), and one would at least need tables for each prime separately if one were to try to use the format used for SO(n).

References

M. Barratt and M. Mahowald, The metastable homotopy of O(n), Bull. Amer. Math Soc. 70 (1964), 758-760.

A. Borel and F. Hirzebruch, Characteristic classes and homogeneous spaces II, Amer. J. Math. 81 (1959), 103-119.

R. Bott, The stable homotopy of the classical groups, Ann. of Math. 70 (1959),313-337.

R. Bott, A note on the Samelson product in the classical groups, Comment. Math. Helv. 34 (1960), 249-256.

M. Crabb and K. Knapp, James numbers and the codegree of vector bundles I, II, (preprint).

M. Crabb and K. Knapp, The Hurewicz map on stunted complex projective spaces, Amer. J. Math 110 (1988), 783–809.

M. Crabb and K. Knapp, James numbers, Math. Ann. 282 (1988), 395–422.

E. Curtis and M. Mahowald, The unstable Adams spectral sequence for the 3-sphere, (preprint).

Y. Hirashima and H. Oshima, A note on stable James numbers of projective spaces, Osaka J. Math. 13 (1976), 157-161.

B. Harris, On the homotopy groups of the classical groups, Ann. of Math. 74 (1961),407-413.

B. Harris, Some calculations of homotopy group of symmetric spaces, Trans. Amer. Math. Soc. 106 (1963), 174-184.

C. Hoo and M. Mahowald, Some homotopy groups of Stiefel manifolds, Bull. Amer. Math. Soc. 71 (1965), 661-667.

H. Imanishi, Unstable homotopy groups of classical groups (odd primary components), J. Math. Kyoto Univ. 7 (1968), 221-243.

M. Imaoka and K. Morisugi, On the stable Hurewicz image of some stunted projective spaces I, Pub. RIMS Kyoto Univ. 39 (1984), 839-852.

M. Imaoka and K. Morisugi, On the stable Hurewicz image of some stunted projective spaces II, Pub. RIMS Kyoto Univ. 39 (1984), 853-866.

M. Imaoka and K. Morisugi, On the stable Hurewicz image of some stunted projective spaces III, Mem. Fac. Sci. Kyushu Univ. 39 (1985), 197-208.

H. Kachi, Homotopy groups of the homogeneous space SU(n)/SO(n), J. Fac. Sci. Shinshu Univ. 13 (1978), 27-34.

H. Kachi, Homotopy groups of the homogeneous space Sp(n)/U(n), J. Fac. Sci. Shinshu Univ. 13 (1978), 36-41.

H. Kachi, Homotopy groups of symmetric spaces Γ_n, J. Fac. Sci. Shinshu Univ. 13 (1978), 103-120.

M. Kervaire, Some nonstable homotopy groups of Lie groups, Ill. J. Math. 4 (1960), 161-169.

K. Knapp, Some applications of K-theory to framed bordism: e-invariant and transfer, Habilitationsschrift, Bonn (1979).

M. Mahowald, The metastable homotopy of S^n, Mem. Amer. Math. Soc. 72 (1968).

M. Mahowald, On the metastable homotopy of SO(n), Proc. Amer. Math. Soc. 19 (1968), 639-641.

H. Matsunaga, The homotopy groups $\pi_{2n+i}(U(n))$ for i=3,4 and 5, Mem. Fac. Sci. Kyushu Univ. 15 (1961), 72-81.

H. Matsunaga, The groups $\pi_{2n+7}(U(n))$, odd primary components, Mem. Fac. Sci. Kyushu 16 (1962), 66-74.

H. Matsunaga, Applications of functional cohomology operations to the calculus of $\pi_{2n+i}(U(n))$ for i=6 and 7, $n \geq 4$, Mem. Fac. Sci. Kyushu Univ. 17 (1963), 29-62.

H. Matsunaga, Unstable homotopy groups of Unitary groups (odd primary components), Osaka J. Math. 1 (1964), 15-24.

M. Mimura, On the generalized Hopf construction and the higher composition II, J. Math. Kyoto Univ. 4 (1965), 301-326.

M. Mimura, Quelques groupes d'homotopie metastables des espaces symetriques Sp(n) et U(2n)/Sp(n), C.R. Acad. Sci. Paris, 262 (1966), 20-21.

M. Mimura, Homotopy groups of Lie groups of low rank, J. Math. Kyoto Univ. 6 (1967), 131-176.

M. Mimura, M. Mori and N. Oda, On the homotopy groups of spheres, Proc. Japan Acad. 50 (1974), 277-280.

M. Mimura and H. Toda, The (n+20)-th homotopy groups of n-spheres, J. Math. Kyoto Univ. 3 (1963), 37-58.

M. Mimura and H. Toda, Homotopy groups of SU(3),SU(4) and Sp(2), J. Math. Kyoto Univ. 3 (1964), 217-250.

M. Mimura and H. Toda, Homotopy groups of symplectic groups, J. Math. Kyoto Univ. 3 (1964), 251-273.

H. Minami, A remark on odd-primary components of special unitary groups, Osaka J. Math. 21 (1984), 457-460.

M. Mori, Applications of secondary e-invariants to unstable homotopy groups of spheres, Mem. Fac. Sci. Kyushu Univ. 29 (1974), 59-87.

K. Morisugi, Homotopy groups of symplectic groups and the quaternionic James numbers, Osaka J. Math. 23 (1986), 867-880.

K. Morisugi, Metastable homotopy groups of Sp(n), J. Math. Kyoto Univ. 27 (1987), 367-380.

K. Morisugi, On the homotopy group $\pi_{4n+16}(Sp(n))$ for $n \geq 4$, Bull. Fac. Ed. Wakayama Univ. (1988), 19–23.

K. Morisugi, On the homotopy group, $\pi_{8n+4}(Sp(n))$ and the Hopf invariant, to appear in J. Math Kyoto Univ.

R. Mosher, Some stable homotopy of complex projective space, Topology 7 (1969), 179-193.

R. Mosher, Some homotopy of stunted complex projective space, Ill. J. Math. 13 (1969), 192-197.

J. Mukai, The S^1-transfer map and homotopy groups of suspended complex projective spaces, Math. J. Okayama Univ. 24 (1982), 179-200.

N. Oda, On the 2-components of the unstable homotopy groups of spheres I, Proc. Japan Acad. 53 (1977), 202-205.

N. Oda, On the 2-components of the unstable homotopy groups of shperes II, Proc. Japan Acad. 53 (1977), 215-218.

N. Oda, Some homotopy groups of SU(3),SU(4) and Sp(2), Fukuoka Univ. Sci. Reports 8 (1978), 77-90.

N. Oda, Periodic families in the homotopy groups of SU(3),SU(4),Sp(2) and G_2 , Mem. Fac. Sci. Kyushu Univ. 32 (1978), 277-290.

S. Oka, On the homotopy groups of sphere bundles over spheres, J. Sci. Hiroshima Univ. 33 (1969), 161-195.

H. Ōshima, On the stable James numbers of complex projective spaces, Osaka J. Math. 11 (1974), 361-366.

H. Ōshima, On stable James numbers of stunted complex or quaternionic projective spaces, Osaka J. Math. 16 (1979), 479-504.

H. Ōshima, On the homotopy group $\pi_{2n+9}(U(n))$ for $n \geq 6$, Osaka J. Math. 17 (1980), 495-511.

H. Ōshima, Some James numbers of Stiefel manifolds, Proc. Camb. Phil. Soc. 92 (1982), 139-161.

H. Ōshima, A homotopy group of the symmetric space SO(2n)/U(n), Osaka J. Math. 21 (1984), 473-475.

H. Ōshima, A remark on James numbers of Stiefel manifolds, Osaka J. Math. 21 (1984), 765-772.

F. Sigrist, Groupes d'homotopie des varietes de Stiefel complexes, Comment. Math. Helv. 43 (1968), 121-131.

H.Toda, A topological proof of theorems of Bott and Borel-Hirzebruch for homotopy groups of unitary groups, Mem. Fac. Sci. Kyoto Univ. 32 (1959), 103-119.

H. Toda, Composition Methods in Homotopy Groups of Spheres, Ann. of Math. Studies 49, Princeton University Press 1962.

H. Toda, On homotopy groups of S^3-bundles over spheres, J. Math. Kyoto Univ. 2 (1963), 193-207.

H. Toda, On iterated suspensions I, J. Math. Kyoto Univ. 5 (1965), 87-142.

H. Toda, On iterated suspensions III, J. Math. Kyoto Univ. 8 (1968), 101-130.

G. Walker, Estimates for the complex and quaternionic James numbers, Quart. J. Math. 32 (1981), 467-489.

G. Walker, The James numbers $b_{n,n-3}$ for n odd, (preprint), 1988.

Department of Mathematics, Box 426
University of Colorado
Boulder, Colorado 80309

1990 Barcelona Conference
on Algebraic Topology.

AN EXAMPLE OF A STABLE SPLITTING: THE CLASSIFYING SPACE OF THE 4–DIM UNIPOTENT GROUP[1]

JOHN R. MARTINO

§0. Introduction.

In a previous paper [**MP**] Stewart Priddy and the author described a method to determine a p–local stable decomposition of a finite group,

$$BG \cong X_1 \vee X_2 \vee \cdots \vee X_N.$$

An alternative method was described by David Benson and Mark Feshbach in [**BF**]. In this paper we apply our method to the classifying space of the unipotent group U_4, the subgroup of upper triangular matrices in $GL_4(\mathsf{F}_2)$. This is an extremely complicated classifying space and is a good illustration of the power of our technique. For a finite group G our method involves the ranks of certain matrices which depend only on the p–subgroups of G and on modular representation theory. Using these matrices we will show that BU_4 has 88 indecomposable summands of 19 different stable homotopy types (see Section 6 for full details). This example is based on a question by John Maginnis.

Splittings of BG are equivalent to idempotent decompositions of the identity in the ring of stable maps $\{BG, BG\}$. Such decompositions, which are usually difficult to obtain, can be partially studied via the ring homomorphism

$$B : \widehat{\mathbf{Z}}_p \operatorname{Out}(G) \to \{BG, BG\}, \tag{1}$$

where $\operatorname{Out}(G)$ is the outer automorphism group of G. If P is a p–group, then $\{BP, BP\}$ is isomorphic to the p–adic completion of the reduced double Burnside ring $\widetilde{A}(P, P)$ [**May**]. This was shown by Lewis, May, and McClure [**LMM**] using [C].

Nishida [N] defined the ideal $J_p \subseteq \{BP, BP\}$ generated by maps of the form $BP \to BQ \to BP$ where $Q \subsetneqq P$. He then showed that (1) induced $\widehat{\mathbf{Z}}_p\operatorname{Out}(P) \approx \{BP, BP\}/J_p$. If X is an indecomposable summand of BP, then it corresponds to a primitive idempotent $e \in \{BP, BP\}$ up to conjugacy. If $e \notin J_p$, then Nishida called X a *dominant* summand of BP. Mod J_p, e is a primitive idempotent of $\widehat{\mathbf{Z}}_p\operatorname{Out}(P)$ and thus corresponds to a simple $\mathsf{F}_p\operatorname{Out}(P)$ module M_x. The multiplicity of X in BP is equal to the dimension of M_x over its endomorphism field. If $e \in J_p$, then by a result of Nishida,

[1] Partially supported by the NSF

X is also a summand of BQ for some proper subgroup $Q \lneqq P$. Thus inductively our problem reduces to considering a summand X of BQ which itself does not come from a proper subgroup of Q and determining its multiplicity in BP. Such an X is a dominant summand of BQ and corresponds to a simple $R = \mathsf{F}_p\mathrm{Out}(Q)$ module $M = M_x$ whose isomorphism class is determined by the homotopy type of X. It is this relationship between summand and module that we exploit in [**MP**].

This paper is organized as follows: Section 1 is a statement of the results in [**MP**]. In Section 2 we analyze the contribution of the unique maximal elementary abelian subgroup of rank four in U_4. Section 3 deals with the other maximal elementary abelian subgroups. Section 4 completes the analysis of elementary abelian subgroups. Section 5 handles the non-abelian subgroups of U_4, and Section 6 gives the complete solution.

For the remainder of this paper all spectra are localized at the prime 2.

Section 1: The matrix $A(Q, M)$.

To state our result, let

$$\mathrm{Split}(Q) = \{q_\alpha : P_\alpha \to Q_\alpha\}$$

be the conjugacy classes of split surjections q_α where $Q_\alpha < P_\alpha < P$ and $Q_\alpha \approx Q$ (we assume chosen a fixed isomorphism). Here q_α is said to be *conjugate* to q_β if there is a commutative diagram

$$\begin{array}{ccc} P_\alpha & \xrightarrow{c_u} & P_\beta \\ {\scriptstyle q_\alpha}\downarrow & & \downarrow{\scriptstyle q_\beta} \\ Q_\alpha & \xrightarrow[c_v]{} & Q_\beta \end{array}$$

for some $u, v \in P$, where c_x is conjugation by x. Let $\overline{W}_{\alpha\beta} = \sum_x q_\beta c_x \bmod J_Q$ for $x \in N(Q_\alpha, P_\beta)/P_\beta$ where $N(Q_\alpha, P_\beta) = \{x \in P \mid Q_\alpha^x < P_\beta\}$. Then $\overline{W}_{\alpha,\beta}$ is a well-defined element of R via the isomorphisms $Q_\alpha \approx Q$ and mod p reduction. Let $n = |\mathrm{Split}(Q)|$, then $A(Q) = (\overline{W}_{\alpha\beta})$ is an $n \times n$ matrix over R. Let $k = \mathrm{End}_R(M)$. Then viewed as a k-linear map of M^n, $A(Q, M) = (\overline{W}_{\alpha\beta}) \in \mathrm{Mat}_{mn}(k)$ where $m = \dim_k M$.

Theorem 1.1. *Let P be a finite p-group. Let X be a dominant summand of BQ, $Q < P$, with corresponding simple $R = \mathsf{F}_p\mathrm{Out}(Q)$ module M. Then the multiplicity of X in BP is* $\mathrm{rank}_k A(Q, M)$.

Corollary 1.2. *The complete stable splitting of BP is given by $BP = \vee \mathrm{rank}_k A(Q, M) X$, where the indecomposable summands $X = X_M$ range over isomorphism classes of simple $\mathsf{F}_p\mathrm{Out}(Q)$ modules M and over isomorphism classes of subgroups $Q < P$.*

Theorem 1.1 can be stated for arbitrary finite groups but that is not needed here.

Definition 1.3. *If* $\mathrm{Split}(Q) = \{id : Q_\alpha \to Q_\alpha\}$, *then we say that* Q *is not a subretract of* P, *in which case the* Q_α*'s range over representatives of the conjugacy classes of subgroups of* P *isomorphic to* Q.

Hypothesis 1.4. *Let* P *be a finite p-group and* Q *a subgroup which is not a subretract. Furthermore, let* X *be a dominant summand of* BQ *with corresponding simple* $R = \mathbb{F}_p\mathrm{Out}(Q)$ *module* M *and* $k = \mathrm{End}_R(M)$.

Corollary 1.5. *Assume Hypothesis 1.4. Let* $Q_1, \ldots, Q_n$ *be a complete set of conjugacy-class representatives of subgroups isomorphic to* Q *and* $\overline{W}_i = \sum_{x \in N_p(Q_i)/Q_i} c_x$. *Then the multiplicity of* X *in* BP *is equal to* $\sum_{i=1}^{n} \dim_k(\overline{W}_i M)$.

Definition 1.6. *Under Hypothesis 1.4, we say the* multiplicity of X in BP from Q_i *is* $\dim_k(\overline{W}_i M)$.

Corollary 1.7. *Assume Hypothesis 1.4. Suppose there exists a normal p-subgroup* N *of* $\mathrm{Out}(Q)$ *such that* $\mathrm{Out}(Q)/N \approx GL_n(\mathbb{F}_p)$ *for some* n. *Let* $W_p(Q) = N_p(Q)/Q$, St_n *be the Steinberg module of* $\mathbb{F}_pGL_n(\mathbb{F}_p)$, *and* X *the corresponding Steinberg summand in* BQ. *If* Q *is self-centralizing and* $W_P(Q)$ *and* $W_P(Q) \cap N = 1$, *then* X *is a summand of* BP *and the multiplicity in* BP *from* Q *is* $p^{\binom{n}{2}}/|W_P(Q)|$.

Definition 1.8. *If* P *is an abelian p-group and* $e \in \mathbb{F}_p\mathrm{Out}(P)$ *is a primitive idempotent,* X' *the dominant summand of* eBP *and* Y' *another summand in* eBP, *then* Y' *is* linked *to* X' *in* BP.

Proposition 1.9. *Assume Hypothesis 1.4. Let* Q *be abelian and* $R \lneqq Q$. *If* Y *is a dominant summand in* BR *linked to* X *in* BQ *and* M_Y *the corresponding simple* $\mathbb{F}_p\mathrm{Out}(R)$ *module. Then the multiplicity of* X *in* BP *from* Q *is equal to the rank of the submatrix of* $A(R, M_Y)$ *where the* (α, β)*-entry is*

$$\overline{W}_{\alpha\beta} : BR_\alpha \xrightarrow{\text{incl}} BP \xrightarrow{tr} BQ \xrightarrow{Bq_\beta} BR_\beta \bmod J_R,$$

tr is the reduced transfer.

All of the above results can be found in [MP].

Section 2: The maximal elementary abelian subgroup of rank 4.

We will refer to the element in U_4 with ones on the diagonal, a one in the (i,j)–position, and zeroes elsewhere as x_{ij}. Then the elements x_{13}, x_{14}, x_{23}, and x_{24} generate a unique maximal elementary abelian subgroup $V = < x_{13}, x_{14}, x_{23}, x_{24} >$ of rank four. In this section we will determine which dominant summands of BV are present in BU_4 and how many copies of each.

A regular partition λ is a sequence $[\lambda_1, \ldots, \lambda_k]$ such that $\lambda_i > \lambda_{i+1} > 0$. Each partition λ describes a diagram with λ_i nodes in the ith column, such a diagram is called a Young diagram. By **[JK]** we know that the simple modules of $\mathbb{F}_2 GL_n(\mathbb{F}_2)$ correspond to Young diagrams with n nodes in the first column (i.e., $\lambda_1 = n$). We will call the simple module corresponding to the partition λ, M_λ. M_n is the trivial module and $M_{n,n-1,\ldots,1}$ is the Steinberg module. The corresponding summand will be called X_λ. $X_{n,n-1,\ldots,1}$ is the symmetric product spectrum $L(n) = \sum^{-n} Sp^{2^n} S^0 / Sp^{2^{n-1}} S^0$ **[HK, MitP]**.

We begin our analysis with the Steinberg summand.

Proposition 2.1. *There are* $2^{\binom{n^2}{2} - \binom{2n}{2} + n^2} = 2^{\frac{n^4 - 3n^2 + 2n}{2}}$ *copies of* $L(n^2)$ *in* BU_{2n}*, and there are*

$$2^{\binom{n(n+1)}{2} - \binom{2n+1}{2} + n(n+1) + 1} = 2^{\frac{n^4 + 2n^3 - 2n^2 - n + 2}{2}}$$

copies of $L(n(n+1))$ *in* BU_{2n+1}.

This follows immediately from Corollary 1.7 and also Corollary 1.5 in the second case. The $+1$ in the second expression of the proposition comes from the fact that there are two copies of $(\mathbb{Z}/2)^{n(n+1)}$ in U_{2n+1}. ■

Thus, there are 16 copies of $X_{4321} = L(4)$ in BU_4.

To determine the number of copies of X_λ we need to calculate the dimension of the image of $\overline{W} : BV \xrightarrow{incl} BU_4 \xrightarrow{tr} BV$ on M_λ. If we define c_{ij} to be conjugation by x_{ij}, then $\overline{W} = 1 + c_{12} + c_{34} + c_{12}c_{34}$. We can find a copy of M_λ as a subquotient of $H^*(BV; \mathbb{F}_2)$. By **[CK,FS]** we know in which dimension of $H^*(BV; \mathbb{F}_2)$ to find the first copy of M_λ.

If $\alpha(n)$ denotes the number of ones in the dyadic expansion of n, then for $I = (i_1, \ldots, i_k)$ define $\alpha(I) = \alpha(i_1) + \cdots + \alpha(i_k)$. For a mononomial $m \in H^*(BV; \mathbb{F}_2)$, let I equal the sequence of exponents of m, then call $\alpha(I)$ the weight of m. By **[FS]** the mononomials which form a basis for the first occurrence of the subquotient M_λ have weight $\lambda_2 + \cdots + \lambda_k$ and are in dimension $\lambda_2 + 2\lambda_3 + \cdots + 2^{k-2}\lambda_k$.

Let $H^*(BV; \mathbb{F}_2) = \mathbb{F}_2[x, y, z, w]$.

M_4 is the trivial module.

M_{41} is 4–dimensional with basis $x,\ y,\ z,\ w$.

M_{42} is 6–dimensional with basis $xy,\ xz,\ xw,\ yz,\ yw,\ zw$.

M_{43} is 4–dimensional with basis $xyz,\ xzw,\ xyw,\ yzw$.

M_{421} is 20–dimensional with basis $x^3y, \ldots, zw^3$, $x^2yz, \ldots$, yz^2w with relations $xyz^2 = x^2yz + xy^2z, \ldots, yzw^2 = y^2zw + yz^2w$.

M_{431} is 14–dimensional with basis $x^3yz, \ldots, yzw^3$ and $x^2yzw + xy^2zw = xyz^2w + xyzw^2$, $x^2yzw + xyz^2w = xy^2zw + xyzw^2$.

M_{432} is 20–dimensional with basis $x^3y^2zw, \ldots, xy^2zw^3$ with relations $x^3yzw^2 = x^3y^2zw + x^3yz^2w, \ldots, xyz^2w^3 = x^2yzw^3 + xy^2zw^3$.

By applying $\overline{W}$ to each of these modules we get the following table

λ	No. of copies of X_λ in BU_4
4	0
41	1
42	0
43	1
421	5
431	3
432	5
4321	16

Section 3: The other maximal elementary abelian subgroups.

The other maximal elementary abelian subgroups are of rank three. They are $< x_{12}, x_{13}, x_{14} >$, $< x_{14}, x_{24}, x_{34} >$, $< x_{12}x_{34}, x_{13}x_{24}, x_{14} >$, $< x_{12}, x_{14}, x_{34} >$, and $< x_{12}x_{13}, x_{14}, x_{24}x_{34} >$. The reader may easily verify this. The first three are normal and the last two are conjugate to each other. In this section we analyse the dominant summands of $B\,\mathbb{Z}/2^3$.

As in the previous section we begin with the Steinberg summand $X_{321} = L(3)$. By Corollary 1.7 each normal subgroup provides one copy of $L(3)$ and the two conjugate subgroups provide an additional two copies. By Corollary 1.5 and Proposition 1.9 there are a total of 21 copies of $L(3)$ since $L(3)$ is linked to $L(4)$ in BV.

The simple modules of $\mathsf{F}_2GL_3(\mathsf{F}_2)$ are M_{321} the Steinberg module, M_{32} the standard module, M_{31} its dual, and M_3 the trivial module. Applying the sum of the elements of the respective Weyl groups leads us to the conclusion that the maximal rank–three elementary abelian subgroups provide no additional copies X_{32}, X_{31}, X_3. Thus all of the copies in BU_4 come from BV. As X_λ is linked to $X_{4,\lambda}$ we have 5 copies of X_{32}, 3 copies of X_{31}, and 1 copy of X_3 in BU_4. By comparing Poincaré series we can easily conclude that $X_3 = BJ_1$ (J_1 is the first Janko group).

Section 4: The non–maximal elementary abelian subgroups.

There are only three more indecomposable summand types which are dominant in classifying spaces of elementary abelian subgroups, namely $X_{21} = L(2)$, $X_2 = BA_4$, and $X_1 = L(1) = \mathsf{R}p^\infty = B\,\mathbb{Z}/2$.

To complete our analysis of their summands we use the cohomology of BU_4 as a short cut [**Mag, TY**].

$$H^*(BU_4;\mathbb{F}_2) = \mathbb{F}_2[x_1,x_2,x_3,\alpha_1,\alpha_2,d,D,T]/\sim$$

where $\sim$ are

$$x_1x_2 = x_2x_3 = 0, \quad x_1d = x_3d = x_1\alpha_2 = x_3\alpha_1\,,$$
$$x_1D = x_3D = x_1^2d, \quad d^2 = x_2D + x_2^2d + \alpha_1\alpha_2\,,$$

and

$$D^2 = x_2^2T + x_2dD + (\alpha_1+\alpha_2+x_1x_3)d^2.$$

Also, $|x_i| = 1$, $|\alpha_j| = 2$, $|d| = 2$, $|D| = 3$, $|T| = 4$; and

$$Sq^1\alpha_1 = (x_1+x_2)\alpha_1,\ Sq^1\alpha_2 = (x_2+x_3)\alpha_2,\ Sq^1d = D$$

$$Sq^2D = x_2T + (x_2^2+d+\alpha_1+\alpha_2)D + x_2(\alpha_1+\alpha_2)d + x_1^3\alpha_2$$
$$Sq^1T = 0,\ Sq^2T = (x_1^2+x_3^2+\alpha_1+\alpha_2+d)T$$

The $\mathbb{Z}/2$'s generated by $< x_{12} >$, $< x_{23} >$, and $< x_{34} >$ are clearly retracts and form independent rows in $A(\mathbb{Z}/2,1)$, so there are at least three copies of $B\,\mathbb{Z}/2$ in BU_4, but $H^1(BU_4)$ is three–dimensional so there are only three copies.

The $\mathbb{Z}/2\times\mathbb{Z}/2$ generated by $< x_{12}, x_{34} >$ is also a retract of U_4, and since $B\,\mathbb{Z}/2\times\mathbb{Z}/2 = BA_4 \vee 2\,L(2)\vee 2B\,\mathbb{Z}/2$ [**HK, MitP**], we have at least one copy of BA_4 and two copies of $L(2)$ in BU_4. In the next section we will show that there is only one copy of BA_4 in BU_4.

We know by linkage (Proposition 1.9) that there are at least 5 copies of $L(2)$ in BU_4 since there are 5 copies of X_{421}. Since $< x_{12}, x_{34} > \not\subseteq V$ the 2 $L(2)$'s from $< x_{12}, x_{34} >$ are independent of the 5 from V. So we have at least 7 $L(2)$'s.

We now want to find 3 additional copies of $L(2)$ in BU_4. First let $E = < x_{ij} \mid i = 1 \text{ or } j = 4 >$, then $< x_{12}, x_{13} >$ and $< x_{24}, x_{34} >$ are both retracts of E. Let i_1, i_2 be the inclusions and r_1, r_2 be the retractions of the two subgroups. Since $r_k \circ i_l = 0$, if $k \neq l$ we have found another two copies of $L(2)$. These two copies of $\mathbb{Z}/2\times\mathbb{Z}/2$ are not contained in either V or $< x_{12}, x_{34} >$. Thus the two new $L(2)$'s are independent of the 7 previous ones.

Let $F_1 = < x_{ij} \mid i = 1 \text{ or } 2 >$ and $T = < x_{13}x_{24}, x_{14} >$. Take $r : F_1 \to (\mathbb{Z}/2)^2$ to be the retraction with kernel $< x_{ij} \mid i = 1 \text{ or } 2,\ i < j \le 3 >$. The map

$$BT \xrightarrow{Bi} BU_4 \xrightarrow{tr} BF_1 \xrightarrow{Br} B(\mathbb{Z}/2)^2$$

gives rise to an $L(2)$. This $L(2)$ is clearly independent of all the $L(2)$'s except possibly the ones from V. But

$$BT \xrightarrow{Bi} BU_4 \xrightarrow{tr} BV$$

is zero mod 2. Thus we have found a total of 10 copies of $L(2)$ in BU_4. Direct inspection of $H^4(BU_4)$ shows that there are only 10 Steenrod–primitive, independent elements with a nonzero Sq^1. Hence there are only 10 copies of $L(2)$ in BU_4.

Section 5: The non-abelian subgroups.

In this section we deal with the non-abelian subgroups.

We introduce three copies of the dihedral group D_8, namely $D_1 = < x_{12}, x_{23} >$, $D_2 = < x_{23}, x_{34} >$, and $D_3 = < x_{12}x_{34}, x_{23} >$. And consider the three maps

$Q_1 : BU_4 \to BD_8$ the quotient by $< x_{14}, x_{24}, x_{34} >$

$Q_2 : BU_4 \to BD_8$ the quotient by $< x_{12}, x_{13}, x_{14} >$

$Q_3 : BU_4 \xrightarrow{tr} BF_1 \xrightarrow{B_\rho} BD_8$ with F_1 as in Section 4 and ρ is the quotient by $< x_{13}, x_{23} >$

Considering the inclusions of D_1, D_2, D_3 and the maps Q_1, Q_2, Q_3 yields the submatrix $\begin{pmatrix} 1 & 0 & 1 \\ 0 & 1 & 0 \\ 1 & 1 & 0 \end{pmatrix}$ of $A(D_8, 1)$, which is of rank three. So there are at least three copies of BA_6, the dominant summand of BD_8 corresponding to the trivial module. Inspecting $H^2(BU_4; \mathbf{F}_2)$ one finds that the three copies of BA_6 exhaust the available space. Therefore, there is only one copy of BA_4 and only three copies of BA_6.

The indecomposable summand $B\ \mathbf{Z}/4$ is not a summand since we have exhausted $H^1(BU_4; \mathbf{F}_2)$.

If R is a subretract of U_4, then $R \times \mathbf{Z}/2$ is a subgroup of U_4; therefore, we have examined every subretract of U_4. The subgroups of U_4 can be determined from [**HS**]. We can now evoke the following:

Proposition 5.1. [M]. *If P is a p-group and $Q < P$ which is not a subretract, then the dominant summand of BQ corresponding to the trivial module is not a summand of BP.*

Thus we need only consider those subgroups whose outer automorphism group have non-trivial simple modules. Those which we have not considered are:

$$(\mathbf{Z}/2)^2 \int \mathbf{Z}/2, \quad D_8 \circ D_8, \quad D_8 \circ \mathbf{Z}/4, \quad Q_8.$$

For BQ_8 (split in [**MitP**]) and $BD_8 \circ \mathbf{Z}/4$ (split in [**FP, M**]) the non-trivial-module dominant summands are 0- and 1-connected respectively; hence they are not summands. There is only one copy of $D_8 \circ D_8$ in U_4, called E in Section 4. $BD_8 \circ D_8$ (split in [**FP**]) has two non-trivial-module dominant summands which we call $X(BD_8 \circ D_8)$ and $eT(\Delta_4)$ each with multiplicity four. Applying the sum of the elements in the Weyl group for E to the corresponding simple modules we find two copies of each summand in BU_4.

There are three copies of $\mathbf{Z}/2^2 \int \mathbf{Z}/2$ in U_4, namely F_1, $F_2 = < x_{ij} \mid j = 3 \text{ or } 4 >$, and $F_3 = < x_{12}x_{34}, V >$. $\mathrm{Aut}(\mathbf{Z}/2^2 \int /2)$ is an extension of $\sum_3$ by a 2-group so there are two simple modules, the trivial module and a two-dimensional Steinberg module. By Corollaries 1.5 and 1.7 we have three copies of the Steinberg summand $\mathrm{St}(\mathbf{Z}/2^2 \int \mathbf{Z}/2)$ in BU_4.

Section 6: The stable splitting of BU_4..

In this section we give a complete splitting of BU_4.

By [Mag] $\mathrm{Out}(U_4) = \Sigma_3 \times \mathbb{Z}/2$ so $\mathbb{F}_2\mathrm{Out}(U_4)$ has two simple modules: the trivial module and a two-dimensional Steinberg module. So there are two types of dominant summands: $\mathrm{Triv}(BU_4)$ and $\mathrm{St}(BU_4)$, with multiplicity one and two respectively. Thus,

Theorem 6.1.

$$\begin{aligned} BU_4 \cong\ & \mathrm{Triv}(BU_4) \vee 2\mathrm{St}(BU_4) \vee 3\mathrm{St}(B\,\mathbb{Z}/2^2 \wr \mathbb{Z}/2) \vee 2eT(\Delta_4) \\ & \vee 2X(BD_8 \circ D_8) \vee 16L(4) \vee 21L(3) \vee 10L(2) \vee 3B\,\mathbb{Z}/2 \vee 5X_{432} \\ & \vee 5X_{32} \vee BA_4 \vee 3X_{431} \vee 3X_{31} \vee 5X_{421} \vee X_{43} \vee BJ_1 \vee X_{41} \vee 3BA_6, \end{aligned}$$

where $\mathrm{Triv}(BU_4)$ *and* $\mathrm{St}(BU_4)$ *are the dominant summands of* BU_4, $\mathrm{St}(B\,\mathbb{Z}/2^2 \wr \mathbb{Z}/2)$ *is the Steinberg summand of* $B\,\mathbb{Z}/2^2 \wr \mathbb{Z}/2$, $eT(\Delta_4)$ *and* $X(BD_8 \circ D_8)$ *are the dominant summands of* $BD_8 \circ D_8$, $L(n)$ *is the Steinberg summand of* $B\,\mathbb{Z}/2^n$, X_λ *is the dominant summand of* $B\,\mathbb{Z}/2^{\lambda_1}$, *corresponding to the Young diagram with partition* λ *(see Section 2).*

References.

[BF] D. Benson and M. Feshbach, *"Stable splittings of classifying spaces of finite groups"*, to appear in Topology.

[C] G. Carlsson, *"Equivariant stable homotopy and Segal's Burnside ring conjecture"*, Annals of Math. **120** (1984), 189–224.

[CK] D. Carlisle and N. Kuhn, *"Subalgebras of the Steenrod algebra and the action of matrices on truncated polynomial algebras"*, J. Alg. **122** (1989), 370–387.

[FP] M. Feshbach and S. Priddy, *"Stable splittings associated with Chevalley groups I, II"*, Comment. Math. Helv. **64** (1989), 474–507.

[FS] V. Franjou and L. Schwartz, *"Reduced unstable A-modules and the modular representation theory of the symmetric groups"*, to appear.

[HS] M. Hall and J. Senior, *"The groups of order 2^n ($n < 6$)"*, MacMillan Co., New York, 1964.

[HK] J. Harris and N. Kuhn, *"Stable decompositions of classifying spaces of finite abelian p-groups"*, Math. Proc. Camb. Phil. Soc. **103** (1988), 427–449.

[JK] G. James and A. Kerber, *"The representation theory of the symmetric group"*, Encyclopedia of Math. **16**, Addison–Wesley, 1981.

[LMM] G. Lewis, J.P. May, and J. McClure, *"Classifying G-spaces and the Segal conjecture"*, Current Trends in Algebraic Topology, CMS Conference Proc. **2** (1982), 165–179.

[Mag] J. Maginnis, Stanford Univ. Ph. D. thesis, 1987.

[M] J. Martino, Northwestern Univ. Ph. D. thesis, 1988.

[MP] J. Martino and S. Priddy, *"The complete stable splitting for the classifying space of a finite group"* to appear in Topology.

[May] J.P. May, *"Stable maps between classifying spaces"*, Contemporary Math. **37**, 1985.

[MitP] S. Mitchell and S. Priddy, *"Symmetric product spectra and splittings of classifying spaces"*, Amer. J. Math. **106** (1984), 219–234.

[N] G. Nishida, *"Stable homotopy type of classifying spaces of finite groups"*, Algebraic and Topological Theories (1985), 391–404.

[TY] M. Tezuka and N. Yagita, *"The cohomology of subgroups of $GL_n(\mathbb{F}_q)$"*, Proc. of the Northwestern Homotopy Theory Conf., Contemporary Math. 19, 1983.

University of Virginia
Department of Mathematics
Mathematics–Astronomy Building
Charlottesville, VA. 22903–3199
U.S.A.

1990 Barcelona Conference
on Algebraic Topology.

ON THE HOMOTOPY UNIQUENESS OF BU(2) AT THE PRIME 2

JAMES E. MCCLURE AND LARRY SMITH

Abstract

The purpose of this note is to prove that a two complete space with the same mod 2 cohomology as $BU(2)$ is homotopy equivalent to the 2–adic completion of $BU(2)$.

This note is an addendum to [2] and a point of depart for [4]. We prove the following result:

THEOREM : *Let X be a 2–complete space such that*

$$H^*(X;\mathbb{Z}/2) \cong H^*(BU(2);\mathbb{Z}/2)$$

as algebras over the mod 2 Steenrod algebra $\mathcal{A}^*(2)$. Then X is homotopy equivalent to $BU(2)^\wedge_2$.

REMARK : Let det : $U(2)\longrightarrow S^1$ be the determinant map and $U(2)\xrightarrow{q}SO(3)$ the map induced by dividing $U(2)$ by its center. It is routine to verify that there is a pullback diagram

$$\begin{array}{ccc} BU(2) & \xrightarrow{q} & BSO(3) \\ \downarrow \det & \bullet\rfloor & \downarrow \\ \mathbb{CP}(\infty) & \longrightarrow & K(\mathbb{Z}/2,2) \end{array}$$

along the two non-trivial homotopy classes

$$BSO(3)\longrightarrow K(\mathbb{Z}/2,2)$$

and

$$\mathbb{CP}(\infty)\longrightarrow K(\mathbb{Z}/2,2).$$

This is the basic idea for the proof that follows.

PROOF : Choose an isomorphism

$$H^*(X,\mathbb{Z}/2) \cong H^*(BU(2);\mathbb{Z}/2) \cong P[c_1,c_2]$$

as algebras over $\mathcal{A}^*(2)$. Form the fibration

$$F\longrightarrow X\xrightarrow{c}\mathbb{CP}(\infty)^\wedge_2$$

where c classifies c_1. An Eilenberg-Moore spectral sequence argument [5] shows that $H^*(F;\mathbb{Z}/2) \cong H^*(\mathbb{HP}(\infty);\mathbb{Z}/2)$ as algebras over $\mathcal{A}^*(2)$. Since X and $\mathbb{CP}(\infty)_{\hat{2}}$ are 2–complete it follows [1] that F is also. Therefore by [2] F is homotopy equivalent to $\mathbb{HP}(\infty)_{\hat{2}}$.

The fibration

$$BS^3_{\hat{2}} \longrightarrow X \xrightarrow{c} \mathbb{CP}(\infty)_{\hat{2}}$$

has a classifying map

$$\vartheta : \mathbb{CP}(\infty)_{\hat{2}} \longrightarrow BHE(BS^3_{\hat{2}})$$

by Stasheff's theorem [6], where $HE(BS^3_{\hat{2}})$ is the monoid of homotopy equivalences of $BS^3_{\hat{2}}$ with itself. By a theorem of Mislin [3] we have

$$HE(S^3_{\hat{2}}) = \coprod_{d\in\Gamma} map(BS^3_{\hat{2}}, BS^3_{\hat{2}})_d$$

where $\Gamma < \mathbb{Z}^*_{\hat{2}}$ is the subgroup of units in the 2–adic integers $\mathbb{Z}^*_{\hat{2}}$. Since $HE(BS^3_{\hat{2}})$ is a loop space any two of its components are homotopy equivalent. Mislin [3] shows that

$$SHE(BS^3_{\hat{2}}) := map(BS^3_{\hat{2}}, BS^3_{\hat{2}})_{id} \cong K(\mathbb{Z}/2, 1).$$

Let

$$\pi_0 : HE(BS^3_{\hat{2}}) \longrightarrow \Gamma$$

denote the homomorphism induced by taking path components. Then we have an exact sequence

$$SHE(B^3_{\hat{2}}) \longrightarrow HE(BS^3_{\hat{2}}) \longrightarrow \Gamma$$

and hence a fibration

$$K(\mathbb{Z}/2, 2) \longrightarrow BHE(S^3_{\hat{2}}) \longrightarrow B\Gamma$$

Consider the classifying map ϑ again. We have the diagram

$$\begin{array}{ccc} & & K(\mathbb{Z}/2,2) \\ & \nearrow^{\bar{\vartheta}} & \downarrow \tau \\ \mathbb{CP}(\infty)_{\hat{2}} & \xrightarrow{\vartheta} & BHE(S^3_{\hat{2}}) \\ & \searrow_{null} & \downarrow \\ & & B\Gamma \end{array}$$

where the lower diagonal map is null-homotopic because $H^1(\mathbb{CP}(\infty)_{\hat{2}};\Gamma) = 0$. Therefore we receive a lift $\bar{\vartheta}$ and hence a fibre square

$$\begin{array}{ccc} X & \longrightarrow & E \\ \downarrow & \bullet\lrcorner & \downarrow \\ \mathbb{CP}(\infty)_{\hat{2}} & \xrightarrow{\bar{\vartheta}} & K(\mathbb{Z}/2,2) \end{array}$$

where E is the total space of the pullback of the universal bundle with fibre $\mathbb{HP}(\infty)_{\hat{2}}$ from $BHE(BS^3_{\hat{2}})$ along τ. There are only 2 possible choices for $\bar{\vartheta}$ up to homotopy : namely $\bar{\vartheta} \sim 0$ and $\bar{\vartheta} \not\sim 0$. The choice $\bar{\vartheta} \sim 0$ does not lead to the correct cohomology for X as an algebra over $\mathcal{A}^*(2)$. Therefore $\bar{\vartheta} \not\sim 0$ independent of which X we started with and so X is determined up to homotopy type by its mod 2–cohomology. □

REFERENCES

1. A. K. Bousfield and D. Kan, Homotopy Limits, Completions and Localizations, Springer Lecture Notes in Math 304 1972.

2a. W. Dwyer, H. Miller, and C.W.Wilkerson, Homotopy Uniqueness of $\mathbb{HP}(\infty)$, Proceedings of the Barcelon Conference on Algebraic Topology 1986, (Editors: J. Aguade and R. Kane), SLNM (1989).

2b. W. Dwyer, H. Miller, and C.W. Wilkerson, Homotopy Uniqueness of BG, private communication.

3. G. Mislin, Self Maps of Infinite Quaternionic Projective Space, Quartely J. of Math. Oxford (3) 38 (1987), 245-257.

4. D. Notbohm and L. Smith, Fake Lie Groups and Maximal Tori II, Göttingen Preprints 1989.

5. L. Smith, Lectures on the Eilenberg-Moore Spectral Sequence, Springer Lecture Notes in Math 134, 1970.

6. Stasheff, J.D., A classification theorem for fibre spaces, Topology 2 (1963) 239-246.

J.E. McClure: Department of Mathematics
University of Kentucky
Lexington, Kentucky
USA 40506

L. Smith: Mathematisches Institut
Bunsenstraße 3/5
D 3400 Göttingen
WEST GERMANY

1990 Barcelona Conference
on Algebraic Topology.

ON INFINITE DIMENSIONAL SPACES THAT ARE RATIONALLY EQUIVALENT TO A BOUQUET OF SPHERES

C.A. McGibbon and J.M. Møller*

Introduction

Let X be a space of the sort mentioned in the title. Suppose that Y is another space whose Postnikov approximations, $Y^{(n)}$, are homotopy equivalent to those of X for each integer n, but not necessarily in any coherent manner. Does it follow that X and Y are homotopy equivalent? This is the problem. In this paper we obtain a fairly general solution to it and we also study some specific examples. To describe our results, we first recall Wilkerson's definition of $SNT(X)$ as the set of all homotopy types $[Y]$ such that $Y^{(n)} \simeq X^{(n)}$ for all integers n,[4]. This is a pointed set with basepoint $* = [X]$. Our main result is the following.

Theorem 1. Let X be a 1-connected space with finite type over some subring of the rationals. Assume that X has the rational homotopy type of a bouquet of spheres. Then the following three conditions are equivalent:
(i) $SNT(X) = *$.
(ii) the map $AutX \xrightarrow{f \mapsto f^{(n)}} AutX^{(n)}$ has a finite cokernel for each n.
(iii) the map $AutX \xrightarrow{f \mapsto f_\#} Aut(\pi_{\leq n}X)$ has a finite cokernel for each n.
Moreover, if $SNT(X) \neq *$, then this set is uncountably large. □

In this theorem, $AutX$ denotes the discrete group of homotopy classes of homotopy self-equivalences of a space X. When we say a homomorphism of groups $\varphi : A \to B$ has a finite cokernel we mean only that its image, $\varphi(A)$, has finite index in B. Since X has finite type, its homotopy groups are finitely generated modules over $\mathbf{Z}_P$, the integers localized at some set of primes P. Finally, $Aut(\pi_{\leq n}X)$ denotes the group of those automorphisms of the graded $\mathbf{Z}_P$ module, $\pi_{\leq n}X$, that preserve the Whitehead product pairing.

Theorem 1 is the Eckmann-Hilton dual of the following result that we proved in [2].

Theorem 2. Let X be a 1-connected, H_0-space with finite type over $\mathbf{Z}_P$ for some set of primes P. Then the following statements are equivalent:
(i) $SNT(X) = *$.

---

* Supported by the Danish Natural Science Research Council

(ii) the map $AutX \xrightarrow{f \mapsto f^{(n)}} AutX^{(n)}$, has a finite cokernel for each n
(iii)the map $AutX \xrightarrow{f \mapsto f^{*}} AutH^{\leq n}(X, \mathbf{Z}_P)$ has a finite cokernel for each n.
Moreover, if $SNT(X) \neq *$, then this set is uncountably large.□

Recall here that an H_0-space is one whose rationalization is homotopy equivalent to a product of Eilenberg-MacLane spaces. Incidently, a key property common to both H_0-spaces and those spaces described by the title is this: each map in the tower, $\cdots \leftarrow AutX^{(n)} \leftarrow AutX^{(n+1)} \leftarrow \cdots$, has a finite cokernel. Of course, by $AutH^{\leq n}(X; \mathbf{Z}_P)$ in Theorem 2, we mean just the group of all automorphisms of the graded cohomology ring indicated.

While Eckmann-Hilton duality suggested what the dual result might be, it didn't give us a proof. The proof here turns out to be quite different from the one we gave for Theorem 2 in [2]. For one thing, in H_0-spaces, the k-invariants have finite order. We exploited this repeatedly in [2]. Here, in the dual case, the k-invariants often rationalize to nontrivial Massey products; things we would just as soon avoid. So instead, we first prove the theorem for the universal example, a bouquet of spheres, and then we relate that result to other spaces within the same rational homotopy type. The second step involves Wilkerson's theorem: "if X and Y are rationally equivalent Postnikov approximations, then $Aut(X)$ and $Aut(Y)$ are commensurable", together with a graded algebra version of his result. The graded algebras in question occur as finite-dimensional quotients of the homotopy algebra ($\pi_* \Omega X$, with Samelson products) and provide the algebraic setting in which to study $Aut(\pi_{\leq n} X)$.

If X is also rationally finite dimensional (eg, the k-fold suspension of a $K(\mathbf{Z}, 2n+1)$), we get a result that is easier to check than that given by the Theorem.

Corollary. Let X be as in the theorem and suppose there exists an integer t such that $H_{>t}(X; \mathbf{Q}) = 0$. Then $SNT(X)$ is trivial if and only if $AutX \to Aut(\pi_{\leq n} X)$ has a finite cokernel in at least one of the following cases:
(i) for $n = t$.
(ii) for some integer $n \geq t$.
(iii) for all integers n. □

We will apply these results to a few examples. Before doing so, it seems natural to ask if the restrictions that the theorem places on the rational homotopy type of X in the theorem are really necessary. Our first example shows that some restrictions are needed.

Example A. Let X denote the localization of complex projective n -space, at a prime p. Then $SNT(X) = *$ when $n < \infty$ because X is finite dimensional. But for $1 < n < \infty$, the cokernel of $Aut(X) \to Aut(\pi_{\leq 2n+1} X)$ is not finite. □.

Of course, $\mathbf{CP}^n$ does not have the rational type of a wedge of spheres since its rational cohomology has nontrivial products. The next three examples involve suspensions of Eilenberg-MacLane spaces. For $n > 1$, the $K(\mathbf{Z}, n)$'s are infinite dimensional and, after suspension, they become somewhat intractible.

Example B. Let $X = K(\mathbf{Z}, 2n+1)$, or the localization of this space at any set of primes P. Then $SNT(\Sigma^k X) = *$, for $k \geq 0$. □

Example C. Let $X = K(\mathbf{Z}, n) \vee S^n$, or the localization of this space at some nonempty set of primes P. Then $SNT(\Sigma^k X) \neq *$ for each $n > 1$ and each $k \geq 1$. □

The next example shows that the suspension map $SNT(X) \to SNT(\Sigma X)$ is, in general, neither surjective nor injective. Examples C and D both use Zabrodsky's result that for $n > 1$, all maps from a $K(\mathbf{Z}, n)$ to a sphere or its loop space are phantom maps.

Example D. Let $X = \Sigma K(\mathbf{Z}, 2n-1) \vee S^n$ for some odd integer $n > 1$ or the localization of this space at some nonempty set of primes P. Then $SNT(X)$ and $SNT(\Sigma^2 X)$ are both trivial but $SNT(\Sigma X)$ is not. □

The next example is not as artificial as it first appears. In it, X could be $\Omega\Sigma K$ where K is a finite complex, or ΩP where P is a finite product of spheres, or an infinite product of finite complexes whose connectivity increases with α.

Example E. Let X be a connected space with finite type over some subring of the rationals and assume that $\Sigma X \simeq \bigvee_\alpha K_\alpha$ where each K_α is finite dimensional. Then $SNT(\Sigma X) = *$. □

All of these examples will be verified after we prove the theorem and its corollary. We end this section with an open question. The reader may have noticed that $K(\mathbf{Z}, 2n)$'s were not mentioned in Example B. The omission was deliberate. We have yet to settle the first interesting case!

Question. Is $SNT(\Sigma\mathbf{CP}^\infty) = *$ or not ?

Proofs

We will first outline the proof of the theorem using the following three lemmas. Let X denote a space that satisfies the hypothesis of the theorem.

Lemma 1. The homomorphism $AutX^{(n)} \xrightarrow{f \mapsto f_\#} Aut(\pi_{\leq n}X^{(n)})$ has a finite kernel and a finite cokernel for each integer $n \geq 1$. □

Lemma 2. The homomorphism $AutX^{(n+1)} \xrightarrow{f \mapsto f^{(n)}} AutX^{(n)}$ has a finite cokernel for each $n \geq 1$. Furthermore, if $H_{n+1}(X, \mathbf{Q}) = 0$, this map also has a finite kernel. □

Lemma 3. Let $G_1 \longleftarrow G_2 \longleftarrow G_3 \longleftarrow \cdots$ be a tower of countable groups in which each map $G_n \longleftarrow G_{n+1}$ has a finite cokernel. Then $\varprojlim{}^1 G_n = *$ if and only if the canonical map

$$\varprojlim G_n \longrightarrow G_k$$

has a finite cokernel for each k. Furthermore, if $\varprojlim{}^1 G_n \neq *$, then it is uncountably large. □

The first lemma implies the equivalence of parts (ii) and (iii) of the theorem. To see this, combine Lemma 1 with the following commutative diagram.

$$\begin{array}{ccc} AutX & ----\longrightarrow & Aut(\pi_{\leq n}X) \\ \downarrow & & \downarrow\approx \\ AutX^{(n)} & ----\longrightarrow & Aut(\pi_{\leq n}X^{(n)}) \end{array}$$

The equivalence of parts (i) and (ii) of the theorem then follows using Wilkerson's formula,[4],

$$SNT(X) \approx \varprojlim{}^1 AutX^{(n)},$$

together with Lemmas 2 and 3. This completes the outline; now we prove the lemmas. Of the three, the first requires the most work. To prove it, we will start with a special case.

Proposition 1.1 Let $W = \bigvee_\alpha S^{n_\alpha}$, a simply connected bouquet of spheres with finite type. Then the map $AutW^{(n)} \longrightarrow Aut(\pi_{\leq n}W^{(n)})$ is injective and has a finite cokernel. □

To relate the automorphism groups associated with W to those of X, we use the following device that we learned about from Wilkerson, [5]: given a map $\varphi : A \longrightarrow B$ in some category, let

$$\Delta(\varphi) = \{(\alpha,\beta) \in Aut(A)\times Aut(B) \,|\, \varphi\alpha = \beta\varphi\}$$

Then $\Delta(\varphi)$ is a group and there are the obvious projections from it to $Aut(A)$ and to $Aut(B)$. Wilkerson shows (Theorem 2.3,ibid.), that if $f : Y \longrightarrow Z$ is a rational equivalence between two simply connected spaces of finite type with only finitely many nonzero homotopy groups, then the projections from $\Delta(f)$ to $Aut(Y)$ and to $Aut(Z)$ have finite kernels and finite cokernels. We will use his theorem in the proof of Lemma 1 along with the following algebraic version of it.

Proposition 1.2 Let $\varphi : A \to B$ be a map in the category of S-algebras with finite type over $\mathbf{Z}_P$. Assume $A_n = B_n = 0$ for $n \leq 0$ and for n sufficiently large. If φ is a rational equivalence, then the projections from $\Delta(\varphi)$ to $Aut(A)$ and to $Aut(B)$ have finite kernels and finite cokernels. □

By an S-algebra over $\mathbf{Z}_P$, (S for Samelson), we mean a graded $\mathbf{Z}_P$ module L, together with bilinear anti-symmetric pairings $L_i \times L_j \to L_{i+j}$ that satisfy a Jacobi identity with signs. So an S -algebra is a graded Lie algebra over $\mathbf{Z}_P$, as defined in [1], page 125, if and only if the primes 2 and 3 are invertible in $\mathbf{Z}_P$. In particular. if 2 and 3 are not in P, then we don't quite have a graded Lie algebra. But for our modest aims, what we have is enough. By [3], p 470, the Samelson product gives $\pi_*\Omega X$ or $\pi_{\leq n}\Omega X$ the structure of an S- algebra over $\mathbf{Z}_P$. Since the Samelson product is the adjoint of the Whitehead product we have $Aut(\pi_{\leq n}X) \approx Aut(\pi_{\leq n-1}\Omega X)$ and the role of Proposition 1.2 becomes apparent.

Let $f : W \longrightarrow X$ be a rational equivalence and consider the following commutative diagram

$$\begin{array}{ccccc} AutW & -\longrightarrow & \Delta(f) & -\longrightarrow & AutX \\ \downarrow & & \downarrow & & \downarrow \\ Aut(\pi_{\leq n}W) & -\longrightarrow & \Delta(f_\#) & -\longrightarrow & Aut(\pi_{\leq n}X) \end{array}$$

Here the horizontal maps are the projections, which all have finite kernels and finite cokernels by Wilkerson's theorem (on top) and by 1.2 (on the bottom). The left side also has this property by 1.1. A diagram chase then confirms that, on the right side, the kernel and cokernel are finite as well.

We now prove 1.1. To see that $AutW^{(n)} \to Aut(\pi_{\leq n}W^{(n)})$ is injective let W_n denote the n-skeleton. Then

$$AutW^{(n)} \approx AutW_n \subseteq [W_n, W_n] \approx \prod_{k_\alpha \leq n} [S^{k_\alpha}, W_n]$$

Moving from left to right, the identity map on $W^{(n)}$ is sent to a certain sequence of inclusions. If $h \in AutW^{(n)}$ induced the identity on $\pi_* W^{(n)}$ then it would be sent (above) to the same sequence of inclusions. Since the above composition is injective, this means $h = 1$.

To show the cokernel is finite requires more work. We do it by induction on n. The initial case is obvious; assume therefore that it is true in dimensions $\leq n$. Let V denote the subset of spheres in W of dimension $n+1$. We have the cofibration

$$W_n \xrightarrow{i} W_{n+1} \xrightarrow{r} V,$$

an equivalence $W_{n+1} \simeq W_n \vee V$, and a short exact sequence,

$$0 \longrightarrow [V, W_n] \longrightarrow AutW_{n+1} \longrightarrow AutW_n \times AutV \longrightarrow 1.$$

Here the map into $AutW_{n+1}$ sends f to $1 + if\pi$. Of course, the co-H structure on W_{n+1} is used to form this sum. The map out of $AutW_{n+1}$ is given by restriction and projection. One can see, using homology, that this indeed works. Exactness is easy to check.

Let L denote the S-algebra defined by Samelson products on the quotient, $\pi_{\leq n}\Omega W_{n+1}/$ torsion. Since the torsion subgroup of $\pi_{\leq n+1}W_{n+1}$ is finite, the kernel of $Aut(\pi_{\leq n+1}W_{n+1}) \to Aut(L)$ is finite. Therefore, to finish the proof of this proposition, it is enough to show that the cokernel of $Aut(W_{n+1}) \to Aut(L)$ is finite. We use a second exact sequence, involving L,

$$0 \longrightarrow Hom(Q_n L, D_n L) \longrightarrow Aut(L) \longrightarrow Aut(L_{<n}) \times Aut(Q_n L)$$

defined in essentially the same way as the first. Here, of course, D and Q indicate decomposables and indecomposables, respectively. Tensor the following diagram with $\mathbf{Q}$. The slanted isomorphisms are then evident and worth noting.

$$\begin{array}{ccccc} & & \pi_n \Omega V & & \\ & & \uparrow \Omega r_* & \approx & \\ D_n L & \longrightarrow & L_n & \longrightarrow & Q_n L \\ & \approx & \uparrow \Omega i_* & & \\ & & \pi_n \Omega W_n & & \end{array}$$

The two exact sequences just described fit together in the following commutative diagram

$$\begin{array}{ccccc} [V, W_n] & \longrightarrow & AutW_{n+1} & \longrightarrow & Aut(W_n) \times Aut(V) \\ \downarrow & & \downarrow & & \downarrow \\ Hom(Q_n L, D_n L) & \longrightarrow & Aut(L) & \longrightarrow & Aut(L_{<n}) \times Aut(Q_n L) \end{array}$$

From left to right, the first two verticals are induced by $f \mapsto \Omega f_*/\text{torsion}$, the third by $(f,g) \mapsto \Omega(f \vee g)_*/\text{torsion}$. To show that the middle map has a finite cokernel, it suffices to prove this for the two sides. The left side has a finite cokernel because it rationalizes to an isomorphism (recall the slanted isomorphisms noted earlier). On the right side, the first factor has a finite cokernel by the inductive assumption. For the second factor, we note that $\pi_n\Omega V$ maps to a direct summand of finite index in Q_nL and from this it follows easily that $Aut(V) \to Aut(Q_nL)$ has a finite cokernel. □

Proof of Proposition 1.2 It is not difficult to see that the projections from $\Delta(\varphi)$ to $Aut(A)$ and to $Aut(B)$ have finite kernels. Indeed, if $L = A$ or B, the kernel of $\Delta(\varphi) \to Aut(L)$ is isomorphic to some subgroup of the kernel induced by rationalization, $Aut(L) \to Aut(L_0)$. Since L is a finitely-generated $\mathbf{Z}_P$-module, this kernel is clearly finite. The cokernels require more work.

Let K, and Q, denote the kernel, and the image, of $\varphi : A \longrightarrow B$. We note that the conditions on A, B, and φ imply that K is finite and that Q has finite index in B. Consider the set, $EXT(Q,K)$, of S-algebra extensions of Q by K. As usual, we identify any two of them, between which there is a S-algebra isomorphism that restricts to the identity on K and induces the identity on the quotient Q. We claim that this set of extensions is finite. The corresponding set of $\mathbf{Z}_P$-module extensions is finite since K is finite and Q is finitely generated. Let E be one such module extension. The number of different brackets on E, compatible with the given one on Q, is finite because (i) each bracket is determined by its values on the generators, (ii) E is finitely generated and (iii) for each pair of generators, x, y, there is only a finite number of possible choices for $[x,y]$. Thus the set $EXT(Q,K)$ is finite. Therefore, if a group G acts on this set, the stabilizer of the extension $K \longrightarrow A \xrightarrow{\varphi} B$ must have finite index in G.

Now let $Aut(B)$ act on the set of all subalgebras of B with the same index (in B) as Q has. This set is finite because B, being a finitely generated $\mathbf{Z}_P$-module, has only a finite number of submodules of a given finite index. Let G be the stabilizer of Q in $Aut(B)$; then G has finite index in $Aut(B)$ and G acts on $EXT(Q,K)$ via the pullback construction, . It follows that the stablizer of the extension defined by φ has finite index in $Aut(B)$. Since this stabilizer is contained in the image of $\Delta(\varphi)$, it too must have finite index.

Thus we have shown that the projection from $\Delta(\varphi)$ to $Aut(B)$ has a finite cokernel. For the other projection, let $Aut(A)$ act on the finite set of those subalgebras of A that have the same order as K. Then let the stabilizer of K act on $EXT(Q,K)$ and argue as before. □

Proof of Lemma 2 Let W again denote a simply-connected bouquet of spheres with finite type. It is easy to see that the lemma is true for W. Indeed, from the commutative diagram,

$$\begin{array}{ccc} AutW^{(n+1)} & \overset{\approx}{\dashrightarrow} & AutW_{n+1} \\ \downarrow & & \downarrow \\ AutW^{(n)} & \overset{\approx}{\dashrightarrow} & AutW_n \end{array}$$

where the right side is given by restriction, it is apparent that the left side is surjective for each n and bijective when $H_{n+1}W = 0$, (for then $W_{n+1} \simeq W_n$).

For a space X, with the rational homotopy type of W, take a rational equivalence

$f : W \longrightarrow X$, and form the commutative diagram

$$\begin{array}{ccccc} AutW^{(n+1)} & \dashrightarrow & \Delta(f^{(n+1)}) & \dashrightarrow & AutX^{(n+1)} \\ \downarrow & & \downarrow & & \downarrow \\ AutW^{(n)} & \dashrightarrow & \Delta(f^{(n)}) & \dashrightarrow & AutX^{(n)} \end{array}$$

Since the four horizontal maps have finite kernels and finite cokernels, the result for X follows from a diagram chase; first in the left square, then in the right.□

Proof of Lemma 3 This is a combination of two results we proved in [2]. To be specific, in Theorem 2 of that paper we show that for an inverse tower of countable groups, $\varprojlim^1$ of it is trivial if and only if the tower is Mittag-Leffler. Moreover, if the $\varprojlim^1$ term is nontrivial, then it is uncountable. Then in Lemma 3.2, we add the hypothesis that each map in the tower has a finite cokernel and conclude that the tower is Mittag-Leffler if and only if the canonical projection from the inverse limit to each term in the tower has a finite cokernel. □

Proof of the Corollary If $SNT(X) = *$ then the theorem implies that condition (iii) in the corollary holds. Obviously, then, parts (i) and (ii) of the corollary hold as well. Going the other way, assume that the weakest condition of the three, namely (ii), holds for some $m \geq t$. In the commutative diagram,

$$\begin{array}{ccc} AutX & \dashrightarrow & Aut(\pi_{\leq m}X) \\ \downarrow & & \downarrow\approx \\ AutX^{(m)} & \dashrightarrow & Aut(\pi_{\leq m}X^{(m)}) \end{array}$$

it follows, using Lemma 1 on the bottom, that the left side has a finite cokernel. Assume a and b are integers with $a > m > b$ Then, in the diagram,

$$\begin{array}{ccccc} & & AutX & & \\ & & \downarrow & & \\ AutX^{(a)} & \longrightarrow & AutX^{(m)} & \longrightarrow & AutX^{(b)} \end{array}$$

the vertical map in the middle has a finite cokernel. By Lemma 2, both horizontal maps have finite cokernels and, since $a > m \geq t$, the first one has a finite kernel as well. It then follows that the two slanted maps out of $AutX$ have finite cokernels too. Therefore, by part (ii) of the theorem we conclude that $SNT(X) = *$. □

The Examples

Example A. In dimensions $\leq 2n+1$, we note that

$$\pi_i X = \begin{cases} \mathbf{Z}_{(p)} & if\ i\ = 2\ or\ 2n+1 \\ 0 & otherwise \end{cases}$$

and when $n > 1$, all ordinary Whitehead products in this range are zero. Thus

$$Aut(\pi_{\leq 2n+1}X) \approx \mathcal{U} \times \mathcal{U}$$

where $\mathcal{U}$ denotes the group of multiplicative units in $\mathbf{Z}_{(p)}$. Let $\mathcal{I}$ denote the image of $Aut(X)$ in this group. A simple calculation shows that

$$\mathcal{I} = \{(u, u^{n+1}) \mid u \in \mathcal{U}\}.$$

This subgroup has infinite index in $\mathcal{U} \times \mathcal{U}$. One way to see this is to consider the element $(1, p+1)$. It is in $\mathcal{U} \times \mathcal{U}$ but no nonzero power of it is in $\mathcal{I}$. Thus the cokernel is infinite as claimed. □

Example B. Let Y denote the localization of $\Sigma^k X$ at a set of primes P. Then Y has the rational homotopy type of a sphere of dimension $d = k + 2n + 1$. By the corollary it is enough to consider

$$Aut(Y) \longrightarrow Aut(\pi_d Y) \approx \mathcal{U}.$$

Here $\mathcal{U}$ denotes the group of multiplicative units in $\mathbf{Z}_P$. Since localization commutes with suspension, it is easy to see that Y has enough self-maps for this composition to be epimorphic. The result follows. □

Example C. Let $Y = \Sigma^k X$ localized at a nonempty set of primes P. To apply our theorem we need $k \geq 1$ if n is even, but k could be zero if n is odd. It then suffices to show that the map

$$Aut(Y) \longrightarrow Aut(\pi_{n+k} Y) \xrightarrow{\approx} GL(2, \mathbf{Z}_P)$$

does not have a finite cokernel. Use the wedge decomposition of Y to obtain a basis for $\pi_{n+k} Y$ and hence to realize the isomorphism above. Given $f \in Aut(Y)$, consider the composition

$$\Sigma^k K(\mathbf{Z}, n) \xrightarrow{i_1} Y \xrightarrow{f} Y \xrightarrow{\pi_2} S^{n+k}$$

where all spaces are localized at P. Zabrodsky [6], Theorem D, proves that this composition must be a phantom map as long as P is nonempty. This means that f is sent to a matrix in $GL(2, \mathbf{Z}_P)$ whose lower left entry is zero. Since the subgroup of upper triangular matrices has infinite index in $GL(2, \mathbf{Z}_P)$, the result follows. □

Example D. Notice that X has the rational homotopy type of $S^{2n} \vee S^n$, where n is odd. By the corollary we can restrict our attention to dimensions $\leq 2n$. In this range, the Hurewicz homomorphism is an isomorphism mod torsion. So consider the composition

$$Aut(X) \longrightarrow Aut(\pi_{\leq 2n} X) \longrightarrow Aut(\pi_{\leq 2n} X/\text{torsion}) \approx \mathcal{U} \oplus \mathcal{U}.$$

It is easy to see that this composition is surjective. Since the second map has a finite kernel, the cokernel of the first is likewise finite. The same argument can be adapted to the double suspension to show that $SNT(\Sigma^2 X) = *$. . It doesn't work on the first suspension, however. Notice that $\pi_{2n+1} \Sigma X_0 \approx \mathbf{Q} \oplus \mathbf{Q}$. Use the wedge decomposition of ΣX to obtain a basis for $\pi_{2n+1} \Sigma X)/\text{torsion}$. Then consider the subgroup of $Aut(\pi_{\leq 2n+1} \Sigma X)$ consisting of those automorphisms that have the matrix representation

$$\begin{pmatrix} 1 & 0 \\ \lambda & 1 \end{pmatrix}$$

on $\pi_{2n+1} \Sigma X/\text{torsion}$ and equal the identity map on all other subgroups of $\pi_{\leq 2n+1} \Sigma X$. The intersection of this infinite abelian subgroup with the image of $Aut(\Sigma X)$ must be trivial because of Zabrodsky's theorem that says that every map from $K(\mathbf{Z}_P, 2n-1)$ to $\Omega^2 S^{n+1}$ is a phantom map when $n > 1$ and P is nonempty. Thus the cokernel of $Aut(\Sigma X) \to Aut(\pi_{\leq 2n+1}\Sigma X)$ must be infinite. □

Example E. Take an integer $n > 1$, and choose V to be a finite sub-bouquet of the K_α's such that the pair $(\Sigma X, V)$ is $(n+1)$-connected. Now consider the commutative diagram

$$\begin{array}{rcl} Aut(\Sigma X) & \dashrightarrow & Aut(\Sigma X)^{(n)} \\ f \mapsto f \vee 1 \uparrow & & \uparrow\approx \\ Aut(V) & \dashrightarrow & Aut V^{(n)} \end{array}$$

The bottom map has a finite cokernel since V is finite-dimensional and satisfies the hypothesis of the theorem. The cokernel of the top map must likewise be finite. Since n was arbitrary, the result follows. □

References

1. F.R.Cohen, J.C.Moore, and J.A.Neisendorfer, Torsion in homotopy groups. Ann. of Math. 109 (1979), pp 121-168.

2. C.A.McGibbon and J.M.Møller, On spaces of the same n-type for all n, preprint, May 1990.

3. G.W.Whitehead, Elements of Homotopy Theory, Graduate Texts in Mathematics 61. Springer-Verlag, New York - Heidelberg - Berlin 1978

4. C.W.Wilkerson, Classification of spaces of the same n-type for all n. Proc. Amer. Math. Soc. (60) 1976,pp 279-285

5. ———— Applications of Minimal Simplicial Groups, Topology (15),(1976) pp111-130.

6. A.Zabrodsky, On phantom maps and a theorem of H.Miller, Israel J. Math. 58 (1978), pp129-143.

C.A.McGibbon: Mathematics Department
Wayne State University
Detroit, MI, 48202
U.S.A.

J.M.Møller: Matematisk Institut
Universitetsparken 5
DK-2100 København Ø
Danmark

1990 Barcelona Conference
on Algebraic Topology.

COHOMOLOGICALLY CENTRAL ELEMENTS AND FUSION IN GROUPS

GUIDO MISLIN

Introduction.

Let G be a compact Lie group with center ZG. For p a prime we put

$$_pZG = \{x \in ZG | x^p = 1\},$$

the maximal elementary abelian p-subgroup of ZG. The cohomology algebra $H^*(BG; \mathbf{F}_p)$ is an unstable algebra over A_p, the mod-p Steenrod algebra. For an arbitrary unstable A_p-algebra R the Dwyer-Wilkerson center ZR of R is defined by

$$ZR := \{f : R \to H^*(B\mathbf{Z}/p; \mathbf{F}_p) \mid T_f : R \to T_f R \text{ an isomorphism}\}.$$

Here, T_f denotes the relative version of Lannes' T-functor, which is defined by $T_f R = TR \otimes_{T^\circ R} \mathbf{F}_p$, where $\mathbf{F}_p$ is considered as a module over the degree zero component $T^\circ R$ of TR via the adjoint $TR \to \mathbf{F}_p$ of $f : R \to H^*(B\mathbf{Z}/p; \mathbf{F}_p)$. It is shown in [DW] that ZR has a natural abelian group structure. There is an obvious group homomorphism

$$\tilde{\varphi} : {_pZG} \to ZH^*(BG; \mathbf{F}_p)$$

which arises as follows. Let x be in $_pZG$, with associated map $\phi(x) : \mathbf{Z}/p \to G$ and $f = (B\phi(x))^* : H^*(BG; \mathbf{F}_p) \to H^*(B\mathbf{Z}/p; \mathbf{F}_p)$. Such an f lies in $ZH^*(BG; \mathbf{F}_p)$, because T_f corresponds to the map induced by the inclusion of the centralizer $C_G(x)$ in G, which is an isomorphism; one then defines $\tilde{\varphi}(x) := f$. It is clear that $\tilde{\varphi}$ is injective, because from basic properties of the T-functor [L] one knows that for any compact Lie group, $\tilde{\varphi}(x) = \tilde{\varphi}(y)$ implies that x and y are conjugate in G, thus equal, since they are central. However, in general φ will not be surjective (for instance the inclusion of an element of order two in the symmetric group S_3 corresponds to a non trivial element in $ZH^*(BS_3; \mathbf{F}_2)$, but the center of S_3 is trivial). Note also that $_pZG$ injects into $_pZ(G/H)$, if $H < G$ is a normal p'-group (a torsion group all of whose elements are of order prime to p) and, if G is finite, one has $H^*(BG; \mathbf{F}_p) = H^*(BG/H; \mathbf{F}_p)$ for such an H. It is therefore natural to consider the quotient $G/O_{p'}(G)$, where $O_{p'}(G)$ denotes the largest normal p'-subgroup of G. Note that $O_{p'}(G)$ may be an infinite group and it is totally disconnected in the Lie group G. Of course, one can still define $_pZ(G/O_{p'}(G))$ as before, although $G/O_{p'}(G)$ may fail to be a Lie group (e.g. consider the case of $G = S^1$). We will see that, however, one has for an arbitrary compact Lie group G a well defined map ${}_pZ(G/O_{p'}(G)) \to H^*(BG;\mathbb{F}_p)$, and we denote this map by φ. Our main theorem can then be expressed as follows.

Theorem 1. *Let G be a compact (not necessarily connected) Lie group and p a fixed prime. Then the canonical map*

$$\varphi : {}_pZ(G/O_{p'}(G)) \to ZH^*(BG;\mathbb{F}_p)$$

is an isomorphism of abelian groups.

As we will see in the course of the proof of this theorem, for an element x in ${}_pZ(G/O_p'(G))$ any counter image $\tilde{x}$ in G will have the property that the inclusion $C_G(\tilde{x}_p) \to G$ is a mod-p homology isomorphism on the classifying space level, where $\tilde{x}_p$ denotes the p-part of the torsion element $\tilde{x}$. Obviously, $\tilde{x}_p$ satisfies $(\tilde{x}_p)^p = 1$, and we will show that $\tilde{x}_p$ depends up to conjugation in G only on its projection x in $G/O_{p'}(G)$; it seems natural to call $\tilde{x}_p$ a "p-cohomologically central element of G". Conversely, an element y in G satisfying $y^p = 1$ and such that $C_G(y) \to G$ induces a mod-p cohomology isomorphism on the classifying space level, will necessarily be of the form $\tilde{x}_p$ for some x in ${}_pZ(G/O_p'(G)$, as we will see. Our Theorem 1 can therefore be rephrased as follows.

Theorem 2. *Let G be a compact Lie group and p a prime. Then there is a natural bijection between the set of conjugacy classes of p-cohomologically central elements y in G satisfying $y^p = 1$, and the group ${}_pZ(G/O_p'(G))$.*

It seems natural to widen the scope a little bit and to consider subsets rather than elements of G, which are cohomologically central in the following sense.

Definition: Let G be a (topological) group and S an arbitrary subset of G. We call S p-cohomologically central in G, if the inclusion of (topological) groups $C_G(S) \to G$ induces an isomorphism in mod-p cohomology on the level of classifying spaces.

We will mainly be interested in the case where G is a compact Lie group. In that case, $C_G(S)$ will be a closed subgroup of G, thus a compact Lie group too; note also that $C_G(S) = C_G(S(\alpha))$ for a suitable finite subset $S(\alpha)$ of S. This can be seen as follows. Obviously, $C_G(S)$ is the intersection of groups $C_G(S(\beta))$ where $S(\beta)$ runs over the finite subsets of S, and each $C_G(S(\beta))$ is compact. But the compact subgroups in a compact Lie group satisfy the descending chain condition, and the result follows easily.

In Section 1, we will discuss fusion from a cohomological point of view and prove Theorems 1 and 2. In Section 2, we discuss two applications, one dealing with finite p-groups of exponent p, and one concerning the cohomology of finite solvable groups.

1) Fusion.

We recall first the classical notion of fusion in the setting of finite groups. Let G be a finite group, p a prime and $P < G$ a Sylow p-subgroup. One says that a subgroup $H < G$ containing P *controls the p-fusion* in G if for any subgroup π of P and any element g in G such that $\pi^g = g^{-1}\pi g < P$ one has $g = ch$ for some c in $C_G(\pi)$ and h in H. For instance, if P is abelian then by a classical theorem of Burnside, the normalizer

$N_G(P)$ controls p-fusion in G. Note also that if $H < G$ controls the p-fusion in G, then G and H have the same mod-p cohomology as one deduces immediately from the description of the mod-p cohomology by means of stable elements in the cohomology of P.

To generalize the notion of fusion to arbitrary groups, we proceed as follows. For G any group and p a prime, the Frobenius category $\mathrm{Frob}_p(G)$ is defined as category with objects the finite p-subgroups of G, and morphisms the group homomorphisms of the form $\mathrm{in}(g) : x \mapsto g^{-1}xg$, for some g in G.

Definition ([MT]): Let G be an arbitrary group, H a subgroup, and p a prime. Then H *controls finite p-fusion* in G, if the inclusion $H \to G$ induces an equivalence of Frobenius categories $\mathrm{Frob}_p(H) \to \mathrm{Frob}_p(G)$.

Fusion is tightly linked to cohomology by the following theorem.

Fusion-Theorem ([M]). *Let $f : H \to G$ be a morphism of compact Lie groups, and p a prime. Then the following are equivalent:*
(i) $(Bf)^* : H^*(BG; \mathsf{F}_p) \to H^*(BH; \mathsf{F}_p)$ is an isomorphism
(ii) f induces an equivalence of categories $Frob_p(H) \to Frob_p(G)$.

The basic ingredient of the proof of Theorem 1 is the following theorem, which links fusion with the internal structure of a group.

Z*-Theorem ([MT]). *Let G be a compact Lie group and p a prime. Let $A < G$ be a p-subgroup (not necessarily finite), or a p-toral subgroup. If $C_G(A)$ controls finite p-fusion in G, then*

$$G = C_G(A)O_{p'}(G) = C_G(A)[A, G]$$

and the commutator group $[A, G]$ is a (normal) finite p'-subgroup of G.

Remark: The reader who is primarily interested in the case of finite groups, should consult [B] for a reduction of the Z^*-theorem for finite groups to the case of finite simple groups. The Z^*-theorem for finite simple groups can then be checked case by case. Unfortunately, no proof of the Z^*-theorem not using the classification is known for finite groups and p an odd prime. However, for the applications offered in Section 2 , one only needs the Z^*-theorem for finite solvable groups. In that case, the methode of reduction as described in [B] provides a complete proof, since for finite solvable simple groups (i.e. finite cyclic groups of prime order) the statement is trivial. In [MT], the Z^*-theorem for compact Lie groups is proved by reducing it to the case of finite groups; no new proof is offered in the case of finite groups.

Proof of Theorem 1: We define $\varphi : {}_pZ(G/O_{p'}(G)) \to ZH^*(BG; \mathsf{F}_p)$ as indicated above. For $x \in {}_pZ(G/O_{p'}(G))$ we choose $y \in G$ satisfying $y^p = 1$, and y over x. Then $[y, G] < O_{p'}(G)$, and $[y, G]$ is finite, since $[y, G^0] = \{1\}$ (it is connected and totally disconnected). Thus $G \to G/[y, G]$ is a morphism of compact Lie groups, which induces a mod-p cohomology isomorphism on the level of classifying spaces. Since $\overline{y} \in G/[y, G]$ is central, it follows by the naturality of Lannes' T-functor that $y \in G$ is p-cohomologically central in G (that is, $C_G(\{y\}) \to G$ is a mod-p cohomology isomorphism); we then put

$$\varphi(x) = f(y) : H^*(BG; \mathbb{F}_p) \to H^*(B\mathbb{Z}/p; \mathbb{F}_p),$$

where $f(y)$ is the map induced by the map $\mathbb{Z}/p \to \langle y \rangle \to G$, ($y$ corresponding to the residue class of 1 in $\mathbb{Z}/p$). To see that $\varphi(x)$ is well defined, let z denote another element in G over x satisfying $z^p = 1$. Then z will also be p-cohomologically central in G and both, y and z, map to central elements in $G/([y,G] \cdot [z,G])$. We claim that y and z are conjugate in G, and therefore $f(y) = f(z)$. Namely, by construction, y and z are both in the torsion group $\langle y, O_{p'}(G)\rangle < G$ and, from the short exact sequence

$$O_{p'}(G) \to \langle y, O_{p'}(G)\rangle \to \langle x \rangle$$

we see that $\langle y \rangle$ and $\langle z \rangle$ are Sylow p-subgroups of $\langle y, O_{p'}(G)\rangle$. Since in a linear torsion group any two Sylow p-subgroups are conjugate [W], we conclude that $z^n = g^{-1}yg$ for some $g \in G$ and some n prime to p. Clearly, we can choose $n = 1$, because y and z both project onto x, and x, y, z all have the same order by construction. To see that φ is injective, suppose that $\varphi(x_1) = \varphi(x_2)$, and y_1, y_2 elements in G over x_1, x_2 satisfying $y_i^p = 1$, so that $f(y_1) = f(y_2) : H^*(BG; \mathbb{F}_p) \xrightarrow{i} H^*(\mathbb{Z}/p; \mathbb{F}_p)$. But then y_1 and y_2 are conjugate in G by one of the basic properties of Lannes' T-functor [L]. In particular, x_1 and x_2 will be conjugate, thus equal, because they are central in $G/O_{p'}(G)$; thus φ is injective. Now choose $f \in ZH^*(BG; \mathbb{F}_p)$, represented by $B\mathbb{Z}/p \to BG$ which, as is well known, is induced by a homomorphism $\rho : \mathbb{Z}/p \to G$; we used here the basic fact that for any compact Lie group G, one has natural bijections

$$mor(H^*(BG; \mathbb{F}_p), H^*(\mathbb{Z}/p; \mathbb{F}_p)) \cong [B\mathbb{Z}/p, BG] \cong Rep(\mathbb{Z}/p, G)$$

where $mor(\ ,\ )$ stands for the set of A_p-algebra morphisms, and $Rep(\mathbb{Z}/p, G)$ for the set of conjugacy classes of homomorphism $\mathbb{Z}/p \to G$. Put now $y = \rho(1 \text{ mod-}p)$. Then y will be p-cohomologically central in G since f was chosen in $ZH^*(BG; \mathbb{F}_p)$. By the Z^*-theorem (loc. cit.) we thus have

$$G = C_G(y)O_{p'}(G),$$

which implies that y maps to an element $x \in {}_pZ(G/O_{p'}(G))$. By construction, $\varphi(x) = f$, showing that p is surjective and completing the proof of Theorem 1.

Note that our proof of Theorem 1 also shows that the conjugacy classes of elements $y \in G$ satisfying $y^p = 1$ and for which $C_G(y) \to G$ is a mod-p cohomology isomorphism, correspond bijectively to the elements of ${}_pZ(G/O_{p'}(G))$, as claimed in Theorem 2.

2) Two applications.

For G a finite p-group, Theorem 1 implies that

$$ {}_pZ(G) \cong ZH^*(BG; \mathbb{F}_p). \tag{1}$$

Since $H^*(B\mathbb{Z}/p^2; \mathbb{F}_p) \cong H^*(B\mathbb{Z}/p^3; \mathbb{F}_p)$ as A_p-algebras, we cannot expect to be able to compute the order of the finite p-group G just from the A_p-algebra structure of $H^*(BG; \mathbb{F}_p)$. But if we assume G to be of exponent p, this is in principle possible, as one can see from the proof of the following proposition.

Proposition 1. *Let G and H be finite groups of exponent p, p a prime. Suppose $H^*(BG; \mathbb{F}_p)$ and $H^*(BH; \mathbb{F}_p)$ are isomorphic as A_p-algebras. Then G and H have the same order.*

Proof: The assumption implies that the centers $Z(G)$ and $Z(H)$ are isomorphic since they are isomorphic to $ZH^*(BG; \mathbb{F}_p)$ and $ZH^*(BH; \mathbb{F}_p)$ respectively. Choose an A_p-algebra isomorphism

$$\lambda : H^*(BG; \mathbb{F}_p) \to H^*(BH; \mathbb{F}_p).$$

Then λ maps the set of A_p-algebra homomorphisms $mor(H^*(BH; \mathbb{F}_p), H^*(B\mathbb{Z}/p; \mathbb{F}_p))$ bijectively onto the corresponding set $mor(H^*(BG; \mathbb{F}_p), H^*(B\mathbb{Z}/p; \mathbb{F}_p))$. These sets correspond naturally to the sets of conjugacy classes of elements in H (respectively G), since

$$mor(H^*(BG; \mathbb{F}_p), H^*(B\mathbb{Z}/p; \mathbb{F}_p)) \cong Rep(\mathbb{Z}/p, G)$$

and $Rep(\mathbb{Z}/p, G) \cong \{$ conjugacy classes in $G\}$, as G has exponent p (similarly for H). If we put

$$J(G) = mor(H^*(BG; \mathbb{F}_p), H^*(B\mathbb{Z}/p; \mathbb{F}_p)) \backslash ZH^*(BG; \mathbb{F}_p)$$

then λ induces a bijection $\lambda^* : J(H) \cong J(G)$. Of course, $J(G)$ (respectively $J(H)$) is in natural bijection with the set of conjugacy classes of non-central elements in G (respectively H). To compute the order $|G|$ of G, we consider the class equation

$$|G| = |Z(G)| + \sum_{J(G)} |G|/|C_G(x)|, \tag{2}$$

If the conjugacy class of $x \in G$ corresponds to $f(x) \in J(G)$, so that $f(x) : H^*(BG; \mathbb{F}_p) \to H^*(B\mathbb{Z}/p; \mathbb{F}_p)$, and $\lambda^*(f(y)) = f(x) \in J(G)$, $y \in H$, then by naturality of Lannes' T-functor one has

$$T_{f(x)}H^*(BG; \mathbb{F}_p) \cong T_{f(y)}H^*(BH; \mathbb{F}_p).$$

But

$$T_{f(x)}H^*(BG; \mathbb{F}_p) \cong H^*(BC_G(x); \mathbb{F}_p),$$

and

$$T_{f(y)}H^*(BH; \mathbb{F}_p) \cong H^*(BC_H(x); \mathbb{F}_p),$$

so that λ induces an A_p-algebra isomorphism

$$H^*(BC_G(x); \mathbb{F}_p) \cong H^*(BC_H(y); \mathbb{F}_p).$$

Since x (respectively y) is non-central, $C_G(x)$ is smaller than G (respectively $C_H(y)$ is smaller than H). By induction on the order of the groups involved, we infer $|C_G(x)| = |C_H(y)|$. Comparing now (2) with the corresponding formula

$$|H| = |Z(H)| + \sum_{J(H)} |H|/|C_H(y)|, \tag{3}$$

we see that $|G|$ and $|H|$ satisfy the same recursion formulas, which implies that $|G| = |H|$.

The second application involves the Fusion-Theorem and was suggested to me by Jacques Thévenaz.

Proposition 2. *Let $f : H \to G$ be a homomorphism of finite groups and assume that G is solvable. If for some prime p the induced map*

$$(Bf)^* : H^*(BG; \mathbb{F}_p) \to H^*(BH; \mathbb{F}_p)$$

is an isomorphism, then f maps $O_{p'}(H)$ into $O_{p'}(G)$ and induces an isomorphism

$$H/O_{p'}(H) \cong G/O_{p'}(G).$$

Proof: Clearly, $G \to G/O_{p'}(G)$ induces an isomorphism in mod-p cohomology and therefore our assumption on f implies that

$$\overline{f} : H \to G/O_{p'}(G) = \overline{G}$$

is a mod-p cohomology isomorphism too. Thus $ker(\overline{f})$ is a p'-group. If $\overline{f}$ is surjective, we conclude that $ker(\overline{f}) = O_{p'}(H)$ and we are done. Put $\overline{H} = H/ker(\overline{f})$. Then the inclusion $\overline{H} < \overline{G}$ will induce a mod-p cohomology isomorphism too and, by the Fusion-Theorem (loc. cit.), we conclude that $\overline{H}$ controls the p-fusion in $\overline{G}$. Now choose $\pi = O_p(\overline{G})$,the largest normal p-subgroup of $\overline{G}$. Since $\overline{G}$ is solvable,

$$C_{\overline{G}}(\pi) < O_p(\overline{G}) = \pi$$

(see [G, 8.1.1]) and, because $\overline{H}$ controls finite p-fusion, π is conjugate to a subgroup of $\overline{H}$, thus $\pi < \overline{H}$ since π is normal in $\overline{G}$. From the definition of the control of p-fusion we infer

$$N_{\overline{G}}(\pi) = N_{\overline{H}}(\pi) \cdot C_{\overline{G}}(\pi),$$

which implies $\overline{G} = \overline{H} \cdot C_{\overline{G}}(\pi)$ because π is normal in $\overline{G}$. But we observed already that $C_{\overline{G}}(\pi) < \pi$ and $\pi < \overline{H}$, thus $\overline{H} = \overline{G}$, completing the proof.

Remark: We only used that the finite group G satisfies

$$C_{\overline{G}}(O_p(\overline{G})) < O_p(\overline{G})$$

where $\overline{G} = G/O_{p'}(G)$; such a group G is called *p-constrained* (cf. [G]). For instance, p-solvable groups are p-constrained (G is called p-solvable, if G admits a subnormal series with quotients either p-groups or p'-groups [G]).

References.

[B] M. Broué: La Z^*-conjecture de Glauberman, Séminaire sur les groupes finis I, Publications Mathématiques de l'Université de Paris VII (1983), 99-103.

[DW] W. G. Dwyer and C. W. Wilkerson: A cohomology decomposition theorem; preprint 1989.

[G] D. Gorenstein: *Finite groups*; Harper's series in modern mathematics, Harper and Row, New York, Evanston and London 1968.

[L] J. Lannes: Sur la cohomologie modulo p des p-groupes Abéliens élémentaires, in "Homotopy Theory, Proc. Durham Symp. 1985" edited by E. Rees and J. D. S. Jones, Cambridge University Press, Cambridge 1987.

[M] G. Mislin: On group homomorphisms inducing mod-p cohomology isomorphisms, Comment. Math. Helv. 65 (1990) 454-461.

[MT] G. Mislin and J. Thévenaz: The Z^*-theorem for compact Lie groups; preprint 1990.

[W] B. A. F. Wehrfritz, *Infinite Linear Groups*; Ergebnisse der Mathematik Vol. 76, Springer Verlag, Berlin-Heidelberg-New York, 1973.

Mathematik
ETH-Zentrum
8092 - Zürich
Schweiz
and
Department of Mathematics
Ohio State University
Columbus, Ohio 43210
USA

1990 Barcelona Conference
on Algebraic Topology.

RATIONAL HOMOTOPY OF THE SPACE OF HOMOTOPY EQUIVALENCES OF A FLAG MANIFOLD

DIETRICH NOTBOHM AND LARRY SMITH

For a space X denote by $HE(X)$ the topological monoid of homotopy equivalences of X with itself. The component of the identity of $HE(X)$ will be denoted by $SHE(X)$. The monoid $HE(X)$ and its classifying space appear in Stasheff's classification theorem [14] for fibrations with fibre X, and in Cooke's work [1] on homotopy actions of finite groups. (See also [12].) For a compact connected Lie group G with maximal torus $T \longrightarrow G$ the space $HE(G/T)$ has via Stasheff's theorem played an important role in our study of fake Lie groups [6],[7],[8]. In this note we show how to use the lovely paper of R. Thom [16] to analyze the rational homotopy of $SHE(G/T)$ and its classifying space $BSHE(G/T)$. These results are needed in [8] to complete our study of the relation between fake Lie groups and maximal torii.

This paper is organized as follows. In section 1 we review Thom's paper [16]. Section 2 is devoted to applications of these results to the rational homotopy of $SHE(G/T)$, and the third section contains some finiteness results for $\pi_0(HE(G/T))$.

1. A Review of Thom Theory.

In this section we review the results of Thom [16] on the topology of function spaces. Let X and Y be spaces and $f : X \longrightarrow Y$ a map. The space of all continuous maps from X to Y suitably topologized will be denoted by $map(X, Y)$, and the component of f by $map(X, Y)_f$. If Y is a topological group, then so is $map(X, Y)$ and hence the components of $map(X, Y)$ are all homotopy equivalent: a homotopy equivalence being given between components by left translation.

THEOREM 1.1 (R. Thom): *Let X be a connected space, π an abelian group and n a positive integer. Then for any*

$$f : X \longrightarrow K(\pi, n)$$

there is a homotopy equivalence

$$map(X, K(\pi, n))_f \simeq \mathop{\times}_{1 \leq i \leq n} K(H^{n-i}(X, \pi), i).$$

PROOF: Since $K(\pi, n)$ is an abelian group, all components of $map(X, K(\pi, n))$ are homotopy equivalent. Let $c : X \longrightarrow K(\pi, n)$ be the constant map at the unit. Then by a theorem of J.C. Moore [5] [2] $map(X, K(\pi, n))_c$ being a *connected* abelian group is a product of Eilenberg-MacLane spaces. Hence its homotopy type is determined by its homotopy groups, so it suffices to compute these. To this end let

$$\phi : S^i \longrightarrow map(X, K(\pi, n))_f$$

represent a homotopy class in $\pi_i(map(X, K(\pi, n))_f, f)$, where $f \in map(X(, K(\pi, n))_f$ has been chosen as a basepoint. The adjoint of ϕ is a map

$$\Phi : S^i \times X \longrightarrow K(\pi, n)$$

where for the basepoint $* \in S^i$ we have:

$$(*) \qquad \Phi(*, x) = f(x) \quad \forall x \in X.$$

The homotopy class of ϕ is completely determined by Φ. The homotopy class of Φ in turn corresponds to a cohomology class $\Phi^*(\iota_n) \in H^*(S^i \times X; \pi)$. By the Künneth theorem $H^n(S^i \times X; \pi) = H^{n-i}(X; \pi) \oplus H^n(X; \pi)$ so $\Phi^*(\iota_n) = u(\phi) + v(\phi)$, $u(\phi) \in H^{n-i}(X; \pi)$ and $v(\phi) \in H^n(X; \pi)$. By (*) it follows $v(\phi) = f^*(\iota_n)$ for all ϕ. Unravelling the definitions shows that the correspondence

$$u : \pi_i(map(X, K(\pi, n))_f, f) \longrightarrow H^{n-i}(X; \pi) \ : \ \phi \mapsto u(\phi)$$

is an isomorphism of abelian groups, yielding the result. □

We come next to Thom's second result. We suppose that

$$\pi : Y \longrightarrow B$$

is a Serre fibration and

$$f : X \longrightarrow Y$$

a fixed map. Then

$$p : map(X, Y)_f \longrightarrow map(X, B)_{\pi f}$$

where $p = map(X, -)$ is also a fibration. What is the fibre? The fibre over $\pi \cdot f$ is the space of lifts g in the diagram

$$\begin{array}{ccc} & \nearrow^{g} & Y \\ & & \downarrow \pi \\ X & \xrightarrow{\pi \cdot f} & B \end{array}$$

with the property that $g \in map(X, Y)_f$. To analyze this situation we convert such a lift to a cross-section by forming the pullback of $Y \downarrow^{\pi} B$ along f, giving the commutative diagram

$$\begin{array}{ccc} E & \longrightarrow & Y \\ s \uparrow\downarrow q & \nearrow^{g} & \downarrow \pi \\ X & \xrightarrow{\pi \cdot f} & B . \end{array}$$

It is elementary to show that the liftings g are in bijective correspondence with the cross-sections s. Next supose that the pullback fibration $E \downarrow^{\pi} X$ is trivial. Let F be the fibre of π, and hence also of q. Choosing a trivialization

$$\begin{array}{ccc} E & \longrightarrow & X \times F \\ & {\scriptstyle q}\searrow \quad \swarrow{\scriptstyle p_X} & \\ & X & \end{array}$$

one sees that the sections of $q : E \downarrow^{\pi} X$ are in bijective correspondence with a union of components of $map(X, F)$. When will the pullback fibration be trivial? Suppose for example that the original fibration $\pi : Y \longrightarrow B$ is in fact a principal G-bundle for a topological group G. Then $E \downarrow^{\pi} X$ is also a principal G-bundle. The lift f of $\pi \cdot f$ corrsponds to a section of $E \downarrow^{\pi} X$, which therefore is trivial, and hence we arrive at a form of Thom's second theorem.

THEOREM 1.2 (R. Thom): *Let X be a connected space, $\pi : Y \longrightarrow B$ a principal fibration with group G, and $f : X \longrightarrow Y$ a fixed map. Then*

$$map(X, Y)_f \xrightarrow{p} map(X, B)_{\pi \cdot f}$$

is a principal fibration, with fibre a union of components of $map(X, G)$. □

2. Analysis of $SHE(G/T)$.

Let G be a compact connected Lie group and $T \overset{\rho}{\hookrightarrow} G$ a maximal torus. We denote by $SHE(G/T)$ the component of the identity in $map(G/T, G/T)$. Let

$$G \longrightarrow G/T \xrightarrow{\eta} BT$$

be the usual principal bundle. Then by (1.2) Thom's second theorem

$$map(G/T, G)_c \longrightarrow SHE(G/T) \longrightarrow map(G/T, BT)_\eta$$

is a fibration, where $c : G/T \to G$ is the constant map. The space BT is an Eilenberg-MacLane space of type $K(\bigoplus_r \mathbb{Z}, 2)$ where r is the rank of T. Therefore by Thom's first theorem (1.1) we get a homotopy equivalence

$$map(G/T, BT)_\eta \cong K(H^0(G/T; H^2(BT; \mathbb{Z}), 2)) = BT$$

(since $H^1((G/T; \mathbb{Z}) = 0$). In particular $map(G/T, BT)_\eta$ is simply connected. This in turn implies from the homotopy exact sequence of the fibration that the fibre is connected. Since G is a topological group and since G/T is finite we have

$$(map(G/T, G)_c)_{\mathbf{Q}} = map(G/T, G_{\mathbf{Q}})_c$$

where $-_{\mathbf{Q}}$ denotes the Bousfield-Kan localization. By a theorem of H. Hopf

$$G_{\mathbf{Q}} = \times K(\mathbf{Q}, m_i)$$

with $r = rank(T)$ factors in the product, where the integers m_i are determined for example from

$$H^*(BG;\mathbf{Q}) = \mathbf{Q}[\rho_1,\ldots,\rho_r]$$

where $deg(\rho_i) = 2d_i = m_i + 1$. Since $H^{odd}(G/T;\mathbf{Q}) = 0$ any map

$$g : G/T \longrightarrow G_{\mathbf{Q}}$$

is null homotopic, so $map(G/T, G_{\mathbf{Q}})$ is connected. We may again apply (1.1) Thom's first theorem to conclude

$$map(G/T, G_{\mathbf{Q}}) = \mathop{\times}_{1\leq i\leq r}\ \mathop{\times}_{1\leq k\leq d_i} K(H^{2d_i-2k}(G/T;\mathbf{Q}), 2k-1).$$

To summarize we have:

PROPOSITION 2.1: *Let G be a compact connected Lie group and $T \hookrightarrow G$ a maximal torus. Then there is a principal fibration* $(*)$

$$\begin{array}{c} map(G/T, G_{\mathbf{Q}}) = \mathop{\times}_{1\leq i\leq r}\ \mathop{\times}_{1\leq k\leq d_i} K(H^{2d_i-2k}(G/T;\mathbf{Q}), 2k_i - 1). \\ \downarrow \\ SHE(G/T)_{\mathbf{Q}} \\ \downarrow \\ map(G/T, BT_{\mathbf{Q}})_\eta = K(H^0(G/T; H^2(BT;\mathbf{Q}), 2)) \end{array},$$

where $H^*(BG;\mathbf{Q}) = \mathbf{Q}[\rho_1,\ldots\rho_n]$ and $deg(\rho_i) = 2d_i$ □

Thus rationally $SHE(G/T)_{\mathbf{Q}}$ is completely determined by the homotopy exact sequence in the fibration (2.1), since being an $H-$space $SHE(G/T)_{\mathbf{Q}}$ is a product of Eilenberg-MacLane spaces. To determine the boundary map

$$\partial_* : \pi_2(map(G/T, BT_{\mathbf{Q}})_\eta) \longrightarrow \pi_1(map(G/T, G_{\mathbf{Q}}))$$

we note that it is equivalent to determine the map

$$\theta_* : \pi_2(map(G/T, BT_{\mathbf{Q}})_\eta) \longrightarrow \pi_2(Bmap(G/T, G_{\mathbf{Q}}))$$

where

$$\theta : map(G/T, BT_{\mathbf{Q}})_\eta \longrightarrow Bmap(G/T, G_{\mathbf{Q}})$$

is a classifying map for the bundle (*). We are gratefull to Michael Crabb for pointing out the following fact:

$$Bmap(-, G) = map(-, BG)_c$$

for any space - , where $map(-, BG)_c$ denotes the component of the constant map. To see this let

$$G \longrightarrow EG \longrightarrow BG$$

be a universal principal G–bundle (so EG is contractible). Then

$$map(-,EG)_c \longrightarrow map(-,BG)_c$$

is a fibration, and Thom's second theorem tells us that in fact the fibre is $map(-,G)$. Since $map(-,EG)$ is contractible we have shown

$$Bmap(-,G) = map(-,BG)_c$$

The classifying map for (*) is now easily seen to be the map

$$\theta : map(G/T,BT) \longrightarrow map(G/T,BG)_c$$

induced by $B\rho$, where $\rho : T \hookrightarrow G$ is the inclusion of a maximal torus. To compute θ_* on $\pi_2(-)$ we apply Thom theory as above to identify

$$\pi_2(map((G/T,BT))) = \mathbf{Z} \oplus \cdots \oplus \mathbf{Z}$$

with generators given as follows: Let

$$t_i : S^2 \longrightarrow BT \qquad i = 1,\dots,r$$

be generators of $\pi_2(BT)$ corresponding to a decomposition $T = S^1 \times \cdots \times S^1$. Then the maps

$$\tau_i : S^2 \times G/T \xrightarrow{t_i \times \pi} BT \times BT \xrightarrow{\mu} BT \qquad i = 1,\dots,r$$

are the adjoints of generators for $\pi_2(map(G/T,BT)_\eta)$, where μ is the H–space product in BT. Composing τ_i with $B\rho$ gives

$$B\rho \cdot \tau_i : S^2 \times G/T \longrightarrow BG \qquad i = 1,\dots,r$$

which is the adjoint of θ_* applied to the corresponding generator of $\pi_2(map(G/T,BT)_\eta)$. $BG_{\mathbf{Q}}$ is a generalized Eilenberg-MacLane space of type $\times_{i=1}^{r} K(\mathbf{Q},2d_i)$ with fundamental classes $\rho_1,\dots,\rho_r$. Thus the homotopy class of $B\rho \cdot \tau_i$ is completely determined by

$$(B\rho \cdot \tau_i)^*|_{G/T}(\rho_j) \in H^{2d_i-2}(G/T;\mathbf{Q})$$

where $i,j = 1,\dots r$ independently. To obtain an explicit formula we identify t_i with the corresponding dual cohomology class and we have:

LEMMA 2.2: *With the preceeding notations*

$$(B\rho \cdot \tau_i)^*(\rho_j) = \frac{\partial \rho_j}{\partial t_i}$$

PROOF: Let $\rho \in P[t_1,\dots,t_r]$ where the t_i are in degree 2. Give $P[t_1,\dots,t_r]$ the structure of a Hopf algebra by declaring the t_i to be primitive. This is the coalgebra structure on $H^*(BT)$ induced by μ. Let the comultiplication be denoted by Δ. Write

$$\Delta(\rho) = \sum_{r+s=d} \rho(r) \otimes \rho(s) \qquad deg\rho = 2d$$

where $\rho(j)$ has degree $2j$. Then we claim:

$$\rho(1) \otimes \rho(d-1) = \sum t_i \otimes \frac{\partial \rho}{\partial t_i}$$

To see this note it is enough by linearity of Δ to prove the formula for a monomial $t_1^{e_1} \cdot \cdot t_r^{e_r}$. In this case one computes:

$$\Delta\rho = \Delta(t_1^{e_1} \cdot \cdot t_r^{e_r}) = \Delta(t_1^{e_1}) \cdot \cdot \Delta(t_r^{e_r}) = (t_1 \otimes 1 + 1 \otimes t_1)^{e_1} \cdot \cdot (t_r \otimes 1 + 1 \otimes t_r)^{e_r}.$$

Expanding with the binomial theorem and collecting terms where the first component is of degree 2 yields:

$$\sum t_j \otimes e_j t_1^{e_1} \cdot \cdot t_j^{e_j - 1} \cdot \cdot t_r^{e_r} = \sum t_j \otimes \frac{\partial \rho}{\partial t_j}$$

as required. □

To put this discussion together into a concrete result we introduce the identifications:

$$\pi_2(map(G/T, BT)_\eta) = H^2(BT, \mathbb{Z})$$

$$\pi_2(map(G/T, BG_{\mathbb{Q}})_c) = \bigoplus_{i=1}^{r} H^{2d_i - 2}(G/T; \mathbb{Q})$$

obtained from Thom theory and have:

PROPOSITION 2.3: *With the precceding identifications the map*

$$\theta_* : \pi_2(map(G/T, BT)_\eta) \longrightarrow \pi_2(map(G/T, BG_{\mathbb{Q}})_c)$$

is given by

$$J : t_i \mapsto \left(\frac{\partial \rho_i}{\partial t_1}, \ldots, \frac{\partial \rho_i}{\partial t_r}\right).$$

Moreover this map is monic and

$$SHE(G/T)_{\mathbb{Q}} \cong \times_{i=1}^{r} \times_{k_i=1}^{d_i} K(H^{2d_i - 2k_i}(G/T; \mathbb{Q}), 2k_i - 1) \times K(H^{2d_i - 2}(G/T; \mathbb{Q})/ImJ, 1).$$

PROOF: We need only remark that [16]

$$det\left(\frac{\partial \rho_i}{\partial t_j}\right) \neq 0$$

and in fact represents the fundamental class of $H^*(G/T; \mathbb{Q})$. By Cramer's rule this implies that the vectors $(\frac{\partial \rho_i}{\partial t_1}, \ldots, \frac{\partial \rho_i}{\partial t_r}) \in \oplus_{i=1}^{r} H^{2d_i - 2}(G/T; \mathbb{Q})$ are linearly independent over **Q**. $\square$

Continuing our analysis of $SHE(G/T)$ we note that there is a map

$$\lambda : G \longrightarrow map(G/T, G/T)$$

given by left translation:

$$\lambda(g)(hT) = ghT.$$

Since G is connected and $\lambda(1) = 1$ it follows that

$$Im\lambda \subset map(G/T, G/T)_{id} =: SHE(G/T).$$

Consider the diagram

$$\begin{array}{ccc} & & map(G/T, G) \\ & & \downarrow \nu \\ G & \longrightarrow & map(G/T, G/T)_{id} \\ & & \downarrow p \\ & & map(G/T, BT)_{\eta} \end{array}$$

where the vertical sequence is a fibration up to homotopy. We claim that the image of λ lies in the image of the inclusion of the homotopy fibre

$$\nu : map(G/T, G) \longrightarrow map(G/T, G/T)_{id}.$$

To see this we return to Thom's analysis of the fibre map p and note that ν is given by

$$\nu(f)(gT) = f(gT)gT$$

where $f \in map(G/T, G)$ and $g \in G$. In other words: if $\chi_g : G/T \longrightarrow G$ denotes the constant function at $g \in G$ we have

$$\lambda(g) = \nu(\chi_g)$$

and so if we define

$$\chi : G \longrightarrow map(G/T, G)$$

by

$$\chi(g) = \chi_g$$

we obtain a commutative diagram

$$\begin{array}{ccc} & \overset{\chi}{\nearrow} \underset{e}{\swarrow} & map(G/T, G) \\ & & \downarrow \nu \\ G & \longleftarrow\!\!\!\longrightarrow & map(G/T, G/T)_{id} \\ & & \downarrow p \\ & & map(G/T, BT)_{\eta} \end{array}$$

where $e : map(G/T, G) \longrightarrow G$ is defined by $e(f) = f(1T)$. The following is now not hard to prove.

PROPOSITION 2.4: *Let G be a compact simply connected Lie group, then the map*

$$\lambda : G \longrightarrow SHE(G/T)$$

induces an epimorphism in rational cohomology.

PROOF: To show that λ^* is epic in rational cohomology, it is equivalent to show that λ_* is monic in homology. Since all the spaces and maps in the preceeding diagram are H-spaces and H-maps it follows from a theorem of Milnor and Moore [4] that it is equivalent to show λ_* is monic in rational homotopy. From (2.3) it follows that

$$\nu_* : \pi_i(map(G/T, G)_c, c) \otimes \mathbf{Q} \longrightarrow \pi_i(SHE(G/T), id) \otimes \mathbf{Q}$$

is monic for $i > 1$. By hypothesis G is simply connected so λ_* will be monic in rational homotopy if χ_* is. But clearly $e \cdot \chi = 1$ so χ_* is in fact split monic integrally. □

For a compact connected Lie group G with maximal torus $T \hookrightarrow G$, the map

$$\lambda : G \longrightarrow SHE(G/T)$$

may fail to be a monomorphism, because one has $\lambda(z) = id$ for elements z that satisfy

$$g^{-1} z g \in T \qquad \forall g \in G$$

But this says

$$z \in \bigcap_{g \in G} gTg^{-1} = \bigcap_{T' < G : T' max} T'$$

and hence $z \in Z_G$, where Z_G is the center of G. Thus the map λ factors

$$\lambda : G \xrightarrow{q} PG \xrightarrow{\tau} SHE(G/T)$$

where $PG := G/Z_G$ is the corresponding projective Lie group and τ is injective. If furthermore G is simply connected then Z_G is finite so q is a rational homotopy equivalence, and hence:

COROLLARY 2.5: *Let G be a compact simply connected Lie group with maximal torus $T \hookrightarrow G$, then*

$$\tau : PG \longrightarrow SHE(G/T)$$

induces an epimorphism in rational cohomology. □

The results (2.3) - (2.5) remain true upon passing to classifying spaces. Specifically one easily sees using the fact that these spaces are rational Eilenberg-MacLane spaces:

PROPOSITION 2.6: *Let G be a simply connected Lie group and $T \hookrightarrow G$ a maximal torus. Then the map*

$$BPG \longrightarrow BSHE(G/T)$$

induce epimorphisms in rational cohomology. If G is semisimple then

$$BG \longrightarrow BSHE(G/T)$$

induce epimorphisms in rational cohomology.

3. Finiteness Results.

DEFINITION: For a space X denote by $HE_0(X)$ the submonoid of $HE(X)$ of those homotopy equivalences of X that induce the identity map in integral homology.

PROPOSITION 3.1: *Let G be a compact connected Lie group and $T \hookrightarrow G$ a maximal torus. Then $\pi_0(HE_0(G/T))$ is finite.*

PROOF: Introduce the fibration

$$G \longrightarrow G/T \xrightarrow{\pi} BT$$

and consider the commutative diagram of Thom theory

$$\begin{array}{ccc} E & \longrightarrow & G/T \\ q\downarrow & \nearrow\phi & \downarrow\pi \\ G/T & \xrightarrow{\pi} & BT \end{array}.$$

Since π_* is epic in $\mathbb{Z}$ cohomology, any map ϕ must induce the identity map in $\mathbb{Z}$ cohomology. Thom theory then yields a fibration

$$map(G/T, G) \longrightarrow HE_0(G/T) \longrightarrow map(G/T, BT)_\pi$$

from which it follows that

$$\pi_0(HE_0(G/T)) = [G/T, G]$$

and thus we must show that $[G/T, G]$ is finite. To this end introduce the notation $X(0, \ldots, s)$ for the sth Postnikov section of a space X, i.e. one has

$$p_i : X \longrightarrow X(0, \ldots, s)$$

such that

$$(p_s)_* : \pi_i(X) \simeq \pi_i(X(0, \ldots, s)) \quad i \leq s$$

and

$$\pi_j(X(0, \ldots, s)) = 0 \quad j > s.$$

Since G/T is a finite complex it follows that

$$[G/T, G] \cong [G/T, G(0, \ldots, s)] \qquad for\ s >> dim(G/T).$$

Let us now proceed to show inductively that

$$[G/T, G(0, \ldots, s)]$$

is finite. The induction starts with $s = 1$. Since G is connected

$$G(0, 1) = S^1 \times \cdots \times S^1$$

and

$$[G/T, S^1 \times \cdots \times S^1] = H^1(G/T; \mathbf{Z} \oplus \cdots \oplus \mathbf{Z}) = 0$$

as G/T is simply connected. Consider next the inductive step. From the fibration

$$\begin{array}{c} G(0,\dots,s+1) \\ \downarrow \\ G(0,\dots,s) \end{array}$$

whose fibre is $K(\pi_{s+1}(G), s+1)$, we obtain an exact sequence

$$\begin{array}{ccc} [G/T, K(\pi_{s+1}, s+1)] & \cong & H^{s+1}(G/T; \pi_{s+1}) \\ \downarrow & & \\ [G/T, G(0,\dots,s+1)] & & . \\ \downarrow & & \\ [G/T, G(0,\dots,s)] & & \end{array}$$

By induction the bottom group is finite. For the top group there are two cases to consider.
- s+1 odd : if s + 1 is odd, then $H^{s+1}(G/T; -) = 0$, so certainly finite.
- s+1 even : then by the theorems of Hopf [3] and Serre [10] $\pi_{s+1} = \pi_{s+1}(G)$ is finite, so the upper group is finite.

Thus the top and bottom groups are always finite, and hence the same holds for the group in the middle. □

COROLLARY 3.2: *Suppose $\phi \in HE(G/T)$, then there are only finitely many homotopy classes $\theta \in HE(G/T)$ such that $\phi_* = \theta_*$ in integral homology.*

PROOF: This follows from the exact sequence

$$1 \longrightarrow HE_0(G/T) \longrightarrow HE(G/T) \longrightarrow Aut(H_*(G/T; \mathbf{Z})) \longrightarrow 1$$

and (3.1). □

Analagous to (3.1) we have:

PROPOSITION 3.3: *Let G be a compact connected Lie group and $T \hookrightarrow G$ a maximal torus. Then $\pi_0(HE_0((G/T)^\wedge_p)$ is finite.*

PROPOSITION 3.4: *For a compact connected Lie group G with a maximal torus $T \hookrightarrow G$, the natural map*

$$\pi : BSHE(G/T) \longrightarrow BHE_0(G/T)$$

is a rational equivalence.

PROOF: Note that we have an exact sequence

$$1 \longrightarrow SHE(G/T) \longrightarrow HE_0(G/T) \longrightarrow \pi_0(HE_0(G/T)) \longrightarrow 1$$

giving a covering

$$\begin{array}{c} BSHE(G/T) \\ \downarrow \pi \\ BHE_0(G/T) \end{array}$$

with covering transformations $\pi_0(HE_0(G/T))$. Note that $\pi_0(HE_0(G/T))$ acts trivially on $H^*(BSHE(G/T);\mathbf{Q})$ because:
- $\pi_*(BSHE(G/T))\otimes\mathbf{Q}$ is a functor of $H^*(G/T;\mathbf{Q})$ by Thom theory and $\pi_0(HE_0(G/T))$ acts trivially on $H^*(G/T;\mathbf{Q})$ so by functorality it therefore acts trivially on $\pi_*(BSHE(G/T))\otimes\mathbf{Q}$
- $H^*(BSHE(G/T);\mathbf{Q})$ is a functor of $\pi_*(BSHE(G/T))\otimes\mathbf{Q}$ and therefore $\pi_0(HE_0(G/T))$ acts trivially on $H^*(BSHE(G/T);\mathbf{Q})$.
By (3.1) $\pi_0(HE_0(G/T))$ is finite, so

$$\tilde{H}^*(\pi_0(HE_0(G/T);\mathbf{Q})) = 0.$$

Thus the Leray-Serre spectral sequence for π collapses to an isomorphism

$$H^*(BHE_0(G/T);\mathbf{Q}) \cong H^*(BSHE(G/T);\mathbf{Q})$$

as required. □

To complete the picture we mention the following proposition, which is due to S. Papadima [9; 1.3].

PROPOSITION 3.5: *Let G be a compact connected Lie group and $T\hookrightarrow G$ a maximal torus. Then $\pi_0(HE(G/T)$ is finite.*

PROOF: Consider the map

$$s : \pi_0(HE(G/T)) \longrightarrow Aut(H^*(G/T;\mathbf{Z}))$$

given by

$$[f] \mapsto f^*.$$

The fundamental class of $H^*(G/T;\mathbf{Z})$ may be written as the product over all the reflections in the Weyl group of G which span $H^2(G/T;\mathbf{Z})$ [15]. $H^2(G/T;\mathbf{Z})$ generates $H^*(G/T;\mathbf{Z})$ as an algebra. Therefore any automorphism of $H^*(G/T;\mathbf{Z})$ is a signed permutation of the reflections, and hence the group $Aut(H^*(G/T;\mathbf{Z})$ is finite. (In fact the case where G is simple the subgroup of orientation preserving automorphisms is isomorphic to the Weyl group of G : see [9] or [13].) The kernel of s is by definition $\pi_0(HE_0(G/T))$ which is finite by (3.1) and the result follows. □

REFERENCES.

1. Cooke, G.E., Lifting Homotopy Actions to Topological Actions, TAMS 237 (1978), 391-406

2. Dold, A. and R. Thom, Unendliche symmetrischen Potenzen und Quasifaserungen, Ann. of Math 58 (1960) 234 - 256.

3. Hopf, H. Über die Topologie der Gruppen-Mannigfaltigkeiten und ihre Verallgemeinerungen, Ann. of Math. 42 (1941), 22 - 52.

4. Milnor, J.W. and J.C. Moore, The Structure of Hopf Algebras, Annals of Math 81 (1964),

5. Moore, J.C. ,in Seminar Cartan 1954/55 Algèbres d'Eilenberg-MacLane et homotopie, Exposé 19 Theorem 6., W.A. Benjamin Inc. New York 1967.

6. Notbohm, D. and L. Smith, Fake Lie Groups and Maximal Tori I, preprint.

7. Notbohm, D. and L. Smith, Fake Lie Groups and Maximal Tori II, preprint.

8. Notbohm, D. and L. Smith, Fake Lie Groups and Maximal Tori III, preprint.

9. Papadima, S., Rigidity properties of compact Lie groups modulo maximal Tori, Math. Ann. 275 (1986), 637-652.

10. Serre, J.-P., Groupes d'homotopie et classes de groupes abéliens, Annals of Math 58 (1953) 258 - 294.

11. Smith, L., A Note on Realizing Complete Intersection Algebras as the Cohomology of a Space, Quaterly J. of Math. Oxford 33 (1982), 379-384.

12. Smith, L., Realizing Homotopy Actions by Topological Actions II, TAMS (to appear).

13. Speerlich, T., Automorpmismen von Ringen von Koinvarianten kristalographischen Gruppen, Staatsexamensarbeit Göttingen 1190.

14. Stasheff, J.D., A classification theorem for fibre spaces, Topology 2 (1963) 239-246.

15. Steinberg, R. , Invariants of Finite Reflection Groups, Cand. J.of Math. 12 (1960), 616 - 618.

16. Thom, R. L'Homologie des Espaces Fonctionels, Colloque de Topologie algébrique Louvain 1957.

Mathematisches Institut
Georg August Universität
Bunsenstraße 3/5
D 3400 Göttingen
WEST GERMANY

1990 Barcelona Conference
on Algebraic Topology.

RATIONAL COHOMOLOGY AND HOMOTOPY OF SPACES WITH CIRCLE ACTION

MARTIN RAUßEN

Abstract

This is the first of a series of papers with the aim to follow up the program initiated in [12], i.e., to investigate which classes of manifolds are symmetric in the sense that they allow some non-trivial group action. The general approach for constructing cyclic group actions on a given manifold M is to start with an action of the circle group $T = S^1$ on some manifold in the rational homotopy type of M and to "propagate" most of the restricted cyclic group actions to M itself. In [12], we only considered free actions, whereas the focus of this and the following paper [15] is on the non-free case.

This note is primarily concerned with a closer look at T-actions on Poincaré complexes and in particular with the interpretation of the Borel localization theorem in the language of deformations of algebras [14,2]. In the subsequent paper [15], we will show how to realize an abstractly given deformation in the rational homotopy category by a circle action on a manifold of the given rational homotopy type.

The first two sections of this paper should be of independent interest, whereas the short final ones are mainly to be considered as background material for [15].

1 Circle actions and cohomology

In this note, we make an explicit investigation of the rational cohomology and homotopy of finite CW-spaces with a circle action. We are mainly interested in manifolds (or Poincaré duality spaces) with circle action and the relations between rational cohomology and homotopy of the total space and the fixed point space. We begin by describing a hierarchy of actions with free actions as one extreme and the "totally non-homologeous-to-zero"case on the other. Cohomology will always be considered with rational coefficients.

Let X be a finite CW-complex upon which the circle $T = S^1$ acts. Let x denote the total cohomological dimension of X,

$$x := \dim_{\mathbf{Q}} H^*(X;\mathbf{Q}) = \sum_{i=0}^{\infty} \dim_{\mathbf{Q}} H^i(X;\mathbf{Q}).$$

The Borel construction associated with this action is defined as $X_T = X \times_T ET$ [4], where ET is a free contractable T-space and $BT = ET/T$ is the classifying T-space. Then $H^*(BT) \cong \mathbf{Q}[e]$, where e is a generator in $H^2(BT)$. Let $c : X_T \to B_T$ be the natural classifying map, such that c^* provides $H^*(X_T)$ with a graded $\mathbb{Q}[e]$-algebra structure. Furthermore, the T-fibration classified by c will be denoted by $p_X : X \simeq X \times ET \to X_T$.

Algebraic information about that fibration is contained in the Serre spectral sequence associated to it, and the latter boils down to the Gysin sequence

$$\cdots \to H^{*-1}(X) \xrightarrow{t_X} H^{*-2}(X_T) \xrightarrow{\cdot e} H^*(X_T) \xrightarrow{p_X^*} H^*(X) \xrightarrow{t_X} H^{*-1}(X_T) \cdots, \tag{1.1}$$

where p_X^* is a ring homomorphism, and t_X is an $H^*(X)$-module homomorphism via p_X^*,

$$t_X(y \cup p_X^* x) = t_X(y) \cup x,\ y \in H^*(X),\ x \in H^*(X_T). \tag{1.2}$$

Any graded $\mathbb{Q}[e]$-module M contains a graded torsion ideal $Tor(M) := \{u \in M \mid Ann(u) \neq 0\}$ and has the free graded $\mathbb{Q}[e]$-algebra $NT(M) := M/Tor(M)$ as a quotient.

The homomorphisms p_X^* and t_X from the Gysin sequence (1.1) give rise to the definition of the following algebraic objects associated with the action:

- $R(X) := p_X^*(H^*(X_T)) \subseteq H^*(X)$;
- $I(X) := p_X^*(Tor(H^*(X_T))) \subseteq R(X)$;
- $R(X)/I(X)$, and
- $H^*(X)/R(X)$.

Let r, resp. i denote the total $\mathbb{Q}$-vector space dimension of $R(X)$, resp. $I(X)$.

Proposition 1.3 *1. $R(X)$ is a graded subring of $H^*(X)$, $I(X)$ is a graded $R(X)$-ideal, $R(X)/I(X)$ inherits a graded ring structure, and $H^*(X)/R(X)$ is a graded $\mathbb{Q}$- vector space.*

2. The map p_X^* induces a ring epimorphism

$$NT(H^*(X_T)) \to R(X)/I(X).$$

3. The map t induces a $\mathbb{Q}$-linear isomorphism of odd negative degree

$$\tilde{t} : H^*(X)/R(X) \to I(X).$$

4. The dimensions satisfy the relation $x = r + i$.

Proof: The first two statements are immediate from the definitions. To prove (3), of which (4) is a corollary, consider the nil-selfmap "multiplication by e" on $Tor := Tor(H^*(X_T))$. Its Jordan- normal form gives rise to an explicit isomorphism of even non-negative degree

$$\Phi : coker(\cdot e \mid Tor) \xrightarrow{\cong} ker(\cdot e).$$

On a suitable (not necessarily graded) basis, this iso will be given as multiplication with various powers of e. The map $\tilde{t}$ above is defined as the composition

$$H^*(X)/R(X) \xrightarrow{t_X \cong} ker(\cdot e) \xleftarrow{\phi \cong} coker(\cdot e \mid Tor) \xrightarrow{p_X^* \cong} I(X).$$

□

Remark 1.4 One might define a T-action on X to be "of type j" if the dimensions of $I(X) \subseteq R(X)$ satisfy: $j = r - i$. Later, we shall see that j is just the total cohomological dimension of the fixed point set of the action. Hence, an action of type 0 is a (rationally) free action, whereas an action of type x is totally non-homologeous to zero. Examples of actions with types different from these extreme cases may be found in [7] and [1]. Observe that the type of the action is congruent to x mod 2. A refinement would be to regard $R(X)/I(X)$ as a $\mathbb{Z}/2$-graded $\mathbb{Q}$-vectorspace, i.e., to split j up into an even and an odd part. □

The information contained in (1.3) suffices to reconstruct $H^*(X_T)$ as a graded $\mathbb{Q}[e]$-*module* via a short free resolution of torsionfree $\mathbb{Q}[e]$-modules. Choose a $\mathbb{Q}$-linear section

$$s_X : R(X) \to H^*(X_T)$$

of p_X^*, such that $s_X(I(X)) \subseteq Tor(H^*(X_T))$. Since $H^*(X_T)$ is finitely generated as a $\mathbb{Q}[e]$-module (cf. Gysin sequence), one can show inductively, that there is an extension of s_X to a linear epimorphism

$$S_X : R(X)[e] \to H^*(X_T),$$

restricting to an epimorphism from $I(X)[e]$ to $Tor(H^*(X_T))$. It projects to an isomorphism of free $\mathbb{Q}[e]$-modules

$$\overline{S_X} : (R(X)/I(X))[e] \to NTH^*(X_T),$$

- which will be used in (1.8). Let $\bar{t}$ denote a lift of t in the left-hand side of the diagram

$$\begin{array}{ccccc} & & I(X)[e] & & \\ & & S_X \downarrow & & \\ & & Tor(H^*(X_T)) & & \\ & & \cup & & \\ \bar{t} : H^*(X)/R(X) \xrightarrow{t} & & ker(\cdot e) & \xleftarrow{\Phi \cong} coker(\cdot e \mid Tor) & \xrightarrow{p_X^* \cong} I(X). \end{array}$$

An inductive diagram argument as in [12], p.564, yields the resolution

$$0 \to \{H^*(X)/R(X)\}[e] \xrightarrow{e \cdot \bar{t}} R(X)[e] \to H^*(X_T) \to 0. \tag{1.5}$$

Instead, one may use the composition $\tau = \bar{t} \circ \tilde{t}^{-1} : I(X) \to I(X)[e]$, which is a lift of the right-hand side of the diagram, to get a resolution of the form

$$0 \to I(X)[e] \xrightarrow{e \cdot \tau} R(X)[e] \to H^*(X_T) \to 0. \tag{1.6}$$

Remark 1.7 More generally, an analysis as above can be carried out in the presence of an "algebraic Gysin sequence"

$$\cdots \mathcal{A}^{*-2} \xrightarrow{\cdot e} \mathcal{A}^* \to \mathcal{B}^* \xrightarrow{t} \mathcal{A}^{*-1} \xrightarrow{\cdot e} \mathcal{A}^{*+1} \to \cdots$$

connecting a graded $\mathbb{Q}[e]$-algebra $\mathcal{A}^*$ and a graded $\mathbb{Q}$-algebra $\mathcal{B}^*$. As a graded $\mathbb{Q}[e]$-*module*, $\mathcal{A}^*$ can be reconstructed from $I(\mathcal{B}^*), R(\mathcal{B}^*)$ and $\tilde{t} : \mathcal{B}^*/R(\mathcal{B}^*) \to I(\mathcal{B}^*)$. On the other hand, the self-map $\cdot e$ on $\mathcal{A}^*$ determines $\mathcal{B}^*$ as a *vector space*, including the ring-structure on the subspace $R(\mathcal{B}^*)$. The considerations on maps between $\mathbb{Q}[e]$-modules below could be carried out in this general framework, too.□

Next, we analyse what happens, when the selfmap $\cdot e$ on $H^*(X_T)$ is perturbed to $\cdot e - c$ for $c \in \mathbb{Q}^*$, which no longer has a $\mathbb{Z}$-, but still a $\mathbb{Z}/2$-graded significance. Since 0 is the only eigenvalue of the nilmap $(\cdot e \mid Tor)$, this new self-map is injective and hence an isomorphism on $Tor(H^*(X_T))$. Dividing out torsion yields, for any $c \in \mathbb{Q}^*$, an injection on the torsionfree quotient $NTH^*(X_T)$:

$$\begin{array}{ccccc} NTH^*(X_T) & \xrightarrow{\cdot e - c} & NTH^*(X_T) & & \\ \bar{S}_X \uparrow\cong & & \bar{S}_X \uparrow\cong & & \\ (R(X)/I(X))[e] & \xrightarrow{\cdot e - c} & (R(X)/(I(X))[e] & \xrightarrow{ev_c} & R(X)/I(X). \end{array} \tag{1.8}$$

Hence, for $c \neq 0$, the cokernel of $\cdot e - c$ on $H^*(X_T)$ is isomorphic to $R(X)/I(X)$ as a $\mathbb{Z}/2$-graded ring via evaluation of polynomials in e at c. Remark, that we have verified the preservation of the *index* under a deformation.

Given an equivariant map $f : X \to Y$ inducing $f_T : X_T \to Y_T$. The map $f_T^* : H^*(Y_T) \to H^*(X_T)$ commutes with both $\cdot e$ and $\cdot e - 1$, and hence there are commutative diagrams

$$\begin{array}{ccc} (R(Y)/I(Y))[e] & \xrightarrow{NT(f_T^*)} & (R(X)/I(X))[e] \\ \downarrow ev_i & & \downarrow ev_i \\ R(Y)/I(Y) & \xrightarrow{f_i} & R(X)/I(X) \end{array} \quad ,i = 0, 1. \tag{1.9}$$

Here, the lower f_0 is the graded ring homomorphism induced by $f^* : H^*(Y) \to H^*(X)$ on the subquotients. The quotient map f_1 obtained by dividing out the ideal generated by $e - 1$ is a linear map of even non-positive degree. It may be described as a deformation of f_0, i.e., there are $\mathbb{Q}$-linear maps $f_1^i : R(Y)/I(Y) \to R(X)/I(X)$ of degree $-2i$ such that $f_1 = f_0 + f_1^1 + f_1^2 + \cdots$. Remark that f_1 is not a ring homomorphism. On the other hand, it does fit into exact sequences of pairs and mappings:

Proposition 1.10 *Let $f : X \to Y$ denote a T-equivariant map inducing a long exact sequence*

$$\cdots \xrightarrow{\delta} H^*(f_T) \xrightarrow{i^*} H^*(Y_T) \xrightarrow{f^*} H^*(X_T) \xrightarrow{\delta} H^{*+1}(f_T) \to \cdots.$$

Then, the corresponding $\mathbb{Z}/2$-graded sequence of subquotients

$$\cdots \xrightarrow{\delta_1} R(f)/I(f) \xrightarrow{i_1} R(Y)/I(Y) \xrightarrow{f_1} R(X)/I(X) \xrightarrow{\delta_1} R(f)/I(f) \to \cdots$$

is exact. The maps i_1 and f_1 preserve and δ_1 reverses the $\mathbb{Z}/2$-grading.

Proof: As a subquotient of a complex, the sequence above is a complex, too. Let $z \neq 0$ be an element in the kernel of one of the maps in the sequence. It can be represented by a non-torsion element $\tilde{z}$ in the corresponding cohomology group, that maps to a torsion element, itself. Multipying with a suitable power of e, we obtain a new representant $z' = e^i \cdot \tilde{z}$ that maps trivially. Hence, z' and thus also z has a pre-image. □

Remark 1.11 The corresponding sequence containing i_0, f_0 etc. is in general *not* exact.

As a simple consequence, we have the well-known

Corollary 1.12 *If f^* is an isomorphism mod $\mathbb{Q}[e]$-torsion, then f_1 is a $\mathbb{Q}$-linear isomorphism.*

Proof: Under the conditions of the corollary, the $\mathbf{Q}[e]$-module $H^*(f_T)$ is torsion, whence $R(f)/I(f) = 0$. □

In other terms, the map $NT(f^*) : NTH^*(Y_T) \to NTH^*(X_T)$ is considered as a $\mathbf{Z}/2$-graded deformation between the algebras $R(Y)/I(Y)$ and $R(X)/I(X)$. From the Borel localization theorem [4,10], we have the following particularly important case:

Corollary 1.13 *If T acts on X with fixed point set F, the inclusion $j_T : F_T \hookrightarrow X_T$ induces a $\mathbf{Z}/2$-graded deformation between $R(X)/I(X)$ and $H^*(F)$. In particular, there is a $\mathbf{Z}/2$-graded linear isomorphism of non-positive degree $j_{T,1} : R(X)/I(X) \to H^*(F)$.*

Observe that (1.13) generalizes the well-known case of an action that is totally nonhomologous to zero [14] - or free! Let $f := dim_{\mathbf{Q}} H^*(F;\mathbf{Q})$ denote the total cohomological dimension of F. From (1.3.4) and (1.13), we get immediately:

Corollary 1.14

$$f = r - i; r = \frac{1}{2}(x + f); i = \frac{1}{2}(x - f).$$

2 T-actions on rational Poincaré duality spaces and cohomology

From now on, we assume X to be a rational Poincaré duality space [3], in particular, there has to be a fundamental class $[X] \in H_n(X)$ such that the *Poincaré duality form*

$$<,>: H^*(X) \times H^{n-*}(X) \to \mathbf{Q}, < u, v > = < u \cup v, [X] >$$

is *non-degenerate.*

Lemma 2.1 *Suppose $F = X^T$ is nonempty. Then*

1. *$I(X) \cap H^0(X) = \{0\}; I(X) \cap H^n(X) = \{0\}$.*
2. *$H^0(X) \cup H^n(X) \subset R(X)$.*
3. *Poincaré duality induces in a canonical way a non-degenrate form $<,>_0$ on $R(X)/I(X)$, which is equivalent to $<,>$ in the Witt ring $W(\mathbf{Q})$.*
4. *There is a non-degenerate $\mathbf{Z}/2$-graded bilinear form $\psi : I(X) \times I(X) \to \mathbf{Q}$ with the following properties:*
   - $| x | + | x' | \geq n \Rightarrow \psi(x, x') = 0$.
   - *If n is odd, then ψ is the sum of a non-degenerate form on both the odd and the even part $I^{odd}(X)$ and $I^{even}(X)$.*
   - *If n is even, then ψ is trivial when restricted to both $I^{odd}(X)$ and $I^{even}(X)$, and it pairs $I^{odd}(X)$ with $I^{even}(X)$.*

Proof:

1. If $H^0(X) \subset I(X)$, then $H^*(X_T)$ is a $\mathbf{Q}[e]$-torsion module, which in view of (1.13) contradicts to $F \neq \emptyset$. Let $j : F \hookrightarrow X$ and $j_T : F_T \hookrightarrow X_T$ denote inclusion maps, and consider the diagram

$$\begin{array}{cccccc} \cdots \rightarrow & H^*(X_T, F_T) & \stackrel{k_T^*}{\rightarrow} & H^*(X_T) & \stackrel{j_T^*}{\rightarrow} H^*(F_T) \cong H^*(F)[e] \rightarrow \cdots \\ & \downarrow p_M^* & & \downarrow p_X^* & \\ \cdots \rightarrow & H^*(X, F) & \stackrel{i^*}{\rightarrow} & H^*(X) & \rightarrow \cdots \end{array}$$

The ideal $I(X)$ is equal to $p_X^*(Tor(H^*(X_T))) = p_X^*(k_T^*(H^*(X_T, F_T)))$, since, by (1.13), $H^*(X_T, F_T)$ has to be a $\mathbf{Q}[e]$-torsion module. Hence it has to have finite dimension $n-1$, (compare [12], §2 and 4), which proves the second half of (1).

2. By (1.14), $\dim R(X) = \frac{1}{2}(x+f) > \frac{1}{2}x = \frac{1}{2}\dim H^*(X)$ for $F \neq \emptyset$. Hence, the restriction of $<,>$ to the subring $R(X)$ cannot be trivial, whence the cohomology fundamental class $[X] \in R(X)$.

3. By definition, $I(X) \cdot R(X) \subset I(X)$, and (1) implies: $< I(X), R(X) >= 0$, so that $<,>$ factors over a bilinear form $<,>_0$ on $R(X)/I(X)$. From (1.14), we conclude that $R(X) = I(X)^\perp$ and hence, that $<,>_0$ is non-degenerate. On the other hand, there is a decomposition $H^*(X) = \hat{I}(X) \oplus \hat{I}(X)^\perp$, where $\hat{I}(X)$ is a split inner product space with $I(X)$ as a maximal self-orthogonal subspace, such that $\hat{I}(X)^\perp \subset R(X)$, and such that $<,>|\,\hat{I}(X)^\perp$ is isomorphic to $<,>_0$.

4. Define $\psi : I(X) \rightarrow I(X)^*$ by completing the diagram

$$\begin{array}{ccc} H^*(X)/R(X) & = & H^*(X)/R(X) \\ \downarrow \tilde{t} & & \downarrow <,> \\ I(X) & \stackrel{\psi}{--\rightarrow} & I(X)^* \end{array}$$

where $<,>$ denotes the restriction of the Poincaré-duality form, and $\tilde{t}$ is the isomorphism from (1.3.3). The properties of ψ follow immediately from those of $\tilde{t}$ and $<,>$.

□

Relations between the Poincaré duality forms on X and on the fixed point set $X^T = F = \coprod_{i=1}^k F_i$ consisting of k connected components have been exhibited by Bredon[6], Chang and Skjelbred[8] and Wu-Yi Hsiang[10]. The main result is, that all the components of F have to be Poincaré complexes, too, and was established by Bredon in the "non-cohomologous to zero"-situation by an algebraic proof, whereas the proof of Chang and Skjelbred in the general case uses approximations of classifying spaces by manifolds. In fact, it is easy to check that Bredon's original proof can be copied almost word for word in the general case using $NTH^*(X_T)$ as a deformation from $R(X)/I(X)$ to $H^*(F)$, cf. (1.13) together with (2.1.3). The essential step is to show that $[F_i] \notin j_{T,1}(R(X)/I(X)^{<n})$. This establishes part (1) of

Proposition 2.2 *Let T act on the n-dimensional Poincaré-complex X with fixed point set $F = \coprod F_i \neq \emptyset$.*

1. *The components F_i are Poincaré complexes, too. Their dimensions n_i satisfy: $n_i \equiv n \bmod 2$.*

2. [10],p.65ff: *Let $[F_i]$ denotes a fundamental class of F_i. Then, there are classes $[X_i] \in H^*(X_T)$, such that $p_X^*([X_i]) = [X]$ and $j_T^*[X_i] = e^{\frac{n-n_i}{2}} \cdot [F_i]$.*

3. *The Poincaré form $<,>$ on $H^*(X)$ agrees with the sum of the Poincaré forms on the $H^*(F_i)$ in the Witt ring $W(\mathbf{Q})$. Hence, with properly chosen orientations, the index of X is equal to the sum of the indices of the F_i.*

4. *Let j_T^i denote the i-th component of the linear map $j_{T,1}$ from $H^*(X)$ into $H^*(F_i)$, introduced in (1.13). Then, for all $a \in H^*(X)$ with $j_T^i(a) \neq 0$, the following inequation holds: $| j_T^i(a) | \geq | a | -(n-n_i)$.*

Proof: (3) Let $j_{T,1} : R(X)/I(X)$ denote the linear isomorphism from (1.13). Define linear forms ϕ_1, ϕ_2 on $H^*(F)$ by

$$ker(\phi_1) = j_{T,1}(R(X)/I(X)^{<n}),\ \phi_1(j_{T,1}[X]) = 1, \text{ and}$$
$$(\phi_2)(H^{<n_i}(F)) = 0,\ \phi_2([F_i]) = 1, 1 \leq i \leq k.$$

Those give rise to bilinear forms $<,>_j$ on $H^*(F), j = 1,2$, defined as

$$<,>_j\colon H^*(F) \times H^*(F) \xrightarrow{\cup} H^*(F) \xrightarrow{\phi_i} \mathbf{Q}.$$

Both forms are easily seen to be non-degenerate. They agree to each other on those $(x,y) \in H^*(F) \times H^*(F)$ for which either $x \in H^*(F_i), y \in H^*(F_j), i \neq j$ or $x,y \in H^*(F_i), | x | + | y | \geq n_i$, since $\Phi_1[F_i] = 1$. Hence, they are equivalent by a Gram-Schmidt orthogonalization argument. Furthermore, $<,>_1$ is equivalent to $<,>_0$ on $R(X)/I(X)$, since $< x,y >_0 = < j_{T,1}(x), j_{T,1}(y) >_1$ for $x,y \in R(X)/I(X)$ and $| x | + | y | \leq n$. The result follows now from (2.1).

(4) Under the given assumptions, there is an element $b \in H^*(X)$, such that $j_T^i(a) \cup j_T^i(b) = [F_i]$. Then, (2) shows that $a \cup b = [X]$+terms of lower degree. □

In the following proposition, we shall gather a bunch of numerical conditions that can be extracted from the results obtained so far. They only reflect the "graded linear algebra level", which in general should be considered as a set of primary obstructions to the existence of T-actions with given properties. On the other hand, for cohomologically simple spaces as the LC2-manifolds of [12], these primary obstructions suffice.

Let X be an n-dimensional Poincaré-complex with fixed point set $X^T = F$. Let $I(X) \subset R(X) \subset H^*(X)$ be given as in (1.3). Let $I^k(X) = I(X) \cap H^k(X)$, $R^k(X) = R(X) \cap H^k(X), i_k(X) = dim_{\mathbf{Q}}(I^k(X)), r_k(X) = dim_{\mathbf{Q}}(R^k(X)), b_k(X) = dim_{\mathbf{Q}} H^k(X;\mathbf{Q})$.

Proposition 2.3 *The dimensions above have to satisfy the following (in)equalities:*

1.

$$\sum_{k \equiv \epsilon \bmod 2} r_k(X) - i_k(X) = \sum_{k \equiv \epsilon \bmod 2} b_k(F);$$

$$\sum_{k \leq r, k \equiv \epsilon \bmod 2} b_k(F) \geq \sum_{k \leq r, k \equiv \epsilon \bmod 2} r_k(X) - i_k(X) \geq \sum_{k \leq r, k \equiv \epsilon \bmod 2} b_{n-k}(F).$$

2.

$$i_{n-k}(X) = b_k(X) - r_k(X);\ r_{n-k}(X) = b_k(X) - i_k(X).$$

3.

$$\sum_{k \le r, k \equiv \epsilon \bmod 2} b_k(X) \le \sum_{k < r, k \not\equiv \epsilon \bmod 2} i_k(X) + \sum_{k \le r, k \equiv \epsilon \bmod 2} r_k(X).$$

Proof: (1) Corollary (1.13) implies the equality and the first inequality. There is a reverse inequality summing over $n - k \le r$, which together with Poincaré duality (2.1.3) on $R(X)/I(X)$ yields the last inequality.

(2) is an immediate consequence of the splitting of the Poincaré form on $H^*(X)$ into one on $R(X)/I(X)$ and another one that pairs $H^*(X)/R(X)$ with $I(X)$, see the proof of (2.1.3).

(3) Finally, the properties of the non-degenerate form ψ on $I(X)$ in the proof of (1.13.4) show (as in [12], Satz 2.3) that: $\sum_{k \le r, k \equiv \epsilon \bmod 2} i_{n-k}(X) \le \sum_{k<r, k \not\equiv \epsilon \bmod 2} i_k(X)$. Combine this with (2.3.2). □

3 The rational decomposition of a smooth T-manifold

In a certain sense, a T-action on a smooth manifold X rationally decomposes into a trivial action on the fixed point set F and a free action on the complement, which are pasted together by a free action on the link of F in X. Since we plan to construct actions by pasting a (homotopically) trivial action to a free action in [15], we have to study this decomposition in some detail:

Let again T act smoothly on the smooth n-dimensional manifold X with fixed point set $F = \coprod F_i$, where the component F_i is a smooth manifold of dimension n_i. Let $\nu = \nu(F, X) = \coprod \nu_i = \coprod \nu(F_i, X)$ denote the normal bundle to the fixed point set in X. Furthermore, let $M = X \setminus F$ denote the complement of the fixed point set, and $i : M \to X$ denote its inclusion into X.

Let us start with the action on a tubular neighborhood U of (a component of) the fixed point set F_i, being equivariantly diffeomorphic to a disk bundle (in some equivariant Riemannian metric) $D\nu$ of the normal bundle ν of F_i in X. Remark, that ν splits according to the non-trivial representations of T on the fibres, and hence supports a complex structure (unique up to choice of orientations). The quotient of the fixedpointfree action on the sphere bundle $S\nu$ is a fibre bundle with fibre a twisted complex projective space SV/T in the terminology of [11]. In that paper, T.Kawasaki also shows, that the latter is homeomorphic to the quotient of an honest complex projective space by the action of a finite cyclic group and hence rationally cohomology-equivalent to a such. It is not difficult to see, that a twisted complex projective space is simply-connected, and hence is rationally homotopy equivalent to an honest complex projective space.

On the other hand, there is a maximal finite cyclic group H containing all the isotropy groups of the T-action (on each fibre of ν). The quotient $S\nu/H$ of this action is a fibre bundle with fibre SV/H, which is a rational homotopy sphere again. Now, the circle group acts fibrewise and freely as $T' = T/H$ on $S\nu/H$ with quotient the fibre bundle

$S\nu/T$ over F with fibre SV/T, the twisted complex projective space from above. If $c: S\nu/T \to BT$ denotes the classifying map of the T' fibre bundle $S\nu/H \to S\nu/T$, we still call $e = c^*(e) \in H^*(S\nu/T)$. By the Leray-Hirsch theorem, $H^*(S\nu/T)$ is a free $H^*(F)$-module with base $1, \ldots e^{\frac{n-n_i-2}{2}}$. Furthermore, one may, as in the classical situation, define Chern classes $c_j(\nu_i) \in H^{2j}(F_i)$ by the relation $e^{\frac{n-n_i}{2}} = -\sum_{2(j+k)=n-n_i} c_j(\nu_i) \cdot e^k$. Of course, these Chern classes are just the usual Chern classes $c_j(\nu) \in H^{2j}(F_i)$ when the T-action is semifree.

The T-action on the complement $M = X\backslash F$ has only finite isotropy groups; rationally, it cannot be distinguished from a free action. In particular, $M_T \sim M/T$ is a rational cohomology manifold with boundary $S\nu/T$. As a result, we show that the deformation map $j_T^*: H^*(X_T) \to H^*(F_T)$ has to satisfy certain properties expressed in terms of the Chern classes defined above. Below, we list some properties needed in the constructive part in [15] related to the rational cohomology of M_T in terms of X_T and F_T. Here $H_*(-) = Hom(H^*(-); \mathbf{Q})$.

Proposition 3.1 *1. $H^*(M_T; S\nu/T) \cong H^*(X_T, F_T)$; both have a fundamental class in dimension n-1.*

2. $H^*(M_T) \cong H_{n-1-*}(X_T, F_T)$.

3. $i_T^*: H^*(X_T) \to H^*(M_T)$ is onto.

4. The $H^*(F_T)$-ideals C_i generated by the total Chern classes $\sum_{2(j+k)=n-n_i} c_j(\nu_i) \cdot e^k$ are - for every component F_i - contained in the image of $j_T^*: H^*(X_T) \to H^*(F_T)$.

5. For every $z \in H^*(X_T)$ with $z \cup H^{n-1-*}(X_T, F_T) = 0$, the element $j_{iT}^* z \in H^*(F_{iT})$ is contained in the ideal C_i.

Proof: The first two statements are elementary using excision, resp. Alexander duality; the third uses diagram (4.2) . The last two stem from the cohomology of the homotopy commutative diagram

$$\begin{array}{ccccc} S\nu/T \sim & (S\nu)_T & \hookrightarrow & (D\nu)_T & \sim F_T \\ & \downarrow & & \downarrow j_T & \\ & M_T & \xrightarrow{i_T} & X_T, & \end{array}$$

to wit

$$\begin{array}{ccccccc} 0 \to & H^*(X_T, M_T) & \to & H^*(X_T) & \xrightarrow{i_T^*} & H^*(M_T) & \to 0 \\ & \downarrow\cong exc. & & \downarrow j_T^* & & \downarrow & \\ 0 \to & H^*(F_T, S\nu/T) & \to & H^*(F_T) & \to & H^*(S\nu/T) & \to 0 \end{array}$$

The ideals C_i form precisely the kernel of $H^*(F_T) \to H^*(S\nu/T)$, which is the image of $H^*(F_T, S\nu/T)$ in $H^*(F_T)$. In view of (2), the requirement about z in (5) is equivalent to $i_T^* z = 0$. □

4 Smooth T-manifolds and Pontrjagin classes

When X is a rational cohomology manifold, then Pontrjagin classes might impose more cohomological restrictions to either the existence of a non-trivial circle action on X at all, or to one with specific properties. Here, we will discuss a particular case: We suppose the circle group T to act *semifreely* on a *smooth* manifold X such that the inclusion $j : F \hookrightarrow X$ of the (non-empty) fixed point set F is nul-homotopic. Let M denote the free T-manifold $X \setminus F$. Remark that $i^* : H^*(X) \to H^*(M)$ is injective in every but the top dimension. Let $\mathcal{P}(X) \subset H^*(X)$ denote the Pontrjagin algebra generated by the Pontrjagin classes $p_i(X) \in H^*(X)$.

Proposition 4.1 *1. $\mathcal{P}(X) \subset R(X)$.*

2. $p \in \mathcal{P}(X) \cap I(X) \Rightarrow$ all Pontrjagin numbers involving p vanish.

Proof: As before, let $\nu = \nu(F, X)$ denote the normal bundle of F in X. Since T is supposed to act semifreely on X, it acts freely on M and hence $S\nu$. Inclusion of the sphere bundle into the disk bundle induces a map $c : \mathbf{C}P\nu/T \to (D\nu)_T \sim F_T$ on the Borel space level. Since the action is free on M, $M_T \simeq M/T$, and the tangent bundles satisfy: $TM \cong p_M^*(T(M/T)) \oplus \mathbf{R}$, and hence, $i^*(p_k(X)) = p_k(M) = p_M^*(p_k((M/T)))$. Inspect the following diagram:

$$\begin{array}{cccccc}
 & H^*(F_T) & \xrightarrow{c^*} & H^*(\partial M/T) \cong H^*(\mathbf{C}P\nu) & \xrightarrow{0} & H^{*+1}(F_T, \mathbf{C}P\nu) \\
 & \uparrow j_T^* & & \uparrow k_T^* & & \uparrow \cong \ exc. \\
(4.2) & H^*(X_T) & \xrightarrow{i_T^*} & H^*(M_T) & \longrightarrow & H^{*+1}(X_T, M_T) \\
 & \downarrow p_X^* & & \downarrow p_M^* & & \\
 & H^*(X) & \xrightarrow{i^*} & H^*(M) & &
\end{array}$$

Observe, that i_T^* is surjective, while i^* is injective apart from the top dimension. Together with (2.1.2) this shows, that all Pontrjagin classes are induced from $H^*(X_T)$. Finally, (2) follows immediately from (2.1.1). □

Let us note the numerical consequences of (4.1), which have to be added to those of (2.3). We use the following abbreviations: $\mathcal{P}_k(X) = \mathcal{P}(X) \cap H^k(X), a_k(X) = dim_{\mathbf{Q}} \mathcal{P}_k(X)$, $a'_k(X) = dim_{\mathbf{Q}}(\mathcal{P}_k(X) \cap I(X))$:

Corollary 4.3 *Under the conditions above, the following inequalities hold:*

1. $a_k(X) \leq r_k(X), a'_k(X) \leq i_k(X)$.

2. $a_k(X) - a'_k(X) \leq r_k(X) - i_k(X)$. □

Finally, we shall see that the map j_1 derived from j_T^* as in (1.9) has to map the Pontryagin classes of X into certain combinations of the tangential and normal Chern classes c_j and $\bar{c}_j$ of F : Since $\partial(M/T) = \mathbf{C}P\nu$, the Pontryagin classes satisfy: $p_i(\mathbf{C}P\nu) = k_T^*(p_i(M/T))$. Let $ev_1 : H^*(F_T) \to H^*(F)$ denote evaluation at $e = 1$ and $\bar{C}$ denote the $\mathbf{Q}$-submodule of $H^*(F)$ generated by the normal Chern classes $\bar{c}_j$ of F. Then a diagram chase in (4.2) shows:

Proposition 4.4 $j_1(p_i(X)) - ev_1(p_i(\mathbf{C}P\nu)) \in j_1^*((R(X)/I(X))^{\leq 4i-2}) + \bar{C}^{\leq 4i-2}$. □

If F consists of several components, this result has to be applied to the maps into every component. To apply (4.4), one has to know the Pontryagin classes of $\mathbf{C}P\nu$. While explicit calculations seem to be hard in general, the following might help, in particular in low-dimensional cases:

Remark 4.5 Let $r = dim_{\mathbf{C}}\nu$, let $q : \mathbf{C}P\nu \to F$ denote the bundle projection, and let γ_ν denote the canonical line bundle over $\mathbf{C}P\nu$. Then,

1. $T\mathbf{C}P\nu \oplus \mathbf{C} \cong \gamma_\nu \otimes q^*\nu \oplus q^*TF$.

2. $C(\gamma_\nu \otimes \nu) = \sum_{s=0}^{r} t^s \cdot \sum_{i+j=s} \begin{pmatrix} r-j \\ i \end{pmatrix} \cdot \bar{c}_j$.

3. $C(\mathbf{C}P\nu) = \sum_s t^s \cdot \sum_{i+j+k=s} \begin{pmatrix} r-j \\ i \end{pmatrix} \cdot e^i \cdot \bar{c}_j \cdot c_k$.

Pontryagin classes may be calculated with the aid of Corollary (15.5) in [13]. In the case treated in this section, one calculates for example:

$$\begin{aligned} p_1(\mathbf{C}P\nu) &= r^2 \cdot e^2 \\ p_2(\mathbf{C}P\nu) &= \frac{r^2 \cdot (r^2-1)}{12} \cdot e^4 + (c_1^2 - r \cdot (c_2 - \bar{c}_2)) \cdot e^2. \end{aligned}$$

□

5 Rational homotopy

In rational homotopy theory, relations between models of the total space of a T-action and its fixed point space are even closer and more straightforward. In the language of Sullivan minimal models [16,5,9], they have been studied in detail by C.Allday [2]. For the convenience of the reader, we will briefly summarize some relevant notions and results from that article and note an easy consequence, which will be used in a subsequent paper [15].

Let $(\mathcal{M}(X), d)$ denote the Sullivan differential graded algebra minimal model of (the rationalization $X_{(0)}$ of) a simply-connected space X (nilpotent suffices!). The Sullivan minimal model $\mathcal{M}(BS^1)$ is the polynomial algebra $R = \mathbf{Q}[e]$ in one generator e of degree 2 with trivial differential. Let $K = \mathbf{Q}(e)$ denote its field of fractions.

Let T act on a simply-connected space X. The Sullivan minimal model of the Borel space X_T of the action is intimately related to that of X . From [9], one deduces a graded algebra isomorphism

$$\mathcal{M}(X_T) \cong \mathcal{M}(X) \otimes R.$$

The differential d_T on $\mathcal{M}(X_T)$ is trivial on the second factor, whereas it is "twisted" on the first, i.e., there is a derivation $\tau : \mathcal{M}(X) \to \mathcal{M}(X) \otimes R$ of degree -1 such that $d_T(x) = d(x) + e \cdot \tau(x), x \in \mathcal{M}(X)$. See [12] for details.

Let $F = X^T = F_1 \cup \cdots F_k$ denote the (components of the) fixed point set of the action. Since $F_T = F \times BS^1$, the differential d_T on $\mathcal{M}(F_T)$ is untwisted. Relations between $\mathcal{M}(X_T)$ and $\mathcal{M}(F_T)$ can be formulated after passing to $\mathbf{Z}/2$-*graded* augmented Koszul-Sullivan(KS) differential algebras: In [2], Allday develops a homotopy theory of such objects similar to those in ordinary rational homotopy theory. In this context, the most important example of such a $\mathbf{Z}/2$-graded object is $\mathcal{M}(X_T) \otimes_R K$, which as a graded algebra is equal to $\mathcal{M}(X) \otimes K$.

Let ε_X, resp. ε_i denote the standard augmentations on $\mathcal{M}(X)$, resp. $\mathcal{M}(F_{iT}), 1 \leq i \leq k$, and let ε_F on $\mathcal{M}(F_T)$ be given by $\varepsilon_F(\sum x_i) = \frac{1}{k} \cdot \sum \varepsilon_i(x_i), x_i \in \mathcal{M}(F_{iT})$. Then, $j_T \otimes id : \mathcal{M}(X_T) \otimes K \to \mathcal{M}(F_T) \otimes K$ is an augmentation preserving homomorphism of dgas in Alldays category, and the Borel localization theorem and Prop. 2.3 of [2] imply:

Proposition 5.1 *The map*

$$j_T \otimes id : \mathcal{M}(X_T) \otimes K \to \mathcal{M}(F_T) \otimes K$$

is a weak homotopy equivalence of $\mathbf{Z}/2$*-graded KS differential algebras. Hence, it has a homotopy inverse in that category.*

Remark 5.2 By more sophisticated use of his category, Allday is able to relate the rational homotopy groups of X and of those of the components of F.

References

[1] C.Allday, Torus actions on a cohomology product of three odd spheres, Trans. Amer. Math. Soc. **203**, 343-358(1975)

[2] C.Allday, On the rational homotopy of fixed point sets of torus actions, Topology **17**, 95-100(1978)

[3] J.Barge, Structures différentiables sur les types d'homotopie rationelle simplement connexes, Ann.scient.Éc.Norm.Sup., 4e série, **9**, 469-501(1976)

[4] A.Borel et al.: Seminar on Transformation Groups. Ann. of Math. Studies **46**, Princeton, N.J.: Princeton University Press 1961

[5] A.K.Bousfield, V.K.A.M.Gugenheim, On PL De Rham theory and rational homotopy type, Memoirs of the Amer. Math. Soc. **179**, 1976

[6] G.E.Bredon, The cohomology ring structure of a fixed point set, Ann. of Math. **80**, 524-537(1964)

[7] G.E.Bredon, Introduction to compact transformation groups, New York: Academic Press 1972

[8] T.Chang, T.Skjelbred, The topological Schur lemma and related results, Ann. of Math. **100**, 307-321(1974)

[9] S.Halperin, Lectures on minimal models, Mémoire de la Soc. Math. de France **9/10**, 1983

[10] W.Y.Hsiang, Cohomology Theory of Topological Transformation Groups, Ergebnisse der Mathematik und ihrer Grenzgebiete **85**, Berlin, Heidelberg, New York: Springer-Verlag 1975

[11] T.Kawasaki, Cohomology of twisted projective spaces and lens complexes, Math. Ann. **206**, 243-248(1973)

[12] P.Löffler, M.Raußen, Symmetrien von Mannigfaltigkeiten und rationale Homotopietheorie, Math.Ann. **272**, 549-576(1985)

[13] J.W.Milnor, J.D.Stasheff, Characteristic Classes, Ann. of Math. Studies **76**, Princeton, N.J.: Princeton University Press 1974

[14] V.Puppe, Cohomology of fixed point sets and deformations of algebras, manuscripta math. **23**, 343-354(1978)

[15] M.Raußen, Symmetries on manifolds via rational homotopy theory, in preparation.

[16] D.Sullivan, Infinitesimal computations in topology, Publ. IHES **47**, 269-331(1977)

Institut for Elektron Systemer
Aalborg University
Strandvejen 19,
DK-9000 Aalborg,
Denmark

1990 Barcelona Conference
on Algebraic Topology.

ON THE ACTION OF STEENROD POWERS ON POLYNOMIAL ALGEBRAS

CHEN SHENGMIN AND SHEN XINYAO

In [P] F. Peterson made a conjecture about the degrees d of generators for the polynomial algebra

$$S_2 = Z/2[u_1, \cdots, u_n] \qquad (1)$$

in n variables, viewed as a module over the Steenrod algebra $\mathcal{A}_2$, where S_2 is identified with the cohomology of the product of n copies of $RP^\infty$with coefficients in $Z/2$. To explain the Peterson Conjecture we say that d is a Peterson n-number if

$$d = \sum_{i=1}^{n} (2^{\lambda_i} - 1),$$

where $\lambda_i \geq 0$. Alternatively expressed the condition states that $\alpha(d+n) \leq n$, where the α function counts the number of ones in the binary expansion of its argument.

Peterson Conjecture. As a module over the Steenrod algebra, $Z/2[u_1, \cdots, u_n]$ is generated by monomials whose degrees are Peterson n-numbers.

A number of people have looked at the Peterson Conjecture. But at last it was solved by R.M.W. Wood. For stating his result we write $\mathrm{Im}\mathcal{A}_2^+ \subset S_2$ for the vector subspace of "hit" elements: the subspace of those elements expressible as a finite sum $\sum_{i>0} Sq^i x_i$, for appropriate $x_i \in S_2$. Now the Peterson Conjecture can be stated as follows. Let f be a monomial of degree d. If $\alpha(d+n) > n$ then $f \in \mathrm{Im}\mathcal{A}_2^+$.

In [W] Wood proved a stronger statement

Theorem. Let f be a monomial of degree d and suppose that e of its exponents are odd. If $\alpha(d+e) > e$ then f is a hit element, i.e., $f \in Im\mathcal{A}_2^+$.

In [S] W. M. Singer proved another result which generalizes the Peterson Conjecture and identifies a new class of monomials in $\mathrm{Im}\mathcal{A}_2^+$.

Now we consider the similar problem for odd primes. In this case, instead of (1) we have the polynomial algebra

$$S_p = Z/p[t_1, \cdots, t_n]$$

in n variables, viewed as a module over the Steenrod algebra $\mathcal{A}_p$, where S_p is identified with the cohomology of the product of n copies of CP^∞ with coefficients in Z/p. We write $\mathrm{Im}\mathcal{A}_p^+$ for the vector subspace of "hit" elements: the subspace of those elements expressible as a finite sum $\sum_{i>0} P^i x_i$, for appropriate $x_i \in S_p$. Now we are concerned with the problem of determining this subspace.

If $m \geq 0$ is an integer, write $m = \sum_{i\geq 0} \alpha_i(m)p^i$ for its p-adic expansion. Now we have two functions $\alpha(m)$ and $\alpha[m]$:

$$\alpha(m) = \sum_{i\geq 0} \alpha_i(m),$$

$$\alpha[m] = \text{the number of elements of } \{\alpha_i(m) | \alpha_i(m) > 0\}.$$

Both of them are analogues of $\alpha(m)$ in the odd prime case.

First we discuss the conditions which are expressed by function $\alpha[m]$.

It is easy to see that the Peterson Conjecture in this case is not true in general. But we have

Theorem 1. In the case $n = 1$, i.e., $S_p = Z_p[t]$, then $t^a \in \mathrm{Im}\mathcal{A}_p^+$ if and only if $\alpha[a+1] > 1$. Obviously. Theorem 1 i s equivalent to

Theorem 1′. $t^a \notin \mathrm{Im}\mathcal{A}_p^+$ iff $\alpha[a+1] = 1$.

This theorem can be proved by induction on the exponent of p when $a+1$ is expressed as a power of p.

If $n > 1$, the condition $\alpha[a+n] > n$ is not the characteristic condition for $t_1^{a_1} \cdots t_n^{a_n} \in \mathrm{Im}\ \mathcal{A}_p^+ (a = \sum_{i=1}^{n} a_i)$.

So we ask the following

Problem. What is the characteristic condition for hit elements in the case $n > 1$?

Next we turn to the conditions which are expressed by function $\alpha(m)$.

Here the situation is the same as above. That is the Peterson Conjecture is not valid again. For example, $n = 1, f = t_1^{2p^2-1}$. But we have the following

Theorem 2. Suppose $x = t_1^{a_1} \cdots t_n^{a_n} \in S_p$ is a monomial of degree $|x| = \sum_{i=1}^{n} a_i$. Let $a_i = pq_i + r_i, 0 \leq r_i < p, i = 1, \cdots, n$. Suppose $\alpha(\frac{|x|-\alpha_0(x)}{p}(p-1) + \alpha_0(x)) > \alpha_0(x)$, where $\alpha_0(x) = \sum_{i=1}^{n} \alpha_0(a_i) = \sum_{i=1}^{n} r_i$. Then x is hit.

This theorem is an analogue of Wood's theorem in the odd prime case. The proof is also similar. Of course, instead of

$$\chi(Sq^i)u = \begin{cases} u^{2^m}, & \text{if } i = 2^m - 1 \text{ for some } m \geq 0, \\ 0 & \text{otherwise,} \end{cases}$$

where χ is the canonical antiautomorphism of the Steenrod algebra, we have to use the following formula

$$\chi(P^i)t = \begin{cases} t^{p^m}, & \text{if } i = \gamma(m), \\ 0, & \text{otherwise,} \end{cases}$$

where $\gamma(m) = \frac{p^m - 1}{p-1}$.

It is not difficult to prove

Proposition 3. $\alpha(\delta(p-1)+e) \leq e$ if and only if δ can be written in the form $\delta = \sum_{i=1}^{e} \gamma(n_i)$ for appropriate $n_i \geq 0$.

We say "δ is e-sharp" if $\delta = \sum_{i=1}^{e} \gamma(n_i)$ for appropriate $n_i \geq 0$ and the sequence $\{n_1, \cdots, n_e\}$ a "representation of δ as an e-sharp".

So we ask: For the monomial x, if $\frac{|x| - \alpha_0(x)}{p}$ is $\alpha_0(x)$-sharp, when x is hit?

Definition 4. The sequence $\{m_1, \cdots, m_e\}$ is called the minimal representation of δ as a e-sharp if, in addition to $\delta = \sum_{i=1}^{e} \gamma(m_i)$ we have

$$\begin{aligned} m_1 &= \cdots = m_{i_1} > m_{i_1+1} = \cdots = m_{i_1+i_2} > \cdots > m_{i_1+\cdots+i_{k-1}+1} \\ &= \cdots = m_{i_1+\cdots+i_k} > m_{i_1+\cdots+i_k+1} = \cdots = m_e = 0, \end{aligned}$$

where $\max\{i_1 \cdots, i_{k-1}\} < p, i_k \leq p$.

Note that the minimal representation is "unique".

If $x \in S_p$ is a monomial, say $x = t_1^{a_1} \cdots t_n^{a_n}$, define $\alpha_i(x)$ to be the integer $\alpha_i(x) = \sum_{j=1}^{n} \alpha_i(a_j), i \geq 0$.

Lemma 5. Let δ be e-sharp with minimal representation $\{m_1, \cdots, m_e\}$. Let $\{q_1, \cdots, q_e\}$ also be a representation of δ as an e-sharp, ordered so that $q_1 \geq q_2 \geq \cdots \geq q_e$. Let q_{i_o} be the smallest of the numbers $\{q_i\}$ for which $q_i \neq m_i$ (if there is such a number). Then $q_{i_o} > m_{i_o}$.

The following lemma is a paraphrase of Lemma 5 that does not assume any special ordering of the $\{q_i\}$.

Lemma 6. Let δ be e-sharp, with minimal representation $\{m_1, \cdots, m_e\}$. Let $\{q_1, \cdots, q_e\}$ also be a representation of δ as an e-sharp. Then there is a unique integer

$d, 0 \leq d \leq \infty$, such that

$$\alpha_i(t_1^{p^{q_1}} \cdots t_e^{p^{q_e}}) = \alpha_i(t_1^{p^{m_1}} \cdots t_e^{p^{m_e}}), \qquad i < d,$$
$$\alpha_d(t_1^{p^{q_1}} \cdots t_e^{p^{q_e}}) < \alpha_d(t_1^{p^{m_1}} \cdots t_e^{p^{m_e}}), \qquad \text{if } d < \infty.$$

In fact, if we arrange the q_i's in descending order, as in Lemma 5, then d is the integer p_{i_o} of that lemma.

Finally we point out two results about the function $\alpha_i(x)$.

Lemma 7. Let y and y' be two monomial. Then there is an integer $m, 0 \leq m \leq \infty$, such that

$$\alpha_m(yy') < \alpha_m(y) + \alpha_m(y'), \quad \text{if} \quad m < \infty,$$
$$\alpha_i(yy') = \alpha_i(y) + \alpha_i(y'), \quad 0 \leq i < m.$$

Lemma 8. Suppose $z = t_1^{r(n_1)} \cdots t_e^{r(n_e)}$. Then for every $i \geq 0$,

$$\alpha_{i+1}(z) + \alpha_{i+1}(t_1^{p^{n_1}} \cdots t_e^{p^{n_e}}) = \alpha_i(z)$$

Theorem 9. Suppose $x \in S_p$ is a monomial of degree $|x| = \sum_{i=1}^{n} a_i$. Let $\{m_1, \cdots, m_e\}$ be the minimal representation of $|x|$. $z = t_1^{r(m_1)} \cdots t_e^{r(m_e)}$ (Note that e may be greater then n. In this case, instead of (2), we consider $S'_p = Z/p[t_1, \cdots, t_e]$. And x is considered as a monomial of S'_p). Suppose there is an integer $k \geq 0$ for which

$$\alpha_i(x) = \alpha_i(z), \quad \text{for all} \quad i < k,$$
$$\alpha_k(x) < \alpha_k(z).$$

Then x is hit.

This theorem can be proved by induction on k.

Remark 10. Theorem 2 implies the case $k = 0$ of Theorem 9.

Remark 11. If $\alpha_0(x) = \alpha_0(z)$, then $\alpha(\frac{|x| - \alpha_0(x)}{p}(p-1) + \alpha_0(x)) \leq \alpha_0(x)$.
The details will be published elsewhere.

The second author would like to thank Prof. W. M. Singer for sending a copy of [S].

References

[P] F.P. Peterson: Generators of $H^*(RP^\infty \wedge RP^\infty)$ as a module over the Steenrod Algebra. Abstracts Amer. Math. Soc. 833–55–89, April(1987).

[S] W.M. Singer: On the Action of Steenrod Squares on Polynomial Algebras. preprint.

[W] R.M.W. Wood: Steenrod Squares of Polynomials and the Peterson Conjecture. Math. Proc. Cam. Phil. Soc. 105(1989), 307–309.

Institute of Mathemathics
Academia Sinica
Beijing
P.R. of China